Signals and Communication Technology

This series is devoted to fundamentals and applications of modern methods of signal processing and cutting-edge communication technologies. The main topics are information and signal theory, acoustical signal processing, image processing and multimedia systems, mobile and wireless communications, and computer and communication networks. Volumes in the series address researchers in academia and industrial R&D departments. The series is application-oriented. The level of presentation of each individual volume, however, depends on the subject and can range from practical to scientific.

Indexing: All books in "Signals and Communication Technology" are indexed by Scopus and zbMATH

For general information about this book series, comments or suggestions, please contact Mary James at mary.james@springer.com or Ramesh Nath Premnath at ramesh.premnath@springer.com.

Phuong T. Tran · Tan N. Nguyen · Lam-Thanh Tu

Developing MATLAB Comprehensive Simulation for 5G/6G Wireless Communications

Springer

Phuong T. Tran
Advanced Intelligent Technology Research Group
Faculty of Electrical and Electronics Engineering
Ton Duc Thang University
Ho Chi Minh City, Vietnam

Tan N. Nguyen
Advanced Intelligent Technology Research Group
Faculty of Electrical and Electronics Engineering
Ton Duc Thang University
Ho Chi Minh City, Vietnam

Lam-Thanh Tu
Advanced Intelligent Technology Research Group
Faculty of Electrical and Electronics Engineering
Ton Duc Thang University
Ho Chi Minh City, Vietnam

ISSN 1860-4862 ISSN 1860-4870 (electronic)
Signals and Communication Technology
ISBN 978-981-95-1116-7 ISBN 978-981-95-1117-4 (eBook)
https://doi.org/10.1007/978-981-95-1117-4

This Springer imprint is published by the registered company Springer Nature Singapore Pte Ltd.
The registered company address is: 152 Beach Road, #21-01/04 Gateway East, Singapore 189721, Singapore

For my wife, Hai-Yen, and my children,
Uyen-Nhi and Thanh-Khoi

Preface

With the introduction of fifth-generation (5G) wireless networks, many newly introduced technologies are emerging and outperform those of previous generations. This change extends to include the entire infrastructure of the mobile communications system. With these major changes, many software or simulation tools available today for telecommunication network evaluation do not really suffice to cover the whole 5G network standard as well as the proposed technologies for sixth-generation (6G) wireless networks. For this reason, there is a need to develop a new kind of tool that will be able to reflect the changes brought by this contemporary network standard.

While research results in 5G and 6G wireless communications are generated quickly, the integration of these findings into visual learning materials that are easy for students to understand and follow is lagging behind. The teaching of advanced communication technologies, especially the potential technologies for 5G and 6G networks, in developing countries is mainly based on theoretical lectures together with some distributed simulation codes developed by the lecturers or collected from previously published results. Current simulation tools available for learners are very limited and the majority of available specialized 5G simulation toolboxes provided by companies tend to have relatively high costs, which many higher education institutions cannot handle.

Motivated by these above facts, this book introduces how to develop an integrated simulation framework that is systematic and well-updated, based on the collection of the unsystematic simulation codes that have been published or shared, in order to enhance the quality of teaching the 5G and 6G concepts for both undergraduate and graduate students. This comprehensive simulation tool covers many groups of advanced technologies or solutions for 5G/6G networks, including modulation waveforms, non-orthogonal multiple access techniques, massive MIMO systems, energy saving solutions (such as RF energy harvesting, self-energy recycling, intelligent reflecting surface, etc.), cognitive radios, physical layer security, and artificial intelligence in wireless communications. It has a user-friendly interactive interface with usage instructions and output graphs to evaluate the performance of each method. The goal of the simulations is to achieve results related to performance indicators at the physical layer, such as blocking probability, throughput, bit error rate or frame error rate, and security capacity.

Different from the previous simulation tools for 5G networks, this book can provide more diversity on the advanced technologies, and support the learning-by-doing approach in wireless communication training. This book can serve as a guide for graduate students to learn how to build a simulation framework for evaluating advanced 5G/6G technologies by themselves and apply them to their research. The book also keeps the readers updated with advanced research topics in wireless communications by an illustrative way.

Ho Chi Minh City, Vietnam
January 2025

Phuong T. Tran

Acknowledgements

First of all, I would like to express my heartfelt thanks to all the colleagues and students who contributed to the research projects that led to this monograph.

I would also take this opportunity to express my sincere gratitude to my former research advisors: Prof. James S. Lehnert, who taught me how to think of digital communications clearly when I was fortunate enough to be his Ph.D. students and also his graduate research assistant; Prof. Hong-Lien Pham, Prof. Thuong Le Tien, and Prof. Hoang Dinh Chien, who created my eagerness in communication engineering when being my research advisors during my master and undergraduate studying.

Especially, the dedicated gratitude must go to my Master student, Mr. Trung Q. Pham, who did most of the programming and simulation tasks to implement my idea of developing the simulation tool in the book. Without his contribution, this book may not have been finished. This book also owes much gratitude to our colleagues in the research groups of Faculty of Electrical and Electronics Engineering at Ton Duc Thang University and Post and Telecommunications Institute of Technology, including my co-authors (Prof. Tan N. Nguyen and Dr. Lam-Thanh Tu), and Prof. Tran Trung Duy, for their innovative research works in 5G and 6G wireless networks, which were selected for simulation in this book.

I also would like to send special thanks to my students, Mr. Phuc-Thien Nguyen, a research internship in my research group, for his tireless help in typesetting, proofreading, and collecting references for this book.

I would like to particularly acknowledge our editor as well as all the editorial staff and production team of Springer for their supporting in completing and producing this book.

Last, but not least, I would like to thank my family members for their unconditional love and encouragement, that gave me necessary support and motivation to overcome the difficulties when writing this book.

Ho Chi Minh City, Vietnam — Phuong T. Tran
January 2025

Contents

Acronyms

1G	First Generation
2G	Second Generation
3G	Third Generation
3GPP	3rd Generation Partnership Project
4G	Fourth Generation
5G	Fifth Generation
6G	Sixth Generation
ADC	Analog-to-Digital Converter
AF	Amplify-and-Forward
AFB	Analysis Filter Bank
AI	Artificial Intelligence
AN	Artificial Noise
AoA	Angle of Arrival
ASC	Average Secrecy Capacity
ASD	Angular Standard Deviation
AWGN	Additive White Gaussian Noise
BB	Baseband Unit
BBU	Baseband Units
BER	Bit Error Rate
BPSK	Binary Phase Shift Keying
BS	Base Station
BTS	Base Transceiver Station
CDF	Cumulative Distribution Function
CDMA	Code Division Multiple Access
CJ	Cooperative Jamming
CN	Core Network
C-NOMA	Cooperative Non-Orthogonal Multiple Access
CP	Cyclic Prefix
CPOFDM	Circularly Pulsed Orthogonal Frequency Division Multiplexing
CP-OFDM	Cyclic Prefix Orthogonal Frequency Division Multiplexing

CR	Cooperative Relaying
CRC	Cyclic Redundancy Check
CRN	Cognitive Radio Network
CS	Cyclic Suffix
CSI	Channel State Information
CWnd	Congestion Window
D2D	Device-to-Device
DC	Direct Current
DF	Decode-and-Forward
DFT	Discrete Fourier Transform
DL	Downlink
DNC	Digital Network Coding
DRoF	Digital Radio-over-Fiber
DSM	Differential Spatial Modulation
DSTSK	Differential Space-Time Shift Keying
DTT	Delay-Tolerant Transmission
EC	Energy Consumption/Ergodic Capacities
EE	Energy Efficiency
EH	Energy Harvesting
eNB	evolved Node B
EP	Eavesdropping Probability
EWMMSE	Element-Wise Minimum Mean Square Error
FBE	Fractional Bit Encoding
FBMC	Filter Bank Multicarrier
FD	Full-Duplex
FDMA	Frequency Division Multiple Access
FDR	Full-Duplex Relay
FFT	Fast Fourier Transform
FIR	Finite Impulse Response
FN	Function Node
FRS	Full Relay Selection
GCN	Generalized Channel Network
GFDM	Generalized Frequency Division Multiplexing
GSM	Global System for Mobile Communications
GUI	Graphical User Interface
HAP	Hybrid Access Point
HCRAN	Heterogeneous Cloud Radio Access Network
HD	Half-Duplex
HDR	Half-Duplex Relay
HetNet	Heterogeneous Network
HRN	Hybrid Relay Node
i.i.d	Independently and Identically Distributed
I/Q	In-phase/Quadrature
IC	Interference Cancellation
ICI	Inter-Carrier Interference

IFFT	Inverse Fast Fourier Transform
IoT	Internet of Things
IP	Intercept Probability/Information Processing
IRS	Intelligent Reflecting Surfaces
ISD	Inter Symbol Distance
ISI	Inter Symbol Interference
LDS	Low-Density Signature/Spreading
LOS	Line-of-Sight
LS	Least Squares
LTE	Long-Term Evolution
LTE-A	Long-Term Evolution-Advanced
MAI	Multiple Access Interference
MAP	Maximum A Posteriori
MATLAB	Matrix Laboratory
MC	Monte Carlo/Mother Constellation
MC-CDMA	Multi-Carrier Code Division Multiple Access
MCD	Minimum-Center-Distance
MD	Multiuser Detection
MIMO	Multiple-Input Multiple-Output
ML	Maximum Likelihood
MMD	Minimum-Mean-Distance
MMSE	Minimum Mean Square Error
MO	Mean Only
MPA	Message Passing Algorithm
MRC	Maximal Ratio Combining
MTC-MC-CDMA	Machine-Type Communication Multi-Carrier CDMA
MUD	Multi-User Detection
MUSA	Multi-User Shared Access
MUST	Multi-User Superposition Transmission
MVC	Model-View-Controller
NEF	Noise Enhancement Factor
NF	Noise Forwarding
NLoS	Non-Line-of-Sight
NOMA	Non-Orthogonal Multiple Access
NR	New Radio
NUS	National University of Singapore
OFDM	Orthogonal Frequency Division Multiplexing
OFDMA	Orthogonal Frequency Division Multiple Access
OLLA	Outer Loop Link Adaptation
OLLA-NOMA	OLLA Non-Orthogonal Multiple Access
OMA	Orthogonal Multiple Access
OOB	Out-of-Band
OOC	Optical Orthogonal Codes
OP	Outage Probability
ORS	Optimal Relay Selection

OTDA	Optimal Time Allocation
OTDPA	Optimal Time and Power Allocation
P/S	Parallel-to-Serial
PAPR	Peak-to-Average Power Ratio
PDF	Probability Density Function
PDMA	Pattern Division Multiple Access
PD-NOMA	Power-Domain Non-Orthogonal Multiple Access
PMF	Probability Mass Function
PNC	Physical Layer Network Coding
PRS	Partial Relay Selection
PSD	Power Spectral Density
PSK	Phase Shift Keying
PSR	Power Splitting Relaying
PTDPA	Proposed Time Duration and Power Allocation
QAM	Quadrature Amplitude Modulation
QMF	Quantize-Map-and-Forward
QoS	Quality of Service
QPSK	Quadrature Phase Shift Keying
RAN	Radio Access Network
RAT	Radio Access Technology
RE	Resource Element
RF	Radio Frequency
RN	Resource Node
RRC	Root-Raised Cosine
RRU	Remote Radio Unit
RS	Relay Selection
RSI	Receiver-Reflector Interference
RTT	Round Trip Time
S/P	Serial-to-Parallel
SC	Superposition Coding/Selection Combining
SC-FDE	Single-Carrier Frequency-Domain Equalization
SC-FDMA	Single-Carrier Frequency Division Multiple Access
SCMA	Sparse Code Multiple Access
SDN/NFV	Software-Defined Network/Network Function Virtualization
SDR	Software-Defined Radio
SE	Spectral Efficiency
SEP	Symbol Error Probability
SFB	Synthesis Filter Bank
SI	Self-Interference
SIC	Successive Interference Cancellation
SINR	Signal-to-Interference-plus-Noise Ratio
SM	Spatial Modulation
SNR	Signal-to-Noise Ratio
SOP	Secrecy Outage Probability
SPA	Sum-Product Algorithm

STSK	Space-Time Shift Keying
SWIPT	Simultaneous Wireless Information and Power Transfer
TAS	Transmit Antenna Selection
TAS/SC	Transmit Antenna Selection/Selection Combining
TCP/IP	Transmission Control Protocol/Internet Protocol
TDD	Time Division Duplexing
TDMA	Time Division Multiple Access
TIM	Topology Interference Management
TIM-NOMA	TIM Non-Orthogonal Multiple Access
TPSR	Time-Switching and Power-Splitting Relaying
TSR	Time Switching Relaying
TV	Television
TWCN	Two-Way Communication Networks
TWRC	Two-Way Relay Channel
TWRN	Two-Way Relay Network
UDP	User Datagram Protocol
UE	User Equipment
UERANSIM	User Equipment Radio Access Network Simulator
UI	User Interface
UL	Uplink
ULA	Uniform Linear Array
UN	User Node
URRLC	Ultra-Reliable Low Latency Communication
VBLAST	Vertical Bell Labs Layered Space-Time
VN	Variable Node
WCDMA	Wideband Code Division Multiple Access
WET	Wireless Energy Transfer
WiFi	Wireless Fidelity
WOLA	Weighted Overlap-Add
WPCN	Wireless Powered Communication Network
WPT	Wireless Power Transfer
WPTN	Wireless Power Transfer Network
XOR	Exclusive OR
ZF	Zero Forcing

Chapter 1
Introduction

1.1 Purpose of the Book

Today, the development of mobile information networks is progressing rapidly, especially with the emergence of the fifth-generation mobile network (5G) and the upcoming sixth-generation (6G). The continuation of technological eras and telecommunication network generations from 1G (1979), 2G (1991), 3G (2001), 4G (2009), to 5G (2019) has been and continues to materialize previous scientific ideas, promising to bring about a completely new dimension to the life of humanity.

With the increasing demand for wireless traffic from users due to the widespread deployment of Internet of Things (IoT) devices and the emergence of smart city applications requiring ultra-low latency networks, the fifth generation (5G) of mobile networks has been introduced as a revolutionary broadband technology aimed at enhancing the Quality of Service (QoS) for mobile subscribers.

The rapid development of 5G and 6G mobile information technology has led to a significant demand for learning among students and technicians. However, the current simulation tools available for learners are very limited. Furthermore, telecommunication research results are generated quickly, but the integration of these findings into visual learning materials that are easy for students to understand and follow is lagging behind. Currently, simulation tools serving students and lecturers in telecommunication topics are limited, mainly consisting of theoretical study materials followed by simulations provided by instructors. However, these programs are often difficult to comprehend and are not user-friendly.

Beside that, the majority of specialized telecommunication network simulation tools provided by companies tend to have relatively high costs, making it difficult for both learners and institutions to access such software. In the following, this book will introduce and analyze some of the aforementioned software solutions.

P. T. Tran et al., *Developing MATLAB Comprehensive Simulation for 5G/6G Wireless Communications*, Signals and Communication Technology,
https://doi.org/10.1007/978-981-95-1117-4_1

The most widely used telecommunication system simulation tool today is probably MATLAB[1]/Simulink, provided by MathWorks Inc. This is a commercial software, originating as an application of matrix computation. Today, MATLAB is a software tool for technical simulation that integrates computational algorithms, visualization, and programming. The working environment of MATLAB is user-friendly and intuitive for the mathematical representation of operations. MATLAB offers a low-cost version for students, but specialized 5G, 6G network simulation tools require very high costs.

Some universities have developed their own 5G network simulation tools based on the MATLAB platform. For example, the Vienna 5G System Level Simulation tool [1] is a MATLAB source code shared for educational purposes. Key features of the Vienna 5G System Level Simulation tool include Monte Carlo simulations, which are used to evaluate the average performance of large-scale networks involving numerous users, devices, and base stations. However, this tool lacks a user-friendly interface, making it relatively complex for university students to learn how to use.

SystemVue (provided by Keysight Technologies) is a simulation and physical-level design software, serving as a platform for system architecture, design, and testing [2]. It is based on mathematical models with comprehensive RF design processes and RF device measurement techniques, making it suitable for all wireless system architects. SystemVue provides the most advanced design and prototyping platform for complex RF systems with fast simulation speed and high fidelity, including circuits, libraries for radar systems, satellites, 5G networks, and WiFi.

Besides, some research institutions or universities have developed open-source tools to simulate 5G networks. OMNeT++ is a C++ library simulation platform that is extendable and modular, mainly used for building network simulators. It supports sensor networks, especially wireless networks, Internet protocols, performance evaluation models, and optical information networks. OMNeT++ is not a network simulator itself, but it has become widely popular as a network simulation platform in both the scientific community and industry. Most OMNeT models are open-source, developed as independent projects, and adhere to their own versions [3].

UERANSIM is an open-source simulation tool for mobile devices (UE) and 5G Radio Access Networks (RAN). The project is used for testing 5G core networks and researching 5G systems. The basic functions of UE and the base station (gNodeB) are complete and ready for use. However, some of its features are still under development. Furthermore, UERANSIM does not provide a complete simulation of the physical layer. The 5G-NR radio interface is partially implemented and simulated using the UDP (User Datagram Protocol), one of the core protocols of the TCP/IP protocol suite [4].

Simu5G is a free and open-source tool, short for simulator for 5G New Radio (NR) Networks, which is an extension of the popular SimuLTE 4G network simulator combined with the new access of 5G. Based on the OMNET++ framework, it is written in C++ and fully customizable with a simple interface. Users can also develop new modules to implement new algorithms and protocols. Simu5G supports

[1] Matrix Laboratory.

large-scale models for the mobility of User Equipment (UE), including the ability to model vehicular mobility. This tool is used by both enterprises and educational institutions [5].

However, most open-source tools are difficult to study and lack user-friendly interfaces. The update of techniques and new research results is also slow. If simulation tools are not systematically built, it becomes challenging to update new content. Therefore, it is necessary to develop simulation tools with systematic structures and easy-to-update capabilities for graduate and undergraduate students studying in the telecommunication field.

As mentioned above, nowadays, large corporations and companies often use specialized paid software (such as SystemVue by Keysight Technologies). However, the teaching and research on 5G and 6G networks at universities in developing countries mainly rely on theory and utilization of some MATLAB and Python source codes for simulation. The cost of MATLAB's 5G simulation toolbox is relatively high, making it difficult for universities to afford. Therefore, simulations using MATLAB and Python still remain fragmented and lack systematic approaches.

So, this book undertakes the development of a systematic simulation tool based on the MATLAB platform to simulate the techniques of 5G/6G networks at the physical layer. With the aim of covering most of key technologies in 5G/6G, we construct a comprehensive simulation tool, which is divided into modules ranging from basic modules such as modulation waveforms to advanced and modern technologies like intelligent reflecting surfaces or deep-learning-based communication schemes.

1.2 Objectives of the Book

The outcome of this book is an integrated simulation tool developed on the MATLAB platform to simulate basic and advanced techniques at the physical layer of 5G networks, such as modulation, multiple access, energy saving, and physical layer security solutions in 5G and 6G networks. This simulation tool has a user-friendly interactive interface with users' manual and output graphs to evaluate the performance of each method. The interface runs relatively fast, ensuring convenient data input and output for users and ease of understanding. The goal of the simulations is to achieve results related to performance indicators at the physical layer, such as blocking probability, throughput, bit error rate or frame error rate, and security capacity. The solutions in the book consist of several groups of solutions currently of interest, each of which will be fully presented in terms of system models, formulas, and simulation results.

The following is an overview of the specific objectives of the research:

1. Establishing a general framework and components of the simulation tool to manage input data systematically and to ensure efficient access when simulating scenarios integrating various techniques.

2. Constructing modules to simulate modulation techniques in 5G/6G networks. To meet the requirements of 5G/6G networks, various modulation techniques have been proposed, including pulse shaping modulation such as FBMC, GFDM, filter bank modulation such as UFMC, f-OFDM, etc.
3. Constructing modules to simulate proposed multiple access techniques for 5G and 6G networks, specifically including sparse code multiple access (SCMA) and non-orthogonal multiple access (NOMA).
4. Building modules to simulate massive MIMO (Multiple-Input Multiple-Output) techniques as depicted in Fig. 1.1. Utilizing numerous antennas will enhance the performance of future 5G systems in terms of data rates and spectral efficiency, catering to a large number of users. This is achieved by focusing on radiation

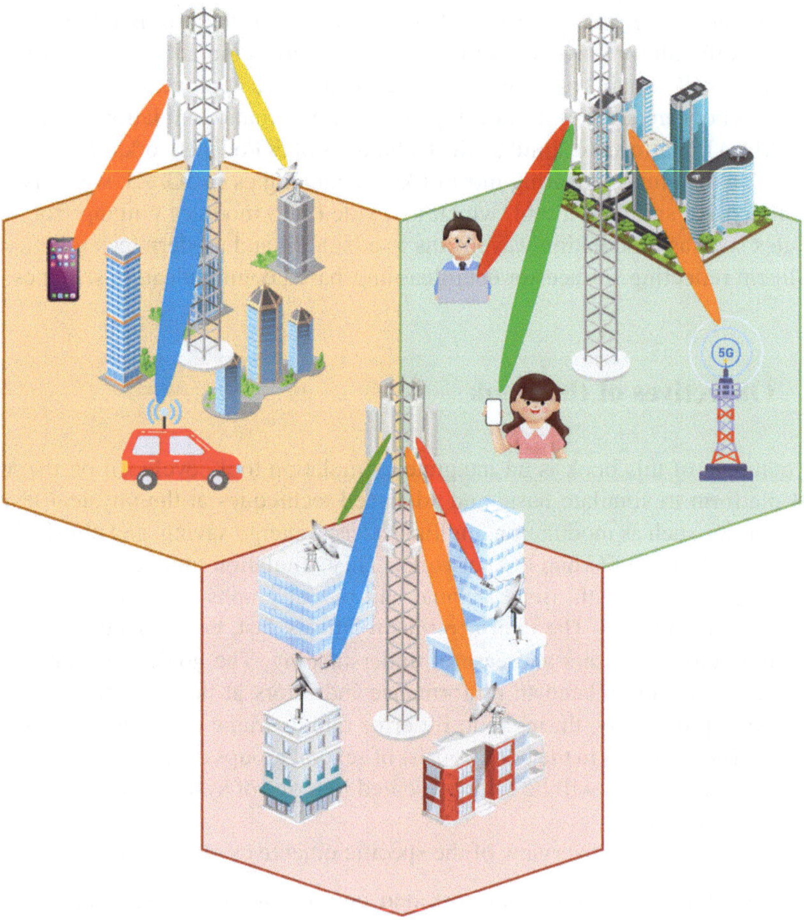

Fig. 1.1 Massive MIMO system

toward intended users, thus enabling simultaneous transmission to multiple users with low interference.

5. Developing simulation tools to evaluate energy-saving solutions in 5G and 6G networks. These solutions include energy harvesting via radio frequency (RF), energy reuse, optimization of energy and power allocation, and the use of energy-efficient devices (e.g., intelligent reflecting surfaces). The book conducts simulations of recent research findings in this area, including solutions from both authors and supervisors currently under development.
6. Developing modules to simulate physical layer security techniques. This technique leverages the characteristics of the wireless channel to implement measures that degrade the signal of eavesdropping receivers without the need for encryption between the transmitter and receiver.
7. Constructing a module to illustrate the using of deep learning algorithms to enhance the performance of wireless communication systems. In particular, we present a deep learning based massive MIMO communication systems that exploit a one-bit quantization to improve the spectral efficiency.
8. Designing interfaces for the aforementioned simulation modules, featuring interactivity and user friendliness. Specifically, the following criteria are considered during interface design: (i) interface logic; (ii) user-friendliness; (iii) aesthetics; (iv) comprehensiveness; and (v) processing speed when accessing results.

1.3 Review of Previous Research Findings

1.3.1 International

Currently, research groups worldwide have developed open-source tools, algorithms, and simulation programs on existing platforms, such as MATLAB, to simulate 5G networks at the physical layer. The research results have been published in scientific papers. However, these tools have almost no user-friendly interfaces. Below, this book summarize some representative results.

The Vienna 5G Link Level Simulator [1], which is introduced earlier, has the advantage of being quite comprehensive and updated with most of the latest 5G technologies. However, its drawback is the lack of a user interface, and the data input and output methods require expertise in the 5G field, making it time-consuming for students and learners to fully understand. Simu5G is an open-source simulation tool available on GitHub with good update capabilities, but it also lacks a user interface, requiring significant time investment to study its coding [6].

Another open-source tool is UERANSIM, capable of simulating emulation on actual hardware using Software-Defined Radio (SDR) devices running on Linux. However, it also lacks a user interface, making it challenging to understand and use [4]. On GitHub, the author Kim-Junseok introduces a 5G simulation platform called 5G K-SimNet [7], utilizing ns-3 and ns-3 mmWave modules. The strength of 5G K-SimNet lies in its inclusion of OpenFlow controllers, OpenFlow switches, modules for evaluating virtualization effects, and mmWave modules for assessing the multi-connectivity of 5G networks. However, it also lacks a user-friendly interface for interaction [8].

Other results have been published in scientific journals. For example, the 5G System Evaluation Tool is a tool that describes functions and presents the evaluation results of 5G based on calculating transmission characteristics using (ray tracing) and SINR/throughput based on Shannon's formula. Additionally, [9] developed a tool to assess the system performance of fifth-generation mobile information systems using multi-antenna technology such as massive MIMO and waveform formats to compensate for increased transmission loss due to high frequencies. Furthermore, [10] discussed the characteristics of 5G simulation, particularly focusing on system-level simulations. Wang et al. [11] analyzed the challenges of building a simulation platform for 5G considering emerging technologies and new network architectures. They proposed a two-tier cloud platform to validate potential 5G technologies. Additionally, 5G K-SimNet, introduced by [7], is a network simulation program designed to evaluate the end-to-end performance of 5G Radio Access Technology (RAT), Software-Defined Network/ Network Function Virtualization (SDN/NFV). Additionally, [12] presented a network simulation framework that addressed key features of 5G, namely, modeling virtual network functions, network slicing, dynamic network parameter adjustments during runtime, and realistic network traffic. In their demo, they deployed various traffic types to demonstrate that their simulation framework can support network management and orchestration applications. In 2021, [13] proposed an mm-wave model and analyzed it using the ns-3 simulation framework. The system's performance was validated through multiple simulation scenarios to evaluate protocols improving mm-wave signal quality, with performance metrics including Congestion Window (CWnd), Round Trip Time (RTT), and throughput. However, most of these results are standalone applications, which may not be convenient for students and learners to accumulate knowledge.

1.3.2 Domestic

According to the authors' knowledge, there is currently no systematically developed 5G/6G simulation tool in the country. Teaching mobile wireless networks such as 5G is mainly conducted using simulation exercise source code collected by instructors or developed and shared with learners.

Previously, in Ton Duc Thang University, a group of authors developed the 2G,3G simulation tool (MobileSim) for teaching purposes [14]. It had the following characteristics: running on the MATLAB programming environment; including simulation tools covering theories from basic to advance in GSM (2G), CDMA (3G), MC-CDMA (3G–4G), and MTC-MC-CDMA (a proposed solution for 4G) mobile networks; users could select parameters for the simulation process; the simulation results were represented through images, waveforms, spectra, and bit error rate (BER) graphs. However, this tool has not been updated to support the 4G and 5G generations.

Based on the research situation mentioned above, this book not only inherits the characteristics of the MobileSim tool, but also further develops it for the 5G and 6G networks to serve teaching purposes for undergraduate and graduate students.

1.4 Practical Significance of the Research

Modern information systems require the ability to meet the demand for high-speed data transmission. To assess the effectiveness of a simple transmission system on a conventional channel, mathematical analytical methods can be highly effective. However, for modern transmission systems operating over complex channels, such as mobile cellular channels, which are heavily influenced by multipath fading and noise, the design and analysis become exceedingly complicated. With their strong development in recent decades, computers have become increasingly powerful and cost-effective. Computer-based design and analysis techniques are becoming more widespread. Complex signal processing at receivers and transmitters, or channel effects, can now be simulated using standard computers.

Simulation is a valuable tool that allows for an in-depth understanding of system operations. A complete simulation system serves as a laboratory, enabling testing at multiple locations within the system. Parameters such as waveforms, signal spectra, constellation diagrams, histograms, or many other charts can be displayed on a computer screen, allowing researchers to analyze, evaluate, and compare with hardware-based results.

Due to the complexity of the 5G standard, 5G network simulations play a significant role in academia, universities, and industry. They can be instrumental in evaluating new network protocols, validating mathematical models, exploring 5G/6G networks, and gaining a better understanding of the relationships between algorithm parameters. Of course, 5G/6G network simulations cannot be the sole tool for research, implementation, and testing. However, network simulations can complement and provide visual insights for researchers studying 5G/6G networks.

These 5G/6G simulations provide standard-compliant functionalities and reference examples for modeling, simulating, and validating 5G NR systems [15]. Therefore, building an efficient simulation tool with systematic features and updates about 5G/6G mobile networks is scientifically significant. Developing such a tool requires understanding and researching 5G/6G network solutions. Organizing and setting up existing tools in a systematic manner is also challenging and demands scientific exploration.

Furthermore, in developing countries, there is currently no comprehensive 5G network simulation tool available. In practice, this topic could address the needs of the majority of society, including students, engineers, technicians, learners, and support for small businesses or startups. To develop 5G solutions, there is a requirement for an intuitive and interactive simulation tool that is easy to understand and use.

1.5 How This Book is Organized

The structure of this book comprises 12 chapters. Chapter 1 provides an overview of the book. Chapter 2 introduces relevant theories regarding the physical layer of 5G networks. Chapter 3 presents the design structure of the simulation tool for the physical layer of 5G networks developed in this book. Chapters 4 to 10 detail the selected solutions for building the simulation models as described in Sect. 1.2. Chapter 11 summarizes the implementation results. Finally, Chap. 12 presents the conclusions of the book and potential areas for further development.

References

1. Pratschner, S., Tahir, B., Nissel, R., Marijanovic, L., Mussbah, M., Kirev, K., Schwarz, S., Rupp, M.: The vienna 5g link level simulator v1. 2. nt. tuwien. ac. at/wpcontent/uploads/2020/05/5gll_usermanual. pdf, Seen In **19**, 25 (2021)
2. Keysight: One platform for system architecture, design, and verification. Online (n.d.). https://www.keysight.com/us/en/products/software/pathwave-designsoftware/pathwave-system-design-software.html. Accessed 2 Oct 2022
3. Varga, A., Hornig, R., Attila, T.: OMNeT++ discrete event simulator (2022). https://omnetpp.org/intro/. GitHub
4. Güngör, A.: Ueransim: Open source 5g ue and ran (gnodeb) implementation. Online (2022). https://github.com/aligungr/UERANSIM. GitHub
5. Giovanni: Simu5g - 5g new radio user plane simulator for omnet++ and inet (2022). URL http://simu5g.org/
6. Giovanni, N.: Simu5g. Online (2022). https://github.com/Unipisa/Simu5G/releases/tag/v1.2.1. GitHub
7. Choi, S., Song, J., Kim, J., Lim, S., Choi, S., Kwon, T.T., Bahk, S.: 5g k-simnet: End-to-end performance evaluation of 5g cellular systems. In: 2019 16th IEEE Annual Consumer Communications & Networking Conference (CCNC), pp. 1–6. IEEE (2019)
8. Siyoung: 5g k-simnet (5g network simulator) (2022). https://github.com/crazysct/kSimNet. GitHub
9. Kitao, K., Benjebbour, A., Imai, T., Kishiyama, Y., Inomata, M., Okumura, Y.: 5g system evaluation tool. In: 2018 IEEE International Workshop on Electromagnetics: Applications and Student Innovation Competition (iWEM), pp. 1–2. IEEE (2018)
10. Cho, S., Chae, S., Rim, M., Kang, C.G.: System level simulation for 5g cellular communication systems. In: 2017 Ninth International Conference on Ubiquitous and Future Networks (ICUFN), pp. 296–299 (2017). https://doi.org/10.1109/ICUFN.2017.7993797
11. Wang, Y., Xu, J., Jiang, L.: Challenges of system-level simulations and performance evaluation for 5g wireless networks. IEEE Access **2**, 1553–1561 (2014)
12. Delgado, O., Jaumard, B., Ding, Z., Bishay, F., Bissonnette, V.: A network simulator for 5g virtualized networks. In: 2022 IEEE 8th International Conference on Network Softwarization (NetSoft), pp. 237–239. IEEE (2022)
13. Nagendram, S., Ahammad, S.H., Nag, M.S.R.K., Satish, K.: Performance assessment of system-level simulation for 5g cellular communication systems. In: 2021 5th International Conference on Electronics, Communication and Aerospace Technology (ICECA), pp. 537–541. IEEE (2021)
14. Tran, P., Duy, N.T.: Simulation models of mobile communication networks. Grin Verlag (2006)
15. Nizzi, F., Pecorella, T., Bastianini, M., Cerboni, C., Buzzigoli, A., Fratini, A.: The role of network simulator in the 5g experimentation. In: Proceedings of the 2019 Workshop on Next-Generation Wireless with ns-3, pp. 13–17 (2019)

Chapter 2
Theoretical Background on 5G Physical Layer

2.1 Components of Cellular Networks

2.1.1 Core Network (CN)

The core network is the system for communication between stations to facilitate the forwarding of calls. As a part of the Internet backbone, the core network provides pathways for data exchange among subordinate nodes. The core network is comprised of a series of routers and the most commonly used circuit-switching switches [1].

2.1.2 Radio Access Network (RAN)

The model combining UE, RAN, and CN is presented in Fig. 2.1. RAN serves as the connection between UE and CN, responsible for wireless access communication. This is the focus of the developed simulation program. RAN consists of nodes, links between nodes, and cells [2].

User Equipment (UE) refers to the devices that end-users utilize to communicate with the mobile network. UE can include computers, mobile phones, or any device equipped with a mobile antenna [1].

2.2 The Architecture of the 5G Networks

Simulation is essential to determine system requirements. Therefore, this section will elucidate the 5G architecture, refer to Fig. 2.2 [2].

P. T. Tran et al., *Developing MATLAB Comprehensive Simulation for 5G/6G Wireless Communications*, Signals and Communication Technology,
https://doi.org/10.1007/978-981-95-1117-4_2

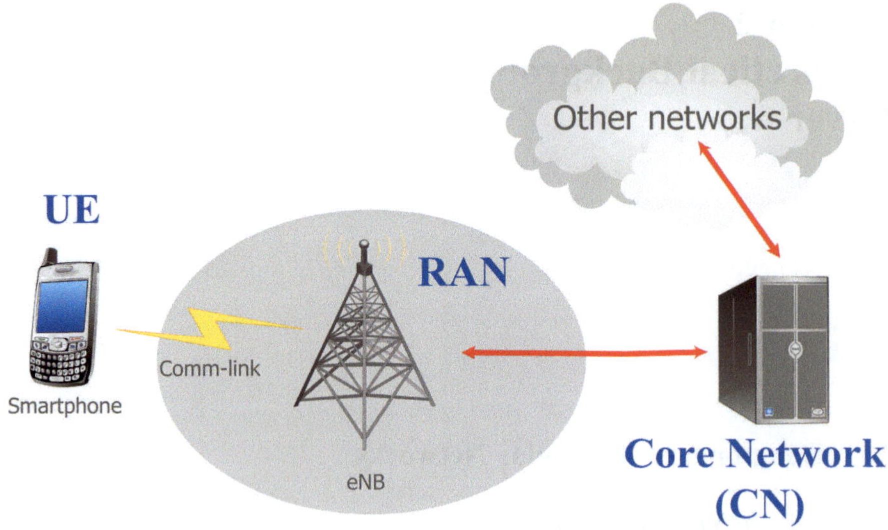

Fig. 2.1 Basic connection between UE, RAN, and CN

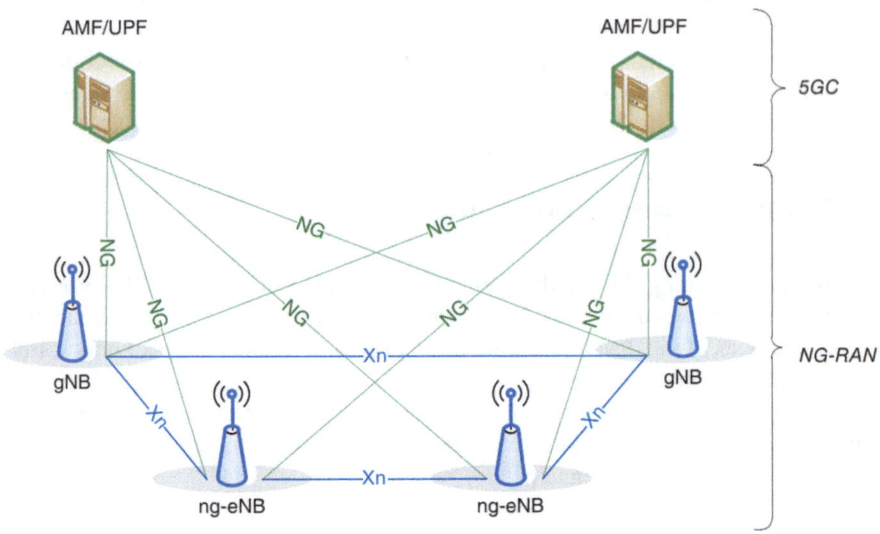

Fig. 2.2 Components of the 5G network (*Source* [3, p. 27])

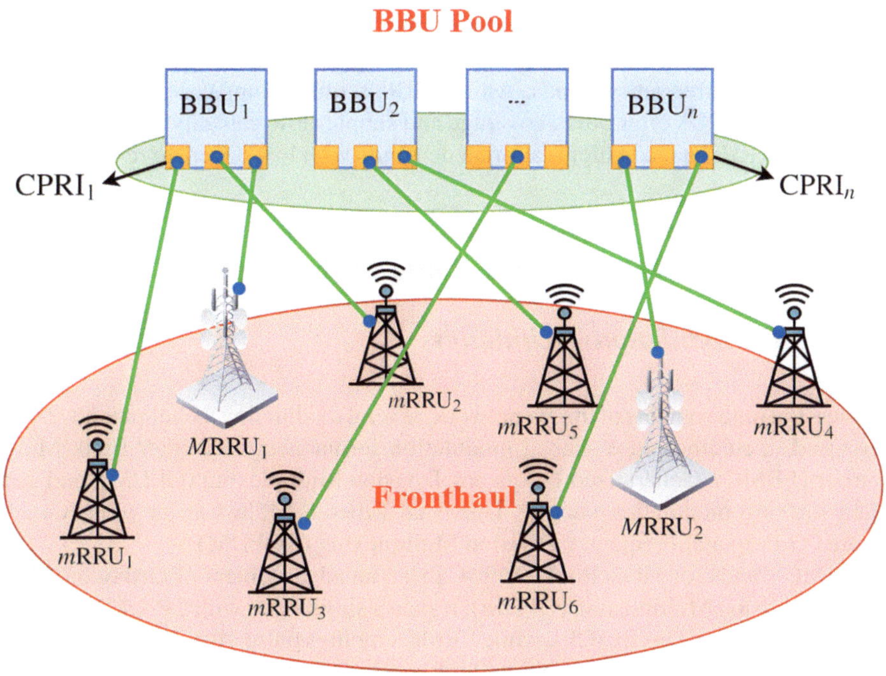

Fig. 2.3 5G network architecture

The 5G network comprises two main components: the Core Network (CN) and the Radio Access Network (RAN). Devices connected to the network, such as smartphones, sensors, or even smart refrigerators, are referred to as User Equipment (UE).

The 5G network generation corresponds to the revolutionary transformation of the telecommunications industry through the incorporation of cutting-edge technologies and architectures such as mmWave, Massive Multiple-Input Multiple-Output (Massive MIMO), Heterogeneous Networks (HetNet), full-duplex systems, Cloud Radio Access Network (C-RAN), and others.

Figure 2.3 illustrates the overall architecture of the 5G network, including macrocells, small cells, Remote Radio Units (RRUs), and Baseband Units (BBUs). As depicted in the figure, the fronthaul corresponds to the connection between RRU and BBU in both uplink and downlink directions. The backhaul model, not shown in the figure, essentially comprises the connection between BBU and the core network in both directions. The original baseband signals are transmitted from BBU to RRU through a digital radio-over-fiber (DRoF) wireless connection over the optical fiber platform. The RRU converts the received radio frequency (RF) signals from the uplink direction of the User Equipment (UE) into the original baseband signals.

The entire coverage area of the network is divided into microcells and macrocells, leading to a multi-layered network architecture. Increasing the station density with small cell Remote Radio Units (mRRUs) allows for reduced load on large cell Remote

Radio Units (MRRUs) while providing better end-to-end latency and increasing energy efficiency, as mRRUs consume less power than MRRUs. In practice, MRRUs use sub-6 GHz frequency bands, while mRRUs utilize mmWave frequencies. In this model, MRRUs offer broad coverage and support low-capacity traffic demands, whereas mRRUs manage high-data-rate demands with low coverage range.

2.3 Key Technologies for 5G Radio Access Network

2.3.1 5G Modulation Techniques

To meet the requirements of the 5G network, several modulation techniques have been proposed, including pulse-shaped modulation formats such as Filter Bank Multi-Carrier (FBMC), Generalized Frequency Division Multiplexing (GFDM), and sub-band filtering modulation such as Universal Filtered Multi-Carrier (UFMC) and filtered Orthogonal Frequency Division Multiplexing (f-OFDM).

FBMC stands for Filter Bank Multi-Carrier, an advancement of Orthogonal Frequency Division Multiplexing (OFDM). It shares similarities with CP-OFDM, where OFDM uses a cyclic prefix (CP), while FBMC employs pulse-shaping filters. FBMC, utilizing efficient filtering, can achieve high spectral efficiency. By considering a flexible waveform for 5G wireless communication, FBMC significantly reduces bandwidth losses compared to OFDM by eliminating the need for a cyclic prefix. FBMC utilizes filter banks with Offset Quadrature Amplitude Modulation (OQAM). The advantages of the FBMC waveform include higher throughput, improved Bit Error Rate (BER), and enhanced Power Spectral Density (PSD) [4].

GFDM (Generalized Frequency Division Multiplexing) is a general form of frequency division multiplexing. It is a viable solution for the physical layer of 5G as it can address various requirements. GFDM is a flexible multicarrier modulation scheme implemented in blocks, where each data block consists of a specific number of subcarriers and certain auxiliary signals. Single-Carrier Frequency-Domain Equalization (SC-FDE) and Circularly Pulsed OFDM (CPOFDM) are special cases of GFDM when the number of subcarriers and auxiliary signals is 1. Furthermore, the choice of pulse-shaping filter in GFDM is a flexible solution to limit out-of-band signal emissions. While FBMC uses linear convolution, GFDM employs circular convolution [5].

2.3.2 Multiple Access Techniques

In mobile communications, it is essential to efficiently manage multiple users over limited frequency resources. In information networks, to conserve resources, a method is often employed where multiple subscribers share a common transmission channel, known as multiple access techniques. Specifically, multiple access techniques in 4G networks include Orthogonal Frequency Division Multiple Access (OFDMA) and Single-Carrier Frequency Division Multiple Access (SC-FDMA). In 3G networks, Code Division Multiple Access (CDMA) and Wideband Code Division Multiple Access (WCDMA) are utilized, while in 2G networks, Frequency Division Multiple Access (FDMA) and Time Division Multiple Access (TDMA) are common. Proposed multiple access techniques for 5G networks include Sparse Code Multiple Access (SCMA), Non-Orthogonal Multiple Access (NOMA), and OFDMA.

Sparse Code Multiple Access (SCMA) is a novel non-orthogonal frequency multiple access technique that can enhance the spectral efficiency of wireless access. With SCMA, different data streams are directly mapped to codewords of various multidimensional codebooks, where each codeword represents a transmission layer. Multiple SCMA layers share the frequency and time resources of Orthogonal Frequency Division Multiple Access (OFDMA) [6].

Non-Orthogonal Multiple Access (NOMA) is categorized into two main approaches: power domain NOMA and code domain NOMA. NOMA primarily relies on the use of Superposition Coding (SC) at the transmitter and Successive Interference Cancelation (IC) techniques at the receiver. In practice, both of these techniques are not entirely new, but the practical application of NOMA is still in its early stages. Currently, Orthogonal Multiple Access (OMA) has been successfully deployed in practical implementations of 5G networks and continues to exhibit good performance. However, the demands of 5G networks are extensive, prompting ongoing development and research into NOMA technology for future performance improvements.

In recent years, NOMA has garnered attention from researchers, who aim to apply it in real-world scenarios to enhance network performance up to the theoretical standards of 5G. Emerging trends include incorporating cooperative communications, Multiple-Input Multiple-Output (MIMO) systems, relay networks, and power-domain NOMA [7].

2.3.3 *Massive MIMO*

Figure 2.4 illustrates the evolution of the transmit technology for mobile networks from 1G to 5G. In 1G and 2G networks, baseband signal processing units were connected to the antenna via coaxial cables and amplifier circuits. Recent 3G and 4G systems employ distributed architectures, where the radio unit, based on baseband processing, is placed directly in front of the antenna. Coaxial cables are replaced with data-carrying optical fibers running up the tower. This architecture achieves a significant advancement as the radio circuitry is positioned alongside the antenna. In 5G systems, the baseband processing unit will be centralized. The radio processing unit will be directly integrated with hundreds of antenna elements, forming a large-scale MIMO array.

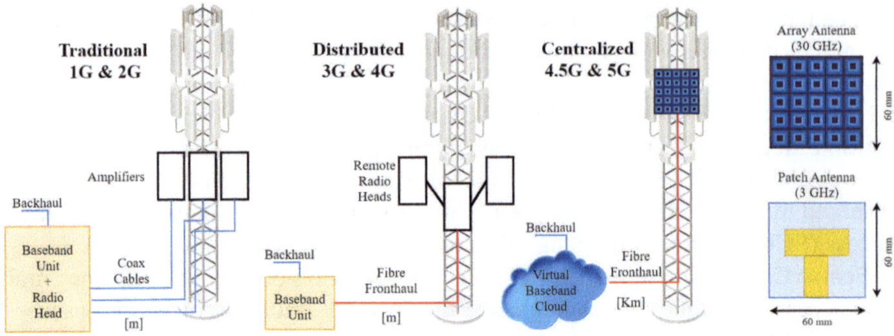

Fig. 2.4 Antenna arrays for millimeter-wave frequencies and the progress of cellular infrastructure

Massive MIMO is poised to enhance the performance of 5G systems in terms of data speed and spectral efficiency, concurrently serving a large number of users. This improvement is achieved by concentrating the radiation toward the intended users, or simultaneously transmitting to multiple users while keeping interference low. This enhances the system's capacity compared to traditional antenna systems, resulting in a spectral efficiency over 50 times greater than practical 4G technology. However, to maximize the potential of this technology and overcome challenges in real-world environments, the implementation of an extremely complex wireless system is required [8].

In general, MIMO technology is employed to transmit data by a protocol such as Wi-Fi in 5G networks to increase bandwidth and capacity. The 802.11n protocol, utilizing MIMO technology, allows achieving speeds of up to 350 Mbits/s. Data transmission quality is improved even in areas with weak signal reception. Additionally, enhanced reliability and the best connection quality are ensured [9].

2.3.4 *Energy-Saving Solutions*

As the demand for energy consumption continues to grow, energy is becoming an increasingly critical issue in the telecommunication industry as a whole and computer networks in particular. Presently, we use a wide variety of devices, so ensuring continuous battery operation with minimal charging is challenging. Therefore, energy conservation is an essential concern in 5G and 6G networks. Recently, numerous energy-saving methods have been proposed, such as optimizing energy utilization efficiency, power allocation optimization, energy reuse, or utilizing energy-efficient devices like smart reflectors.

Wireless Energy Harvesting involves capturing energy from surrounding sources and converting it into a usable form to power any application or store energy for future use. Particularly, harvesting energy from radio frequency (RF) waves is gaining sig-

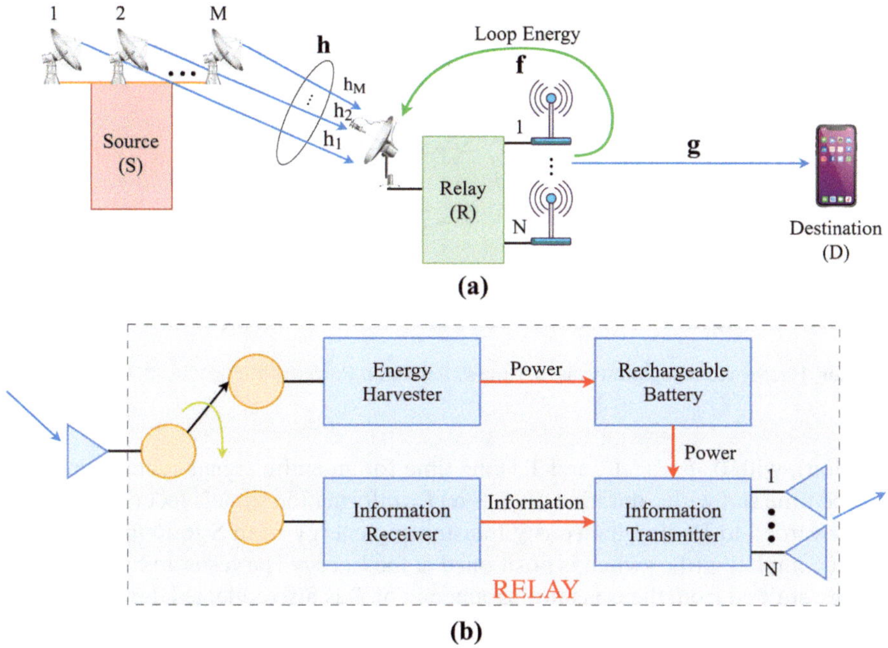

Fig. 2.5 Energy recycling: **a** Full-duplex wireless network energy recycling; **b** Receiver structure

nificant attention. In this method, electrical energy from various sources is captured by antennas, converted into electric energy, and utilized as a power source for other devices. RF energy harvesting technology enables the accumulation of energy from sources such as mobile towers, WiFi networks, and TV/radio transmitters.

Based on the technology of harvesting energy from electromagnetic waves, the concept of *self-energy recycling (S-ER)* in cooperative networks has been proposed. In this approach, the energy from self-interference can be harvested and reused. This idea was first introduced by Prof. Zeng and Prof. Rui Zhang from the National University of Singapore (NUS) in 2015 [10], as illustrated in Fig. 2.5.

In Fig. 2.5, a signal source S communicates with a destination node D with the assistance of a relay node R. The channel gains for $S \rightarrow R$ và $R \rightarrow D$ are denoted by h and g, respectively. Let f be the self-interference channel gain from the transmitting antenna to the receiving antenna of the relay node R. In general, S and R can be equipped with M and $N + 1$ antennas, respectively. Among them, at each time, a maximum of N antennas can be activated to transmit information, and 1 antenna is used to receive information signals and harvest energy. Assuming that the direct link from S to D is of very poor quality and can be neglected.

There are two main energy recycling protocols that have been proposed: time-based protocol and power-based protocol. Figure 2.6a illustrates the time-based protocol. The first time slot has a duration of αT , during which information signals are transmitted from S to R (the switch in Fig. 2.6b is in the Information Receiver

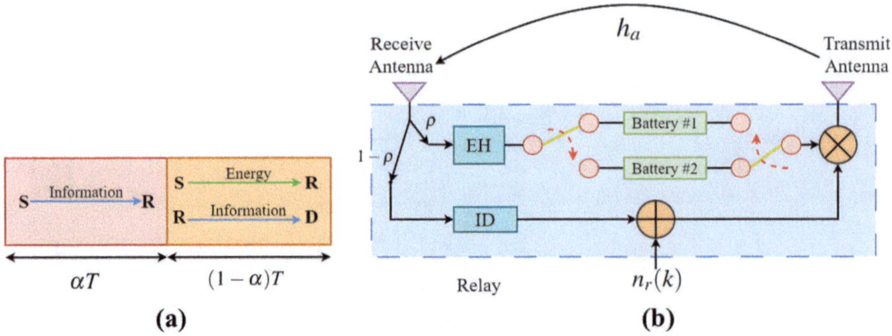

Fig. 2.6 Energy recycling protocols: **a** Time switching; **b** Power-splitting

position), with $0 < \alpha < 1$, and T is the time for an entire signal cycle. During the second time slot with a duration of $(1-\alpha)T$, information signals received by R will be forwarded to D, simultaneously transmitting energy from S to R for harvesting and accumulation(the switch is positioned at the Energy Harvester in Fig. 2.6b [10]. Energy emitted from the transmitting antenna of R is also collected during this time slot.

The total energy harvested at R during this stage is calculated by

$$E_r = \eta(P_S\|h\|^2 + P_R\|f\|^2)\frac{(1-\alpha)T}{2}, \quad (2.1)$$

where $0 < \alpha < 1$ is the energy conversion efficiency from RF to DC and charging for the battery, P_S and P_R are the transmit power of S and R, respectively.

Figure 2.5b illustrates the power-based protocol in which each data signal block is split into two equal time slots, denoted as the odd slot and the even slot, both having a duration of $T/2$. Two battery sources are activated to alternately harvest energy and supply power to the relay node. In each time slot, the power of the signal received by the relay node is divided into two parts: one for energy harvesting and one for decoding the signal in the ratio $\rho : (1-\rho)$, where $\rho \in (0, 1)$ denotes the power division coefficient. The harvested energy at the relay node can be determined by the formula:

$$E_r = \eta\rho(P_S\|h\|^2 + P_R\|f\|^2)\frac{T}{2} \quad (2.2)$$

In summary, a portion of the energy, referred to as loop energy, used to transmit signals from the relay will be harvested and reused along with the energy sent from the signal source, thereby allowing for an enhanced overall energy efficiency of the system.

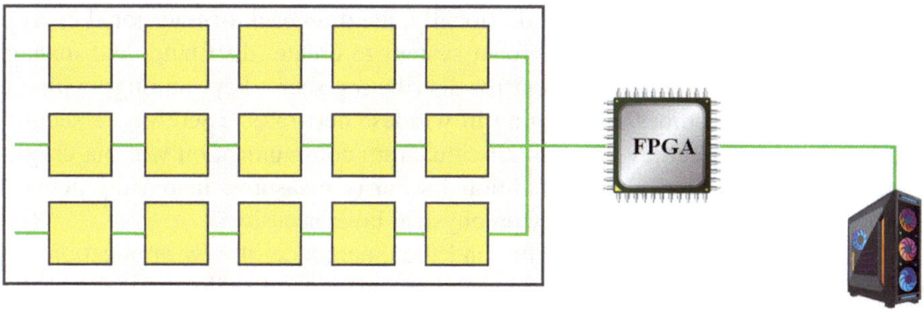

Fig. 2.7 The architecture of the IRS

Another solution for energy saving in wireless communication is to exploit the *zero-energy devices* such as intelligent reflecting surfaces (IRS). IRS is a flat surface consisting of multiple passive reflecting elements, each element can independently adjust a necessary phase shift to the incoming signal [11]. Based on the specific materials of the reflecting elements, IRS can be classified into the structures based on antenna arrays [12] and surface-based structures [13]. By properly adjusting the phase shift of all reflecting elements, the reflected signals can be reconfigured to propagate in the desired direction. Due to the rapid development of electromagnetic metamaterials, the reflection coefficient of each element can be reconfigured in real-time to adapt to the changing of radio transmission environment.

As shown in Fig. 2.7, a typical IRS architecture is fabricated using metamaterials and mainly consists of a flat surface and a controller.The optimal reflectivity of the IRS is calculated at the base station (eNB), then sent to the IRS controller through a predetermined feedback path. The parameters are adjusted depending on CSI. This information is only updated when the CSI changes, usually over a much longer time interval than the symbol period.

In networks with multiple users or multiple relay nodes, adjusting the energy harvesting levels at nodes or altering power allocation for users significantly influences the network performance, particularly energy efficiency. Therefore, there is a demand for *optimizing power or energy allocation* for various nod es or users under system resource constraints and Quality of Service (QoS) requirements. Solving these optimization problems in such scenarios is a research direction of current interest. An example of this solution will be presented in Chap. 7 (Sect. 7.4).

2.3.5 Physical Layer Security

Traditional security methods using cryptographic techniques are always carefully chosen and considered to ensure confidentiality for most of current applications. However, with the rapid development of computational capabilities in processing systems, especially the advent of quantum computers, researchers have had to develop algorithms with higher accuracy. Additionally, the drawback of traditional encryption schemes is the constant need for a system to create, distribute, and manage cryptographic keys. The requirement for distributing secret keys among legitimate components also becomes challenging in wireless networks. Therefore, a security solution at the physical layer or utilizing quantum communication without cryptographic techniques can enhance traditional security measures, increasing network security. This section focuses solely on physical layer security.

The rapid growth of mobile traffic and the emergence of new applications for devices, driven by the inherent broadcast characteristics of radio channels, lead to various security risks. Therefore, it is necessary to strengthen system security because radio channels are highly susceptible to eavesdropping and signal interception [14].

2.3.6 Cognitive Radio Networks

Cognitive radio networks (CRNs) are emerging as a promising cornerstone technology for the advancement of both 5G and the prospective 6G wireless networks due to their adaptability and efficient spectrum management capabilities. The explosive growth in mobile data traffic, coupled with the proliferation of connected devices, demands innovative solutions to effectively utilize the radio spectrum. CRNs provide a dynamic approach by allowing devices to intelligently detect unused spectrum bands and adapt their transmission parameters accordingly [15]. This capacity for spectrum sensing and dynamic spectrum access is crucial for the crowded spectrum in 5G environments and is expected to become even more critical as we advance toward 6G, which is anticipated to introduce even more demanding use cases such as augmented reality, seamless IoT integration, and ultra-reliable low latency communications (URRLC) [16]. Furthermore, cognitive radio's ability to coexist with legacy systems and its application in technologies such as smart grids, broadband cellular networks, and emergency response networks highlight its versatility and potential to support a wide range of services.

In 6G networks, CRNs may crucially underpin new paradigms like symbiotic radio networks, where devices cooperate and share spectrum resources more intimately than ever before [17]. Continued research and development in this area, supported by advancements in artificial intelligence and machine learning, will be essential for overcoming current limitations, such as higher hardware complexity and the need for improved security protocols [18]. Recent studies have explored the integration of machine learning techniques to enhance spectrum prediction and decision-making processes, as well as the development of new protocols to support the ultra-dense network environments foreseen in 6G [19, 20]. Additionally, the exploration of blockchain for secure spectrum transactions adds an innovative layer of trust and transparency to CRN operations [21]. As such, cognitive radio networks hold the potential to drastically enhance spectrum efficiency, network scalability, and the overall performance of next-generation wireless communication systems.

2.3.7 *Artificial Intelligence Applications in Wireless Communications*

Artificial intelligence (AI) is poised to be an integral technology in the evolution of 5G and future 6G wireless networks because it can provide unprecedented capabilities in terms of network optimization, management, and service delivery. In 5G, AI has already been exploited to enhance key performance aspects such as dynamic resource allocation, automated network slicing, and real-time data processing. As we look toward 6G, the role of AI will expand even further. 6G networks are expected to support complex use cases like holographic communications, pervasive intelligence, and massive IoT, all of which demand highly flexible and intelligent network frameworks [22].

Machine learning, a subset of AI, is instrumental in advancing network automation and intelligence. Techniques such as reinforcement learning and federated learning are being utilized to optimize network performance, reduce latency, and manage resources under dynamically changing conditions without human intervention [23]. In 6G, AI is anticipated to play a critical role in the integration of diverse technologies such as terahertz communications, quantum computing, and blockchain, facilitating seamless interoperation and end-to-end security [24]. Furthermore, sustainability is a significant focus for 6G, and AI contributes by optimizing energy consumption and improving spectrum efficiency [25].

2.4 Software Platform

To build the simulation tool, it is necessary to explore various programming languages, libraries, and software platforms to find a suitable language for this research. The criteria to select these languages include two factors: firstly, which simulation libraries are available, and secondly, which tools can be used to develop a graphical user interface (GUI).

Using existing comprehensive systems or software platforms is a feasible idea. Many available simulation platforms can be adjusted or modified to suit the intended purpose. Among the platforms available on the market, MATLAB is selected for this work due to its popularity in the academic community, especially for graduate and undergraduate students.

App Designer is an application design tool that allows you to create applications without needing to be a professional software developer. You can drag and drop visual components to layout your GUI design and use the integrated editor to quickly write a program to describe its behavior. You can share your application through MATLAB Drive or by creating web applications or standalone desktop applications with MATLAB Compiler and Simulink Compiler. MATLAB can be used to build interactive user interfaces that can be integrated into various environments. There are two types of user interfaces you can build [26]: Apps and Live Editor. This book introduces the method of building Apps (with an interface that perform activities based on user interactions). Through integration, MATLAB functions and algorithms can be incorporated into models, and the results can be exported to MATLAB for further analysis [1].

References

1. Asante, J., Olsson, J.: 5g simulation framework (2018). http://www.diva-portal.org/smash/get/diva2:1230756/FULLTEXT01.pdf. Retrieved from DiVA portal
2. Sjöstrand, J.: Extension of a 5g ran simulator by modeling user equipment (2019)
3. Sirotkin, S.: 5G Radio Access Network Architecture: The Dark Side of 5G Book (2020). http://gen.lib.rus.ec/book/index.php?md5=CE87262DF0C7880ABD40AD74CCDC5BE1
4. Aldababseh, M., Jamoos, A.: Estimation of fbmc/oqam fading channels using dual kalman filters. Sci. World J. **2014** (2014)
5. Ali: Gfdm modulation forms for wireless communication (2021). https://www.linkedin.com/pulse/gfdm-modulation-forms-wireless-communication-ali-ahmed/. Retrieved from LinkedIn
6. Nikopour, H., Yi, E., Bayesteh, A., Au, K., Hawryluck, M., Baligh, H., Ma, J.: Scma for downlink multiple access of 5g wireless networks. In: 2014 IEEE Global Communications Conference, pp. 3940–3945. IEEE (2014)
7. Islam, S.R., Avazov, N., Dobre, O.A., Kwak, K.S.: Power-domain non-orthogonal multiple access (noma) in 5g systems: potentials and challenges. IEEE Commun. Surv. & Tutor. **19**(2), 721–742 (2016)
8. Khwandah, S.A., Cosmas, J.P., Lazaridis, P.I., Zaharis, Z.D., Chochliouros, I.P.: Massive mimo systems for 5g communications. Wireless Pers. Commun. **120**(3), 2101–2115 (2021)
9. Biradar, A., Murthy, N.S., Awasthi, P., Srivastava, A.K., Akram, P.S., Lakshminarayana, M., Abidin, S., Vadi, V.R., Sisay, A.: Massive mimo wireless solutions in backhaul for the 5g networks. Wirel. Commun. Mob. Comput. **2022**(1), 3813610 (2022)
10. Zeng, Y., Zhang, R.: Full-duplex wireless-powered relay with self-energy recycling. IEEE Wireless Commun. Lett. **4**(2), 201–204 (2015)
11. Wu, Q., Zhang, R.: Towards smart and reconfigurable environment: Intelligent reflecting surface aided wireless network. IEEE Commun. Mag. **58**(1), 106–112 (2019)
12. Tan, X., Sun, Z., Jornet, J.M., Pados, D.: Increasing indoor spectrum sharing capacity using smart reflect-array. In: 2016 IEEE International Conference on Communications (ICC), pp. 1–6. IEEE (2016)
13. Cui, T.J., Qi, M.Q., Wan, X., Zhao, J., Cheng, Q.: Coding metamaterials, digital metamaterials and programmable metamaterials. Light: Sci. & Appl. **3**(10), e218–e218 (2014)
14. Ha, D.H., Duy, T.T., Son, P.N., Le-Tien, T., Voznak, M.: Security-reliability trade-off analysis for rateless codes-based relaying protocols using noma, cooperative jamming and partial relay selection. IEEE Access **9**, 131087–131108 (2021). https://doi.org/10.1109/ACCESS.2021.3114343
15. Mitola, J.: Cognitive radio an integrated agent architecture for software defined radio (2000). https://api.semanticscholar.org/CorpusID:61207536
16. Zhao, Q., Sadler, B.M.: A survey of dynamic spectrum access. IEEE Signal Process. Mag. **24**(3), 79–89 (2007). https://doi.org/10.1109/MSP.2007.361604
17. Haykin, S.: Cognitive radio: brain-empowered wireless communications. IEEE J. Sel. Areas Commun. **23**(2), 201–220 (2005). https://doi.org/10.1109/JSAC.2004.839380
18. Akyildiz, I.F., Lee, W.Y., Vuran, M.C., Mohanty, S.: Next generation/dynamic spectrum access/cognitive radio wireless networks: a survey. Computer Networks **50**(13), 2127–2159 (2006). https://doi.org/10.1016/j.comnet.2006.05.001. https://www.sciencedirect.com/science/article/pii/S1389128606001009

19. Letaief, K.B., Chen, W., Shi, Y., Zhang, J., Zhang, Y.J.A.: The roadmap to 6g: Ai empowered wireless networks. IEEE Commun. Mag. **57**(8), 84–90 (2019). https://doi.org/10.1109/MCOM.2019.1900271
20. Zong, B., Fan, C., Wang, X., Duan, X., Wang, B., Wang, J.: 6g technologies: Key drivers, core requirements, system architectures, and enabling technologies. IEEE Veh. Technol. Mag. **14**(3), 18–27 (2019). https://doi.org/10.1109/MVT.2019.2921398
21. Liang, Y.C.: Blockchain for Dynamic Spectrum Management, pp. 121–146. Springer Singapore, Singapore (2020). https://doi.org/10.1007/978-981-15-0776-2_5
22. Saad, W., Bennis, M., Chen, M.: A vision of 6g wireless systems: applications, trends, technologies, and open research problems. IEEE Network **34**(3), 134–142 (2020). https://doi.org/10.1109/MNET.001.1900287
23. Zhang, C., Patras, P., Haddadi, H.: Deep learning in mobile and wireless networking: a survey. IEEE Commun. Surv. & Tutor. **21**(3), 2224–2287 (2019). https://doi.org/10.1109/COMST.2019.2904897
24. Dang, S., Amin, O., Shihada, B., Alouini, M.S.: What should 6g be? Nat. Electr. **3**, 20–29 (2019). https://api.semanticscholar.org/CorpusID:211095143
25. Noman, H.M.F., Hanafi, E., Noordin, K.A., Dimyati, K., Hindia, M.N., Abdrabou, A., Qamar, F.: Machine learning empowered emerging wireless networks in 6g: recent advancements, challenges and future trends. IEEE Access **11**, 83017–83051 (2023). https://doi.org/10.1109/ACCESS.2023.3302250
26. MathWorks: Matlab app building (2022). https://www.mathworks.com/help/pdf_doc/matlab/creating_guis.pdf. Retrieved from MathWorks

Chapter 3
Develop a Wireless Network Simulation Tool for 5G/6G Technologies

3.1 Simulation System Architecture

The simulation system can run on a local computer and does not require additional machines to operate. It is designed following the MVC (Model-View-Controller) pattern, as depicted in Fig. 3.1. The main user interface (UI) of the simulation tool is designed as an application on the MATLAB App Designer platform (**main5G.mlapp**). This interface is controlled by object handlers and interface controllers (implemented as a private function **control(app, num, command)**). The functionalities of the controller include: (i) allowing users to vary the simulation model by selecting from the simulation directory tree; (ii) managing input/output parameters on the user interface according to the selected model. The View component manages the graphical output data for users to observe. It can support real-time observation of simulation results or reviewing the stored simulation results. Finally, the simulation core contains the necessary simulation models for techniques in 5G and 6G networks.

Figure 3.2 shows the list of specific modules in the simulation core, which includes the modules for 5G modulation waveforms, multiple access techniques, massive MIMO methods, energy-saving techniques, physical layer security solutions, cognitive radio communication models, and AI applications in 6G networks.

3.2 General Simulation Procedure for a Specific Technical Solution for 5G/6G Networks

To simulate a specific communication model in a 5G/6G network, this work applies the method used in the Vienna 5G Link-Level Simulator. In this method, object-oriented programming is utilized in most modules. The components of the simulation core are divided into separate packages, and network elements such as base stations,

P. T. Tran et al., *Developing MATLAB Comprehensive Simulation for 5G/6G Wireless Communications*, Signals and Communication Technology,
https://doi.org/10.1007/978-981-95-1117-4_3

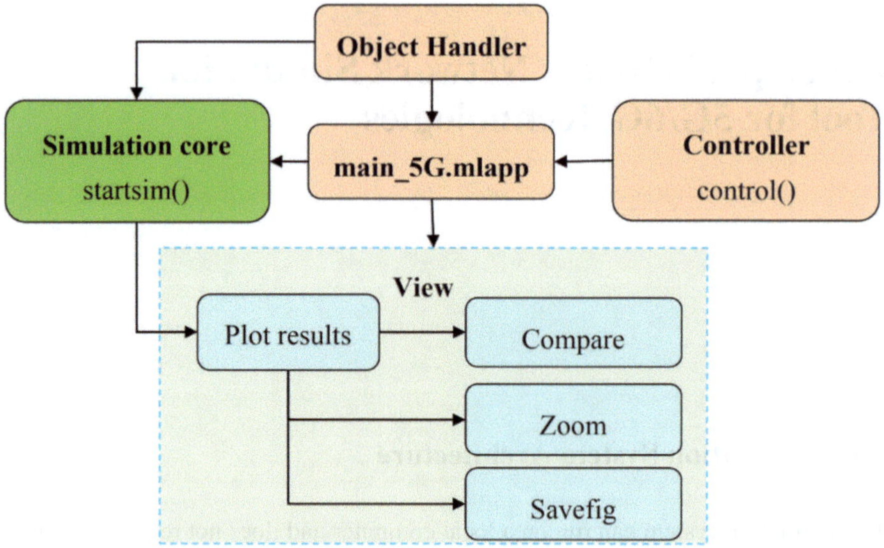

Fig. 3.1 Complete architecture of the simulation system

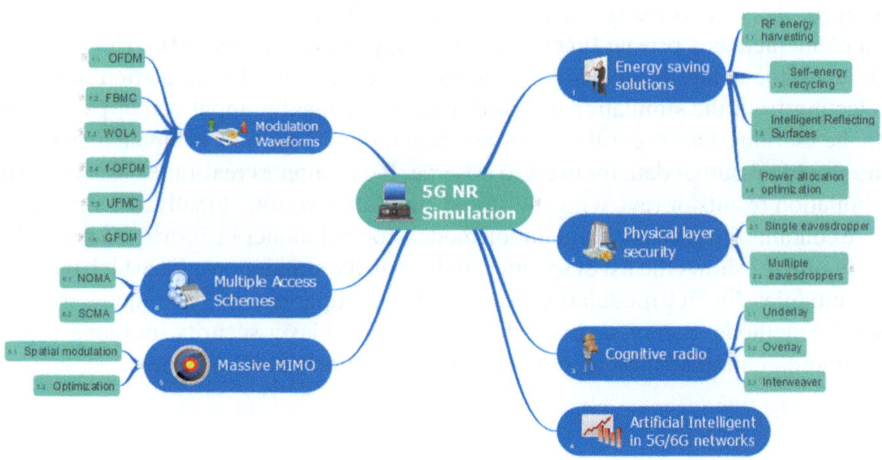

Fig. 3.2 List of simulation modules

mobile subscribers, and transmission channels are defined as classes. The simulation tool is divided into modules corresponding to different techniques, making it easy to update. When adding new simulation modules, programmers do not need to modify parameters in the existing modules. The simulation sequence for a specific module is presented in Fig. 3.3.

The input parameters are defined according to the class structure, which includes simulation configuration parameters, modulation parameters, channel parameters,

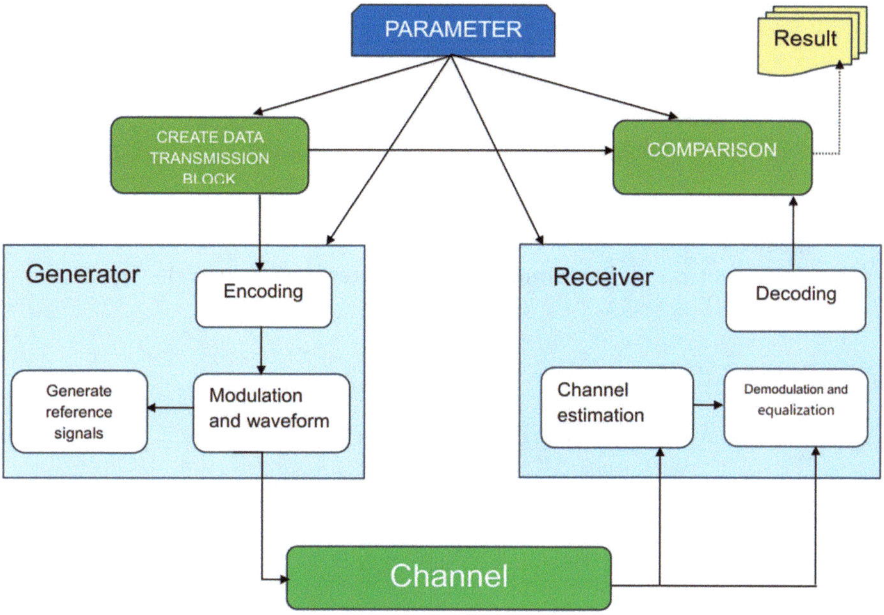

Fig. 3.3 General simulation procedure

multiple access parameters, network topology parameters, channel encoding parameters, feedback channel parameters, etc., as illustrated in Fig. 3.4. In this figure, the blue-colored blocks represent parameter groups that are currently set to default in this initial version, while the green blocks represent parameter groups that users can input from the UI.

The class structure of the input parameters is shown in the following program codes:

```
classdef SimulationParameters < handle
    % this class contains all simulation parameters, including
        user defined
    % parameters and constants

    properties
        simulation      % general simulation parameters
        topology        % network topology, describing which
            nodes are connected to each other
        schedule        % the time constant user allocation
        modulation      % modulation related parameters
        phy             % physical parameters
```

```matlab
        coding          % channel coding parameters
        channel         % parameters related to the channel model
        constants       % physical constants
        feedback        % feedback related parameters
        layerMapping
    end

    methods
        function obj = SimulationParameters( scenario )
            % simulation parameters class constructor
            % ...
        end %function SimulationParameters

        function dependentParameters(obj)
            % this method calculates all parameters that are
                dependent on
            % other parameters
            % ... code is not shown here
        end

        function checkParameters(obj)
            % this method check whether the choice of parameters
                is
            % consistent and allowed
            %...
        end

        function loadScenario(obj, scenario)
            % this method loads parameters from a scenario file
            %...
        end

        function UpdateSweepValue(obj,iSweep)
            % this method overwirtes a specific object property
                with the
            % current sweep variable value
            %...
            end

        function setConstantParameters( obj )
            % defines constant and pre-defined parameters that
                should not
            % be changed
```

```
            %...
        end %setConstantParameters

        function Links = initializeLinks(obj, Links, BS, UE)
            % this function  initializes all links according to some
            % scenario setting
            %...
        end %generateLinks

        function checkLinks(obj, Links)
            % this method checks if parameters of the generated links are
            % consistent and correct
            %...
        end %checkLinks
    end %methods
end %classdef
```

3.3 5G Modulation Waveforms Module

The first module of the simulation program is to simulate the currently proposed modulation waveforms for 5G networks. The input parameters for this module include the following:

- Simulation parameters: transmit power, sampling frequency, and number of iterations.
- Communication model parameters: center frequency of the carrier wave, number of subcarrier waves, spacing between subcarrier waves, filter parameters, modulation scheme, number of transmit and receive antennas.
- Channel parameters: velocity of movement, distance between nodes, exponential decay factor, Doppler model, power delay profile model, channel distribution parameters.

To implement this module, we can follow the flowcharts depicted in Figs. 3.5 and 3.6. After inputting the data, memory is allocated to store the values of transmit power and power spectral density (PSD). Then, the Monte Carlo simulation is conducted as follows. We continuously generates random channel gain values based on the given channel distribution. For each value of the channel gain, we can generate transmit data, perform signal modulation, calculate PSD, create noise corresponding to each SNR value, perform demodulation, and compare it with the original data to update the bit error rate (BER). Finally, the output data is the average result over multiple iterations of the Monte Carlo simulation.

Fig. 3.4 List of simulation parameters

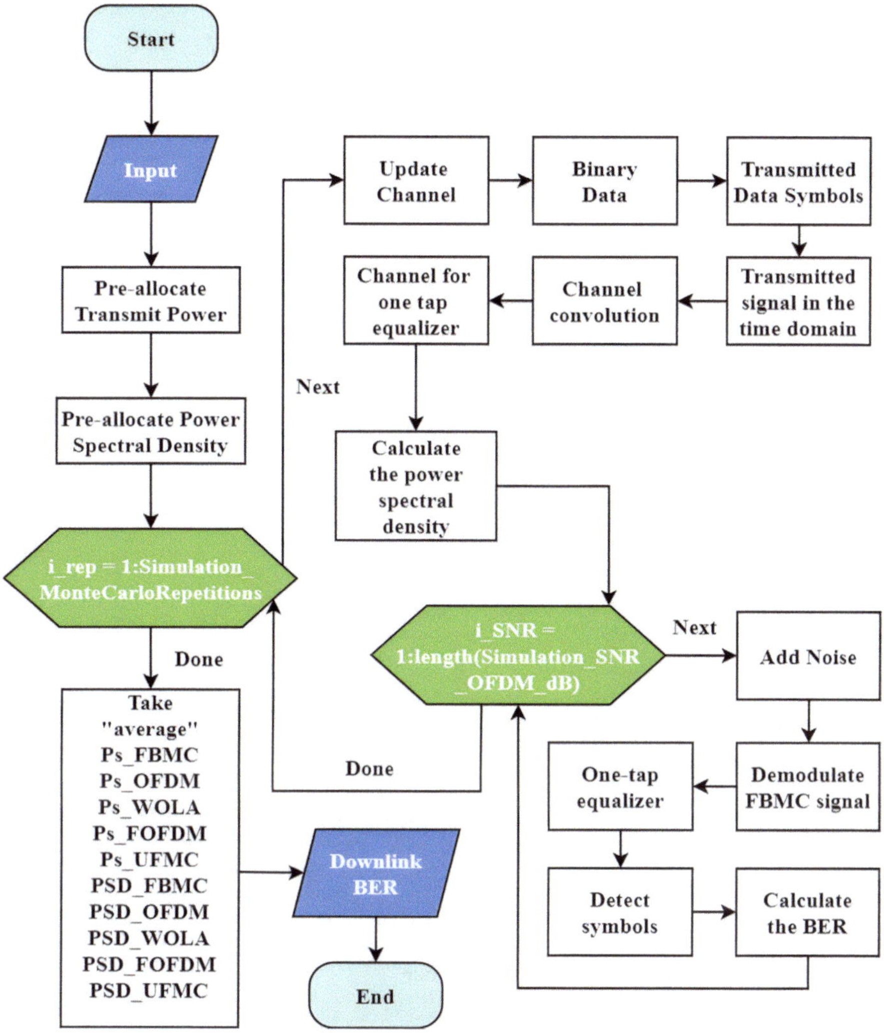

Fig. 3.5 Algorithm flowchart for simulating FBMC/OFDM/WOLA/f-OFDM/UFMC modulation waveforms

3.4 Multiple Access Technology Module

Two multiple access techniques, which are introduced for 5G/6G networks and simulated in this book, are NOMA and SCMA. For the NOMA module, the input data consists of the following:

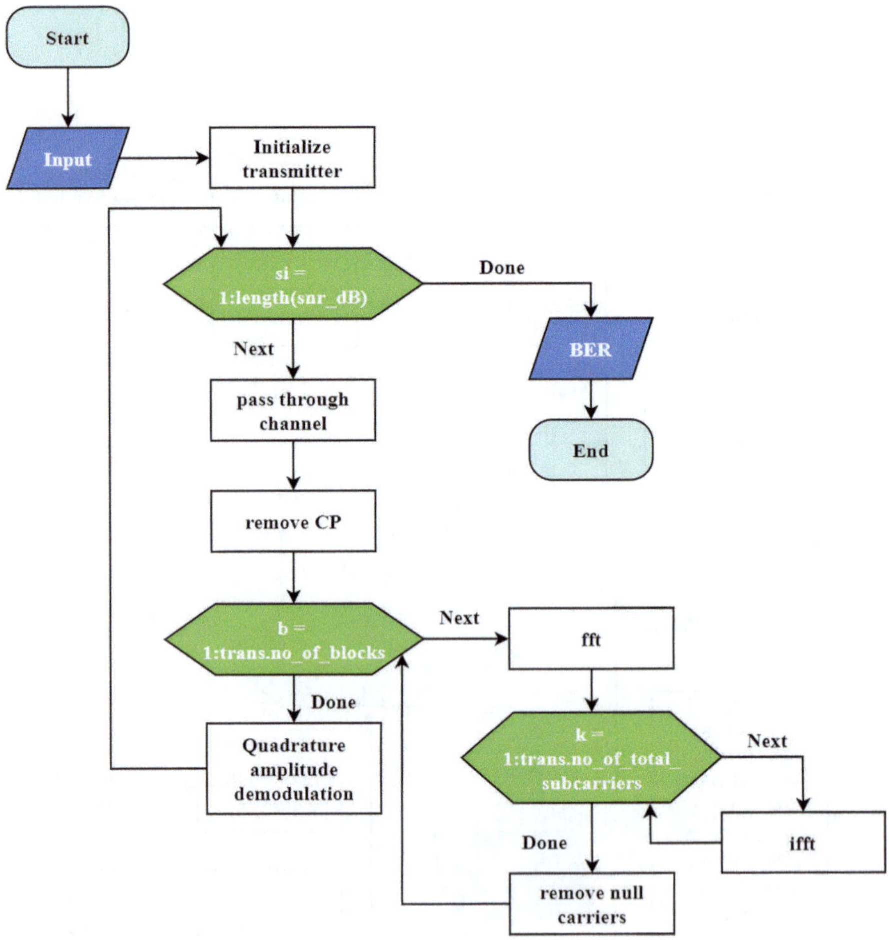

Fig. 3.6 Algorithm flowchart for GFDM modulation

- Simulation parameters: transmit power at the base station (BTS) and user equipment (UE), sampling frequency, number of data frames, and transmission direction (downlink or uplink).
- Communication model parameters: modulation waveform, center frequency of the carrier wave, number of subcarrier waves, spacing between subcarrier waves, filter parameters, modulation scheme, number of transmit and receive antennas at the base station (BTS) and user equipment (UE).
- Channel parameters: similar to the parameters in the 5G modulation waveform module.
- Network configuration parameters: list of nodes and their connections.

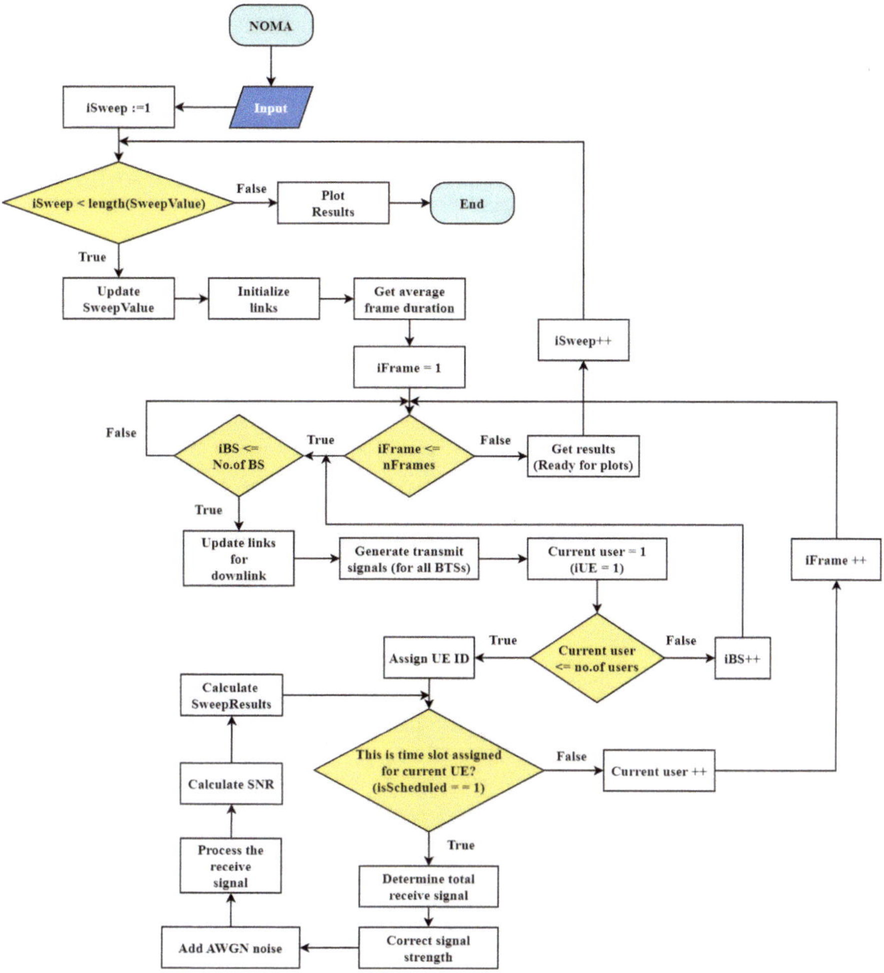

Fig. 3.7 Flowchart for simulating NOMA technique algorithm

This module is programmed according to the flowchart in Fig. 3.7 (example for downlink). After inputting the simulation parameters and the sweep data, the program iterates through each value of the sweep data until the end. For each sweep value, the program initializes the communication links according to the input network topology and processes each data frame sequentially. Transmit data is generated for all BTS stations, then the program runs through each user. Each user is assigned an ID, then the program checks whether the current time slot is assigned to the current user (isScheduled == 1) or not. If yes, the program proceeds to calculate the total received signal power, adjust the received signal power, add AWGN noise to the signal, process the received signal, calculate SNR, and count the number of bit errors. Finally, BER is calculated for each user and the results are plotted in the output graphs.

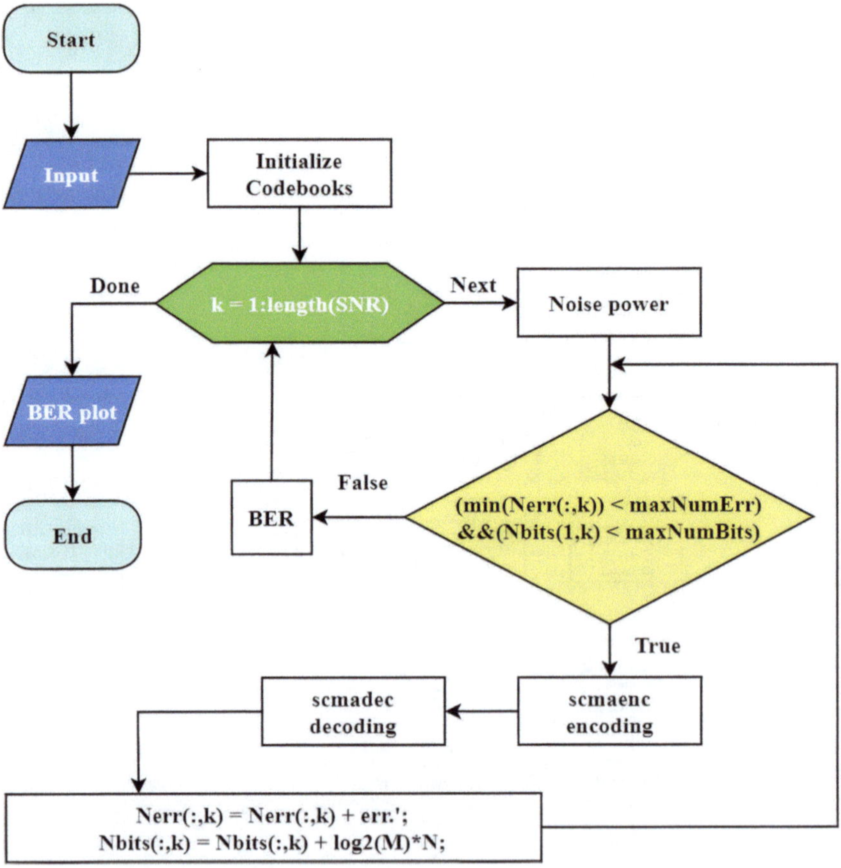

Fig. 3.8 Flowchart for simulating SCMA algorithm

For the SCMA module, the input data consists of parameters similar to the NOMA module, but in this book we only present the simulation for downlink channel and use the default selected modulation scheme. Additionally, users can input the number of SCMA signals in one frame and the maximum number of bit errors before terminating the simulation. This module is programmed according to the flowchart in Fig. 3.8. After inputting the data, the program scans through each input SNR value and performs data generation, SCMA encoding, noise generation and adding to the signal, SCMA decoding, and then counts the number of bit errors. The program runs until the number of transmitted bits or the number of bit errors exceeds the predefined threshold. At that point, the program updates the BER value and outputs it to the screen.

3.5 Massive MIMO Module

For the Massive MIMO module, we introduce two case studies: one is the optimization of spectral efficiency in a massive MIMO system, and the other is the simulation of spatial modulation (SM) technique.

For the Spatial Modulation module, the input data includes the following parameters:

- Simulation parameters: transmit power at the BTS, number of statistical samples, and maximum number of symbol errors.
- Communication model parameters: include modulation scheme, number of transmit and receive antennas at the BTS and UE.
- Channel parameters: similar to the NOMA module but without the mobility speed.

Figure 3.9 presents the algorithm flowchart for implementing this module. The flowchart consists of 2 loops scanning through each SNR value, one for calculating the BER for the SM method and one for calculating the BER for the differential spatial modulation (DSM) method. For each SNR value, Monte Carlo simulation is performed. Binary data is generated continuously, as well as the random fading channel coefficients. The noise power is calculated from the input SNR value, then the program generates AWGN noise with the corresponding power and adds it to the signal. Finally, SM or DSM decoding is performed, and the number of bit errors is recorded until the total number of transmitted bits or the total number of bit errors exceeds the limit. At that point, the program calculates the BER of the two methods and outputs the results visually.

The module for simulating the optimization of spectral efficiency in massive MIMO systems includes the following input parameters:

- Simulation parameters: transmit power at the BTS, number of statistical samples, transmission direction (uplink or downlink), number of channel realizations.
- Communication model parameters: include modulation scheme, number of BTSs and UEs, number of transmit and receive antennas at the BTS and UE, bandwidth, cell size, power control parameters, pilot reuse factor, angle deviation for each transmission path.
- Channel parameters: maximum LOS distance, maximum distance between BTS and user, channel statistical parameters.

Figure 3.10 illustrates the algorithm for simulating the optimization of spectral efficiency in massive MIMO systems. Users can select different channel estimation methods such as LS, MMSE, and EWMMSE, as well as choose the parameters they want to investigate (sweep parameter). The program will generate a large number of random channel gain values, from which it determines the performance metrics such as bit error rate, spectral efficiency according to the proposed optimization formulas, and outputs the corresponding graphs.

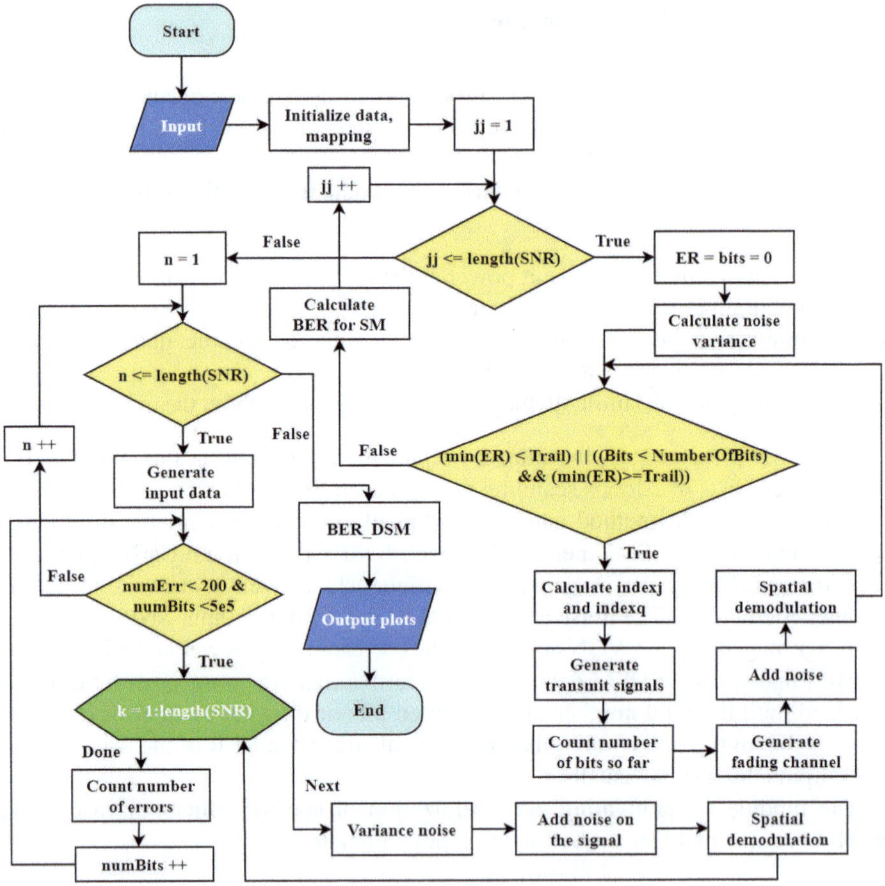

Fig. 3.9 Algorithmic flowchart for Spatial Modulation simulation

3.6 Module for Energy-Saving Solutions

3.6.1 RF Energy Harvesting

For the RF Energy Harvesting module, the input data includes parameters as illustrated in Fig. 3.11. The simulation configuration parameters consist of the number of samples, required data rate, transmitter power, and noise power. The channel parameters encompass the statistical channel model and the path loss exponent. Communication parameters involve modulation and transmission methods. Energy harvesting parameters comprise EH model, EH protocol, time-switching and/or power-splitting factors, and energy conversion efficiency.

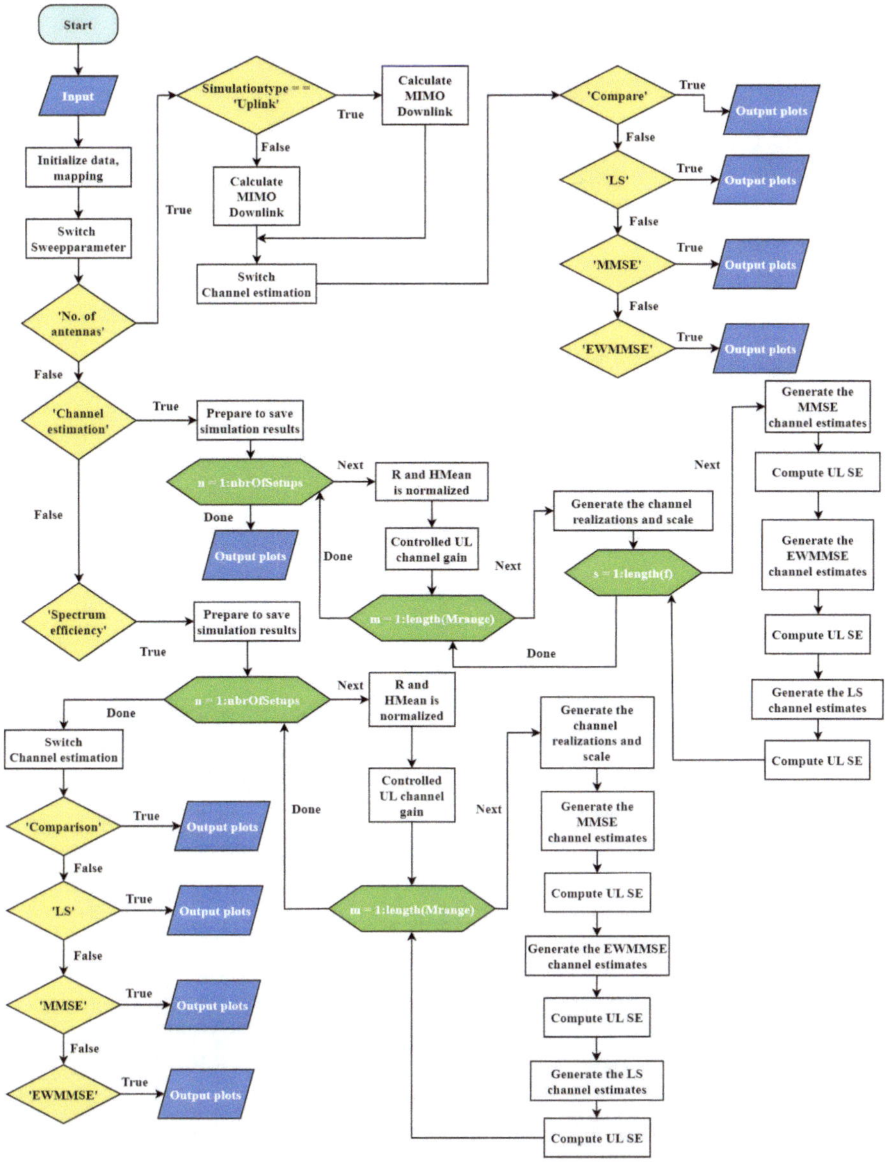

Fig. 3.10 Algorithm for simulating spectral efficiency optimization in Massive MIMO systems

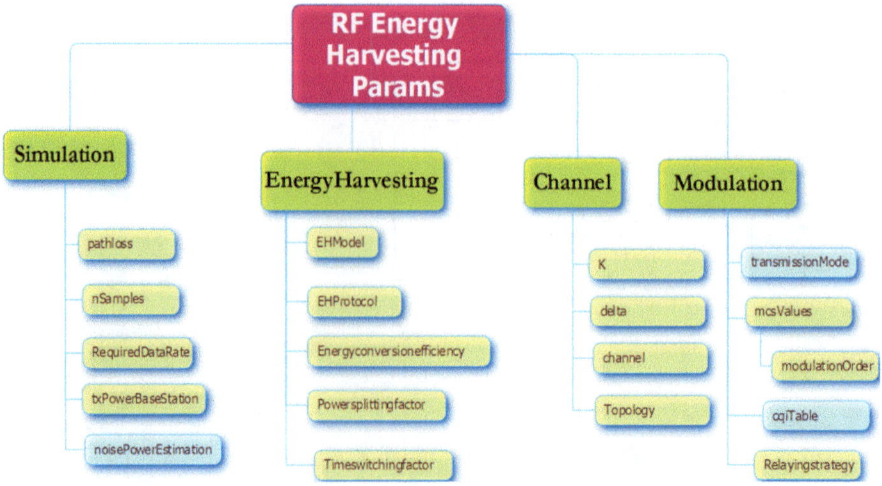

Fig. 3.11 Input parameters for the RF energy harvesting module

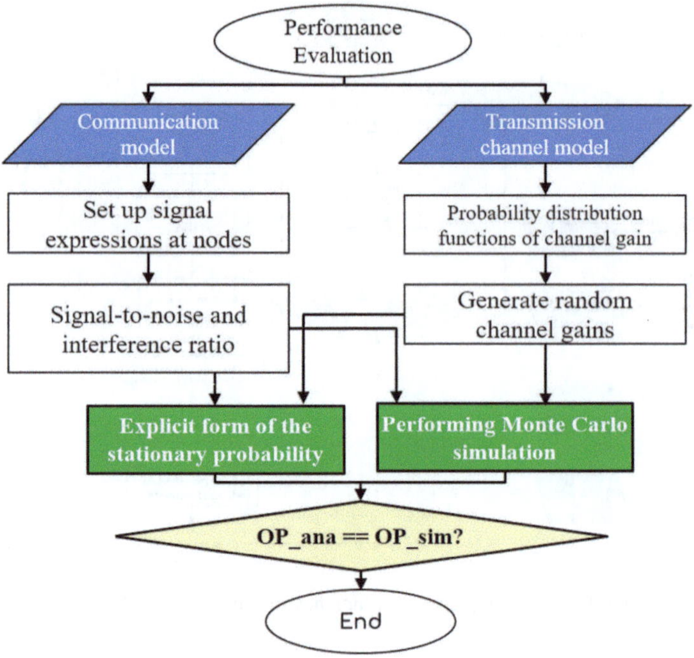

Fig. 3.12 Monte Carlo simulation method

Figure 3.12 presents the flowchart of a general Monte Carlo simulation algorithm for evaluating performance metrics related to probability and statistics. The algorithm begins by collecting user-input data, which includes parameters of the communica-

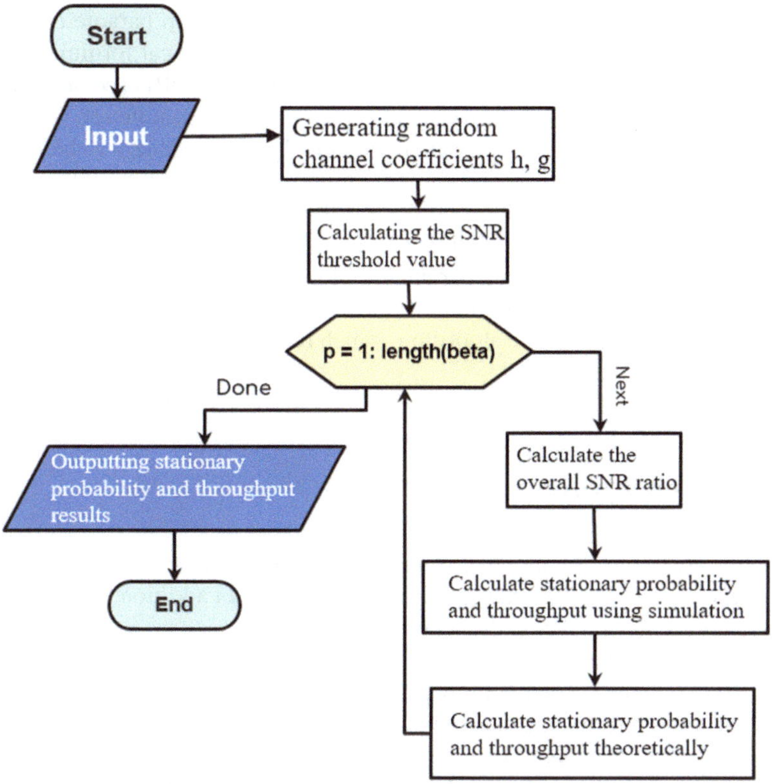

Fig. 3.13 Algorithm flowchart for simulating the RF energy harvesting

tion model and channel model. Communication model parameters encompass transmission protocols (DF, AF, or hybrid) and energy harvesting protocols (TSR, PSR, or combined), energy conversion efficiency, time-switching and/or power-splitting factors, transmit signal-to-noise ratio $\frac{P_S}{N_0}$, and desired data transmission rate. The channel model includes the statistical fading channel model, path loss exponent, distance between nodes, level of hardware noise, etc. Additionally, users can input the number of Monte Carlo simulations.

Figure 3.13 depicts a specific simulation algorithm for the RF energy harvesting model, where the impact of the power-splitting factor is investigated. Based on the channel model parameters, the algorithm generates random channel coefficients according to the given distribution. Subsequently, the algorithm select a sweep parameter (to plot the outage probability versus this parameter) by checking which parameters among the input variables such as $\frac{P_S}{N_0}$, data rate R, power-splitting/time-switching factors (ρ or α) are entered in vector form.

After that, we perform a loop for each value of the sweep parameter. For each iteration, we calculate the OP in two ways: using the theoretical formula (based on Eq. (7.8)) or through Monte Carlo simulation. To estimate the OP based on simulation, we generate t sets of random channel gain values. For each sample set, we calculate the end-to-end SINR ratio and compare it with the threshold value γ (which depends on the data rate R). The ratio of the number of times the end-to-end SINR is below γ to the total number of simulation samples t is the outage probability for the current iteration. From the outage probability, we can also calculate the system throughput.

Finally, the program outputs a parametric plot of either OP or throughput against the scanning parameter. Both simulation and theoretical graphs must closely coincide to ensure the correctness of the algorithm, which requires a sufficiently large number of simulations (denoted as t).

3.6.2 *Self-energy Recycling*

The input parameters for the energy recycling module are presented in Fig. 3.14, which are similar to the RF energy harvesting module. In addition, there is also the self-interference noise level at the relay node.

The self-energy recycling simulation algorithm is introduced in Fig. 3.15a for the i.n.d. channel scenario. The main program consists of data input commands and the calls of four functions to calculate the outage probabilities both by analysis and simulation for the partial relay selection (PRS) and full relay selection (FRS) methods, as illustrated in Fig. 3.15a. Figure 3.15b, c, d, and e depict the flowcharts of each function mentioned above. For example, in Fig. 3.15b, the program is repeated for

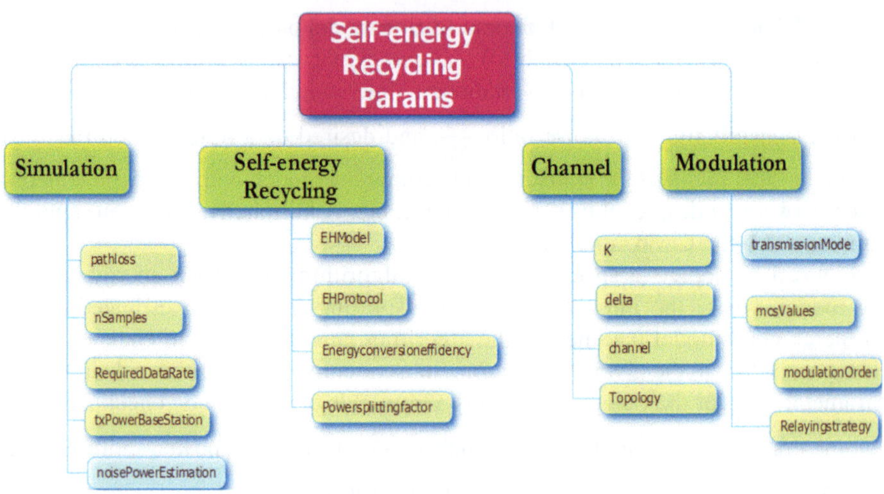

Fig. 3.14 Input parameters for simulating the RF energy harvesting

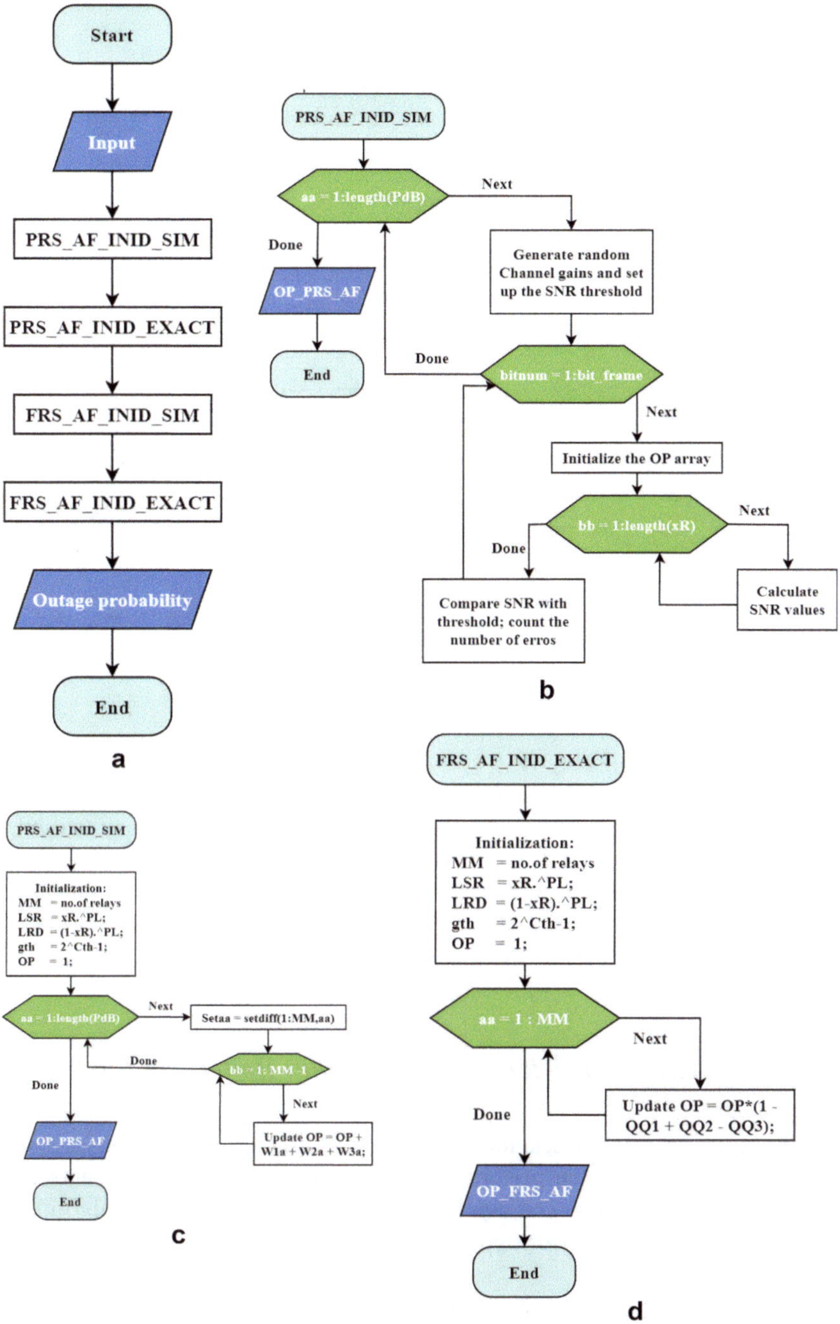

Fig. 3.15 **a** Flowchart for main program. **b** Flowchart for simulating PRS method in i.n.d. scenario. **c** Flowchart for calculating exact OP of PRS method in i.n.d. scenario. **d** Flowchart for calculating exact OP of FRS method in i.n.d. scenario.**e** Flowchart for simulating of FRS method in i.n.d. scenario

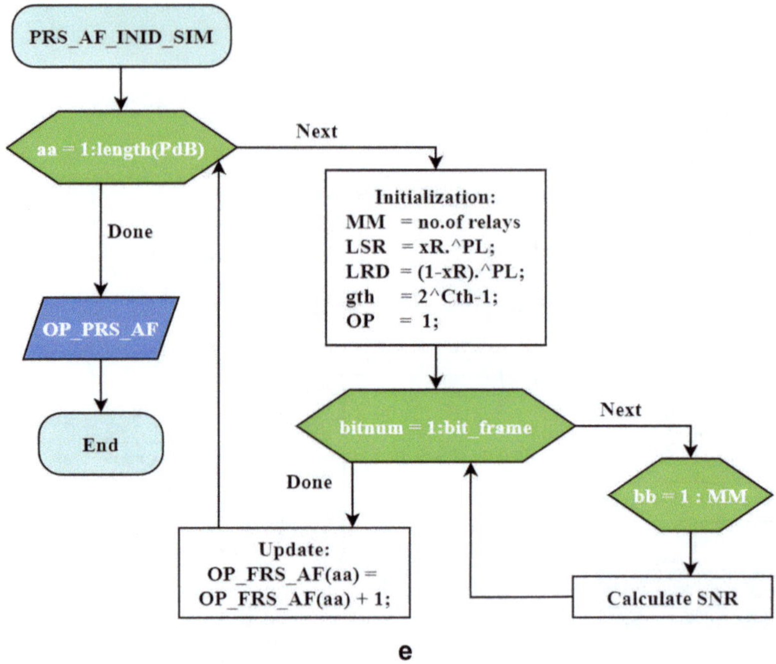

Fig. 3.15 (continued)

each transmission power value. For each PdB value, the program generates random channel samples and determines the SNR threshold. With each transmitted bit, the program calculates the SNR at the relays, selects the best relay, initializes a vector to store outage probabilities, compares the end-to-end SNR with the threshold to calculate the outage probability, and counts the number of bit errors. After sweeping through all PdB values, the resulted outage probability is returned.

3.6.3 Zero-Energy Device: The Intelligent Reflecting Surface

The use of zero-energy devices, including the Intelligent Reflecting Surfaces (IRS), as a solution to enhance the performance of wireless communication networks is investigated in this section. The input parameters and the simulation algorithm for evaluating IRS-assisted network performance are presented in Figs. 3.16 and 3.17, respectively.

Similar to the cases above, Fig. 3.17 illustrates the algorithm flowchart for IRS simulation. The algorithm begins by collecting user-input data. Communication model parameters include modulation method, number of simulation samples, the number of reflecting elements, transmit-power-to-noise ratio $\frac{P_S}{N_0}$, and desired data transmis-

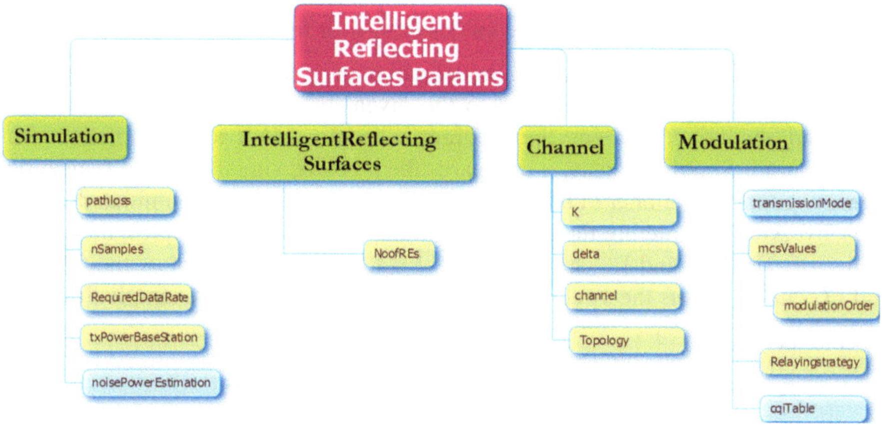

Fig. 3.16 Input parameters for the IRS simulation module

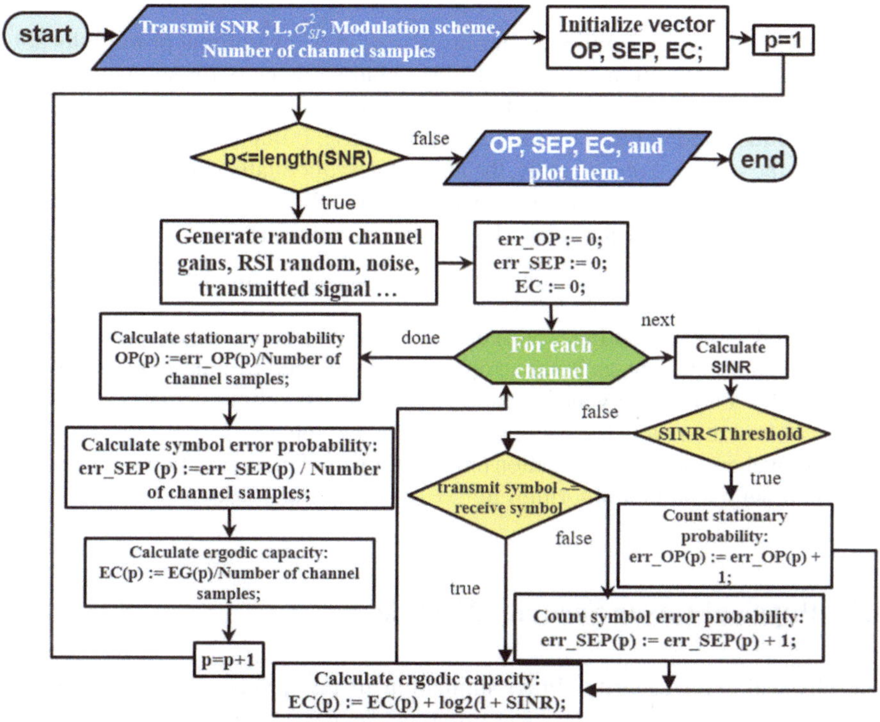

Fig. 3.17 Algorithm flowchart for simulating IRS

sion rate. The channel model comprises the statistical fading channel model, path loss exponent, distance between nodes, and residual self-interference level.

The program sweeps through the values of $\frac{P_S}{N_0}$ (or the sweeping parameter input from the UI). Based on the channel model parameters, the algorithm generates random channel coefficients according to the given distribution functions. For each generated channel coefficient, the program calculates SINR and compares it with the threshold. Simultaneously, it compares the received signal with the transmitted signal to count the number of bit errors, and then calculates the accumulated ergodic capacity. After completing the loop on the number of channel samples, the average values of OP, SEP, and EC are computed. Finally, the program plots the OP, SEP, and EC curves versus the sweeping parameter (in the illustrated example flowchart, it is plotted versus $\frac{P_S}{N_0}$).

3.6.4 *Power Allocation Optimization*

Another energy-saving solution under investigation is the optimal power allocation approach for nodes within the network. The input parameters for simulating a case study for this technique are depicted in Fig. 3.18, with the corresponding algorithmic flowchart illustrated in Fig. 3.19 for an AF relay node. Initially, user-provided data is input into the system. Subsequently, the algorithm sequentially performs calculations for the relay-user distance, channel gain between the BTS and the relay, and between the relay and the user equipment (UE). Then, it calculates $A = \sum \gamma_k$, initializes the value τ_0, τ_k, EE. Let $k = \frac{\tau_0^{AF}}{1 - \tau_0^{AF}}$. Next, we sequentially solve for the optimal values (single-variable optimization) of k, P_u^{AF} and P_d^{AF}, resectively, using a search algorithm. Subsequently, an iterative optimization method to continuously update k, P_u^{AF} and P_d^{AF} until convergence occurs. Finally, the algorithm calculates the resulted τ_0, τ_k, EE, and then plots the energy performance graph.

The flowchart for power allocation optimization in a DF relay network is illustrated in Fi. 3.20, with a similar algorithm for AF shown in Fig. 3.19.

3.7 Physical Layer Security Module

The physical layer security (PLS) simulation in this project is divided into two categories: one with only one eavesdropping device and the other with multiple eavesdropping devices. In this book, we provide a case study for the case with two eavesdropping devices, where the input data includes the following:

- Simulation parameters: transmit power of the BTS, the number of channel samples, the transmit power of the jammer, the number of switching nodes and jammer nodes, and the secrecy capacity threshold.
- Communication model parameters: modulation method, relaying strategy.

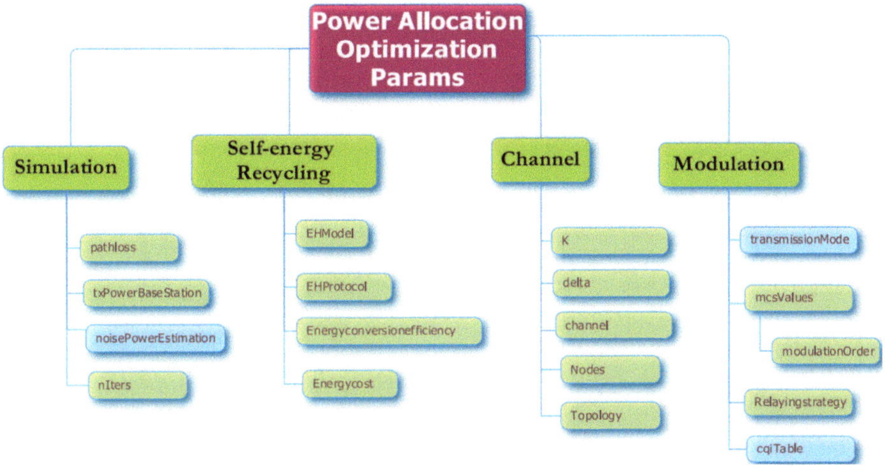

Fig. 3.18 Input parameters for the power allocation optimization module

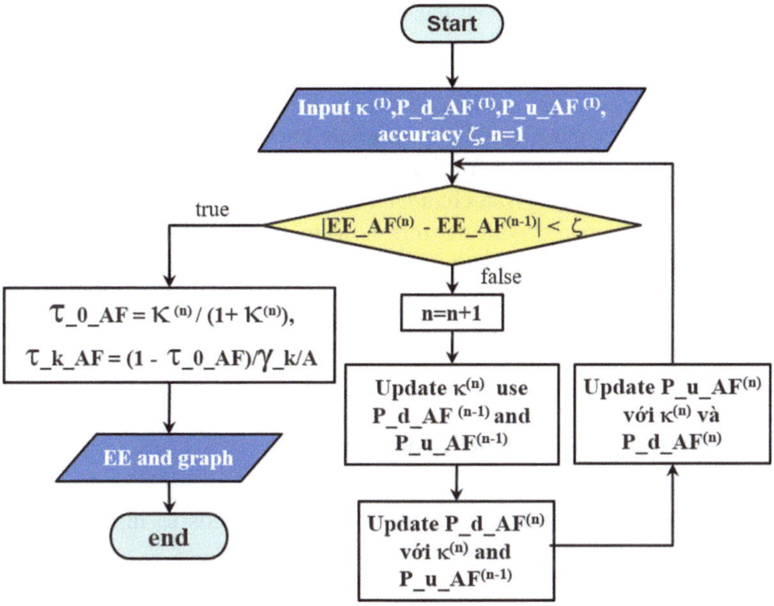

Fig. 3.19 Simulation algorithm flowchart for the power allocation optimization algorithm

- Channel and topology parameters: distance between sources and relay, path loss exponent, and statistical parameters of the channels.
- Energy harvesting model parameters: model type, protocol, power splitting and time-switching factors, energy conversion efficiency.

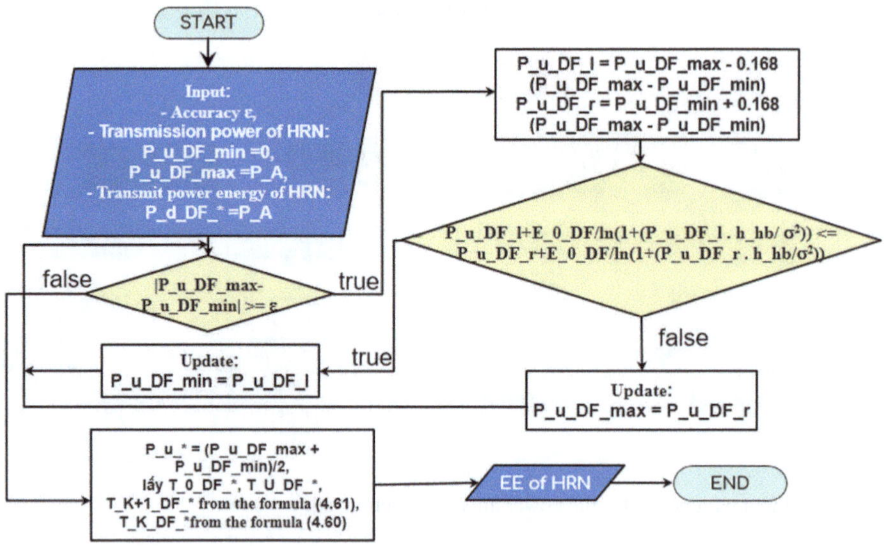

Fig. 3.20 Algorithm flowchart for evaluating the PLS scheme

The flowchart in Fig. 3.21 explains the steps of implementing the PLS simulation algorithm. After data input step, the program iterates through each value of the sweeping parameter. For each iteration, channel gain coefficients h_{ar}, h_{rb}, h_{ae}, and h_{je} are generated from the given channel distribution. Subsequently, the SNR ratio is calculated and compared with the threshold to compute the OP and the average secrecy capacity (ASC) by using the Monte Carlo method. OP and ASC values for each sweeping parameter are also calculated using analytical expressions. Finally, the program plots the OP and ASC versus the sweeping parameter.

3.8 Cognitive Radio Simulation Module

For the cognitive radio simulation module, the user needs to input the following parameters:

- Simulation parameters: base station transmit power, number of simulation samples, and required data rate.
- Channel parameters: network topology, distance between nodes, path loss exponent, and the distribution model of channel gain.
- Communication model parameters: modulation method, cognitive radio model parameters, including the maximum interference level at primary users and the noise levels at secondary transceivers.

The simulation flowchart for this module is illustrated in Fig. 3.22.

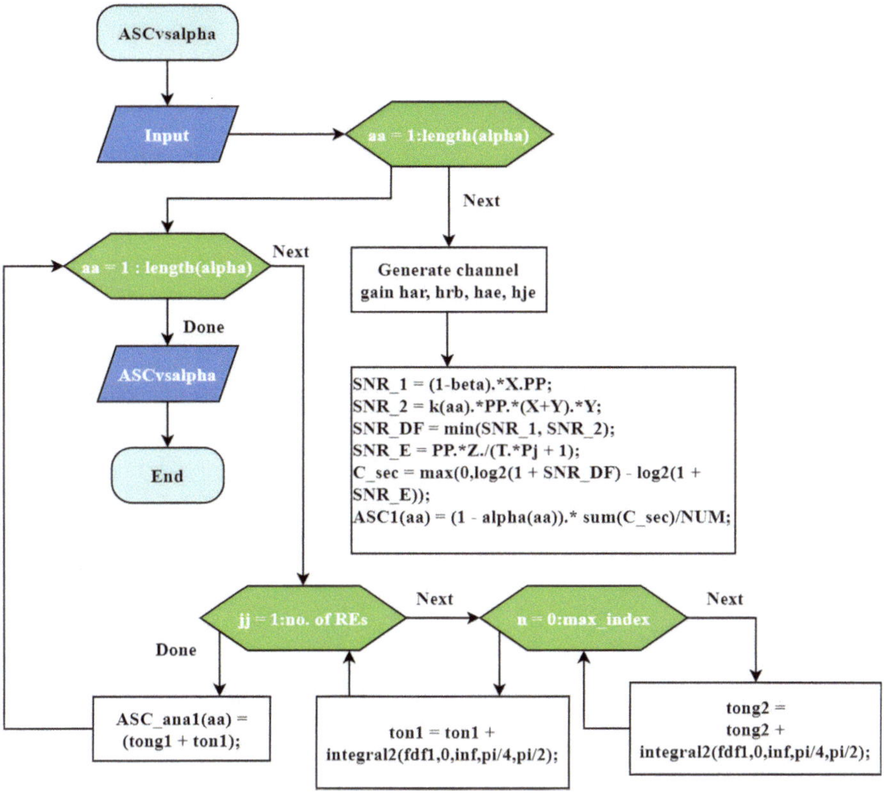

Fig. 3.21 Simulation flowchart for physical layer security

3.9 Artificial Intelligence Application Module

For the AI application in wireless communications, the simulation module requires the following input parameters:

- Simulation parameters: number of data frames, number of simulation samples per frame, and SNR sweep data.
- Channel parameters: number of users. number of mmWave links, number of fading channel realizations, and the distribution model of channel gain.
- Communication model parameters: number of antennas, modulation method, CRC length.
- AI parameters: number of training epochs for each symbol, bias value, variance of dithering.

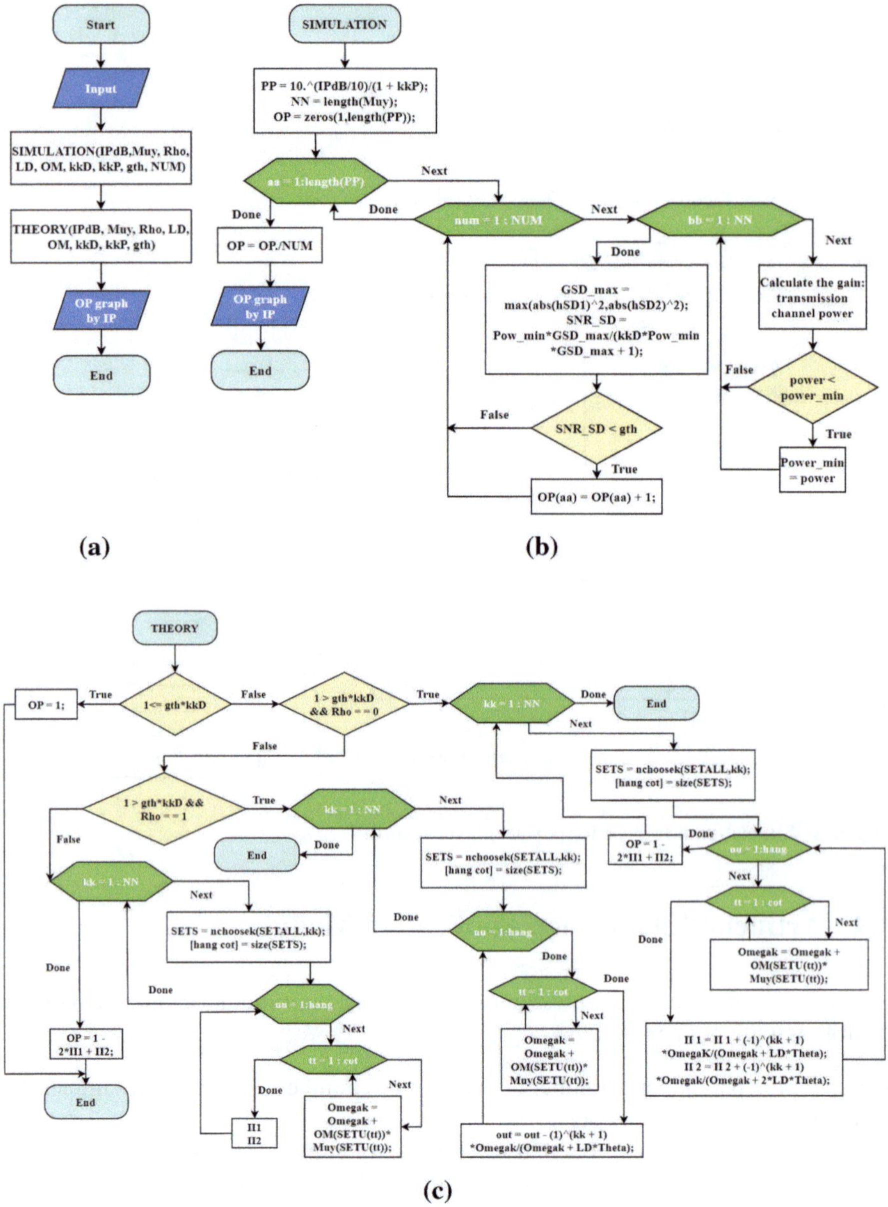

Fig. 3.22 The simulation algorithm flowchart for cognitive radio: **a** Main program flowchart; **b** Monte Carlo simulation flowchart; **c** Theoretical calculation flowchart

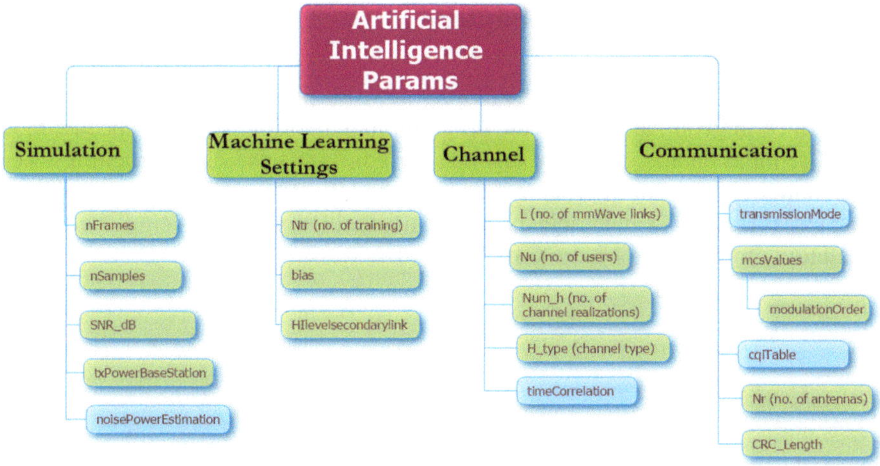

Fig. 3.23 Input parameters for artificial intelligent module

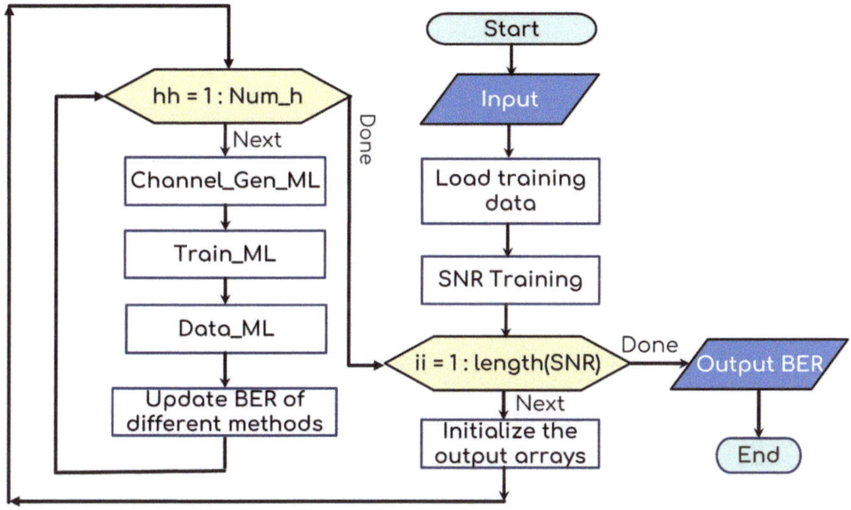

Fig. 3.24 The simulation algorithm flowchart for artificial intelligence applications in wireless communications

Figure 3.23 shows the input parameters for this module. The module's flowchart is presented in Fig. 3.24.

3.10 Conclusion of the Chapter

In this chapter, the structure of the 5G/6G network physical layer simulation tool is described in detail, along with the programming algorithms for the specific modules. The input data is generally organized according to the class structure, where each element of the information system, such as UE, BTS, and the communication channels, is defined as data objects. The main simulation method employed in this simulation project is the Monte Carlo method.

Chapter 4
Modulation Waveforms for 5G Networks

4.1 Theoretical Background

4.1.1 FBMC

4.1.1.1 Concept

The exploration of alternatives to OFDM in the context of 5G networks is an area of significant interest, with Filter Bank Multicarrier (FBMC) emerging as a promising candidate. The elimination of the cyclic prefix (CP), a notable limitation in OFDM, can be effectively achieved through pre-IFFT filtering operations utilizing bandpass filters on data streams. FBMC, akin to OFDM, segments bandwidth into multiple sub-channels and employs Offset Quadrature Amplitude Modulation (OQAM). This strategic integration circumvents the need for CP while maximizing bit rate efficiency.

FBMC represents an evolved iteration of OFDM by substituting traditional multi-carrier systems with those based on bandpass filters; specifically replacing IFFT plus CP with synthesis filter banks (SFB), and FFT plus CP with analysis filter banks (AFB). Table 4.1 delineates primary distinctions between these techniques across various attributes such as CP presence, sidelobe behavior, synchronization requirements, Doppler effect resilience, MIMO compatibility, spectrum sensing capabilities, and computational complexity.

Some disadvantages of OFDM include reduced spectral efficiency caused by CP, high spectral leakage from the effect of rectangular pulses, and interference between asynchronous signals in adjacent bands. The FBMC technique overcomes these limitations of OFDM by adding generalized pulse shaping filters, providing well-contained sub-channels in both time and frequency domains. Consequently, FBMC systems occupy more spectrum, refer to the signals, and offer more efficient utilization of radio resources without the need for CP.

P. T. Tran et al., *Developing MATLAB Comprehensive Simulation for 5G/6G Wireless Communications*, Signals and Communication Technology,
https://doi.org/10.1007/978-981-95-1117-4_4

Table 4.1 Comparison between OFDM and FBMC

Nature	OFDM	FBMC
Dependence on CP	The cyclic prefix is mandatory in OFDM, which consumes bandwidth	The cyclic prefix is not mandatory, preserving bandwidth
Sidelobes	Large sidelobes and interference	Small sidelobes
Synchronization	For accurate detection, multiple access interference (MAI) cancellation should be performed at the receiver	MAI is suppressed due to excellent frequency localization of subcarriers
Doppler Effects	Highly sensitive to carrier frequency offset	Less sensitive, thus performing significantly better with increased user mobility
MIMO Systems	High flexibility in applying MIMO techniques	Limited flexibility
Spectrum Sensing	Low spectrum sensing performance due to spectral leakage effects	High spectrum sensing performance
Computational Complexity	Very low computational complexity	High computational complexity

4.1.1.2 Techniques in FBMC

The system necessitates precise synchronization between the receiver and the transmitter. This requirement arises from the multi-path channel's impulse response, which causes the overlapping of multicarrier symbols at the input of the receiver, thereby prohibiting demodulation using only FFT due to a loss of orthogonality caused by ISI. The OFDM approach to mitigate this issue involves a CP insertion. Alternatively, one can maintain identical symbol time and duration while incorporating additional processing in the IFFT/FFT stages.

In this configuration, the bandpass filter employed at the transmitter is termed a synthesis filter, whereas its counterpart at the receiver is known as an analysis filter. As illustrated in Fig. 4.1, initially, the input signal undergoes serial-to-parallel conversion before being processed by the synthesis filter and subsequently reverted to serial form. On reaching the receiver side after traversing through the synthesis filter and passing through the channel, the signal is converted to parallel form again by using serial-to-parallel conversion followed by analysis filtering. It is ultimately converted back into its original serial format upon reception.

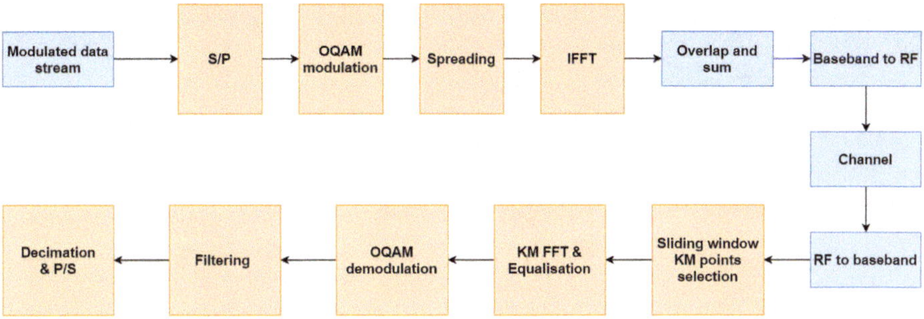

Fig. 4.1 FBMC system diagram

4.1.1.3 OQAM

For FBMC method, the separation of sub-channels allows for flexibility in modulation methods. Specifically, when only even or odd indexed sub-channels are utilized, QAM can be employed due to the absence of overlapping modulation. Nevertheless, this approach does not achieve full transmission rate. The primary objective of FBMC is to leverage all available sub-channels effectively; thus, modulation techniques must address the frequency-domain overlapping between adjacent sub-channels.

Two critical characteristics of FBMC merit attention. First, orthogonality is essential to mitigate the problem of overlapping among neighboring sub-channels. This is achieved through an IFFT module that compensates by processing even-indexed inputs as real components and odd-indexed inputs as imaginary components.

To maximize power efficiency by increasing the symbol rate per sub-channel by two times, an alternating use of real and imaginary parts in the IFFT process is implemented. Consequently, rather than transmitting both components simultaneously, the imaginary part experiences a delay equivalent to half of a symbol duration relative to its real counterpart. This strategic alternation ensures optimal utilization of each complex data symbol's potential within FBMC frameworks.

4.1.1.4 Advantages and Disadvantages of FBMC

One of the most important advantages of FBMC is the avoidance of inserting a CP. This leads to full bit rate achivement and better spectral efficiency. FBMC can provide a more efficient system with better spectral selectivity and robust narrowband interference suppression.

However, one of the disadvantages to note is the increased complexity of implementing FBMC compared to OFDM, due to the need to use bandpass filters in the receiver. The complexity depends on the number of sub-channels M in the system and also on the overlapping factor K. Designing wideband and high dynamic range systems with FBMC presents significant RF development challenges.

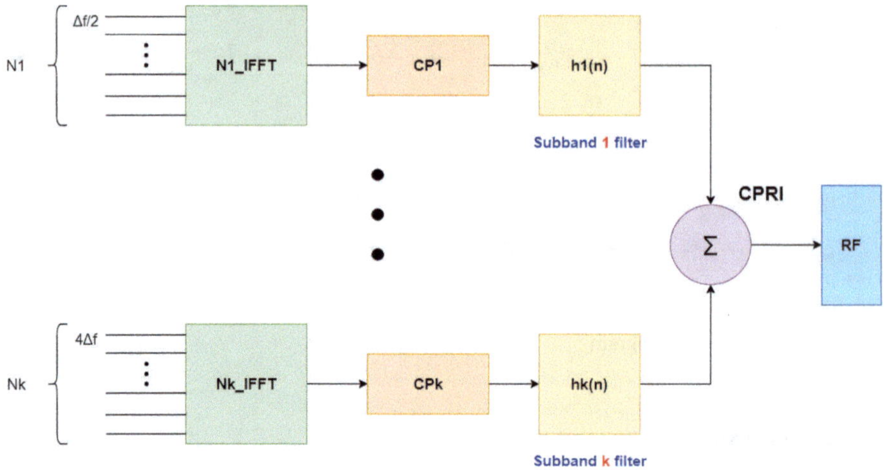

Fig. 4.2 The diagram of the F-OFDM transmitter

4.1.2 f-OFDM

Figure 4.2 depicts the F-OFDM transmitter. In conventional OFDM, the entire band is composed of a single block, and the frequency spacing between each subcarrier is identical (denoted by Δf). In contrast, in F-OFDM, the entire bandwidth is divided into multiple sub-blocks (sub-bands). The subcarrier spacing within each sub-band varies (for example, the subcarrier spacing for sub-band N_1 is $\Delta f/2$, and the subcarrier spacing for sub-band N_k is $4\Delta f$), and each sub-band has its own CP (the length of each CP can also vary). Each sub-band is also applied with its own filter.

Filters that can be used in F-OFDM technology include: soft truncation filters utilizing Hann or root-raised-cosin (RRC) windows, and equi-ripple filters using the Remez exchange algorithm.

F-OFDM offers an increased throughput compared with the conventional OFDM techniques. It significantly reduces the use of guard bands, leading to more efficient spectrum utilization. The main disadvantage is the complexity of the structure and implementation (Table 4.2).

4.1.3 GFDM

Generalized Frequency Division Multiplexing (GFDM) [1] is a promising multicarrier modulation technique gaining traction as a potential solution for next-generation wireless communication systems (5G and beyond). It addresses several shortcomings of the widely adopted OFDM while offering improved flexibility and spectral efficiency. The increased demand for high data rates, efficient spectrum utilization,

Table 4.2 Comparison between OFDM and f-OFDM

Specifications	OFDM	F-OFDM
Method of use	Orthogonal Frequency Division Multiplexing	Orthogonal Frequency Division Filtering
Filter granularity	Full bandwidth	Sub-band
Filter length	$\leq$ CP length	$\leq$ (1/2)*Symbol Duration
Orthogonal in the time domain	Orthogonal	Non-orthogonal
Orthogonal in the frequency domain	Orthogonal	Non-orthogonal

and low-latency communications necessitates advanced modulation schemes like GFDM, which is well-suited to handle the diverse requirements of heterogeneous network environments.

GFDM has garnered significant attention due to its advantageous properties, which are particularly relevant in the context of modern communication systems. As a generalization of OFDM, GFDM retains many beneficial characteristics of OFDM while effectively resolving its inherent drawbacks [2]. Notably, Farhang et al. [3] elucidated that the constraints posed by carrier aggregation in OFDM systems are mitigated through GFDM's implementation, primarily because it exhibits minimal out-of-band radiation.

Furthermore, GFDM employs circular filtering as opposed to the linear filtering utilized in OFDM. This methodological shift eliminates the transient intervals in the prototype filter and consequently reduces its latency—a critical factor for low-latency applications such as the IoT and machine-to-machine (M2M) communications. Additionally, GFDM demonstrates robustness against orthogonality loss and effectively suppresses the intercarrier interference (ICI) by employing a meticulously designed prototype filter for subcarrier processing. These attributes underscore GFDM's potential to enhance performance across various communication scenarios while maintaining rigorous standards of efficiency and reliability.

4.1.3.1 Concept and Methodology

GFDM operates on a block-based structure, with each block composed of multiple subcarriers and sub-symbols, creating a two-dimensional time-frequency representation of the transmitted data. Unlike OFDM's reliance on orthogonal subcarriers and rectangular pulse shaping, GFDM employs flexible pulse shaping and allows for non-orthogonal subcarrier configurations. This fundamental difference provides increased degrees of freedom for system optimization and adaptation. The flexible pulse shaping, achieved through prototype filter design, is a core aspect of GFDM's adaptability, enabling tailored characteristics to meet various bandwidth and latency demands. A tail-biting CP further enhances spectral efficiency and robustness compared to OFDM's fixed CP.

4.1.3.2 GFDM Transceiver

GFDM transceiver system model is shown in Fig. 4.3 (source: [4]). Each GFDM block employs K subcarriers, with each transmitting M complex-valued subsymbols, and therefore a total of $D = KM$ symbols are transmitted within one block.

For a GFDM block $d \in \mathbb{C}^D$, its mth subsymbol on the kth subcarrier $[\mathbf{d}]_{k+mK}$ is pulse-shaped by a vector $\mathbf{g}_{k,m}$, whose nth entry is $\left[\mathbf{g}_{k,m}\right]_n = \left[\mathbf{g}\right]_{<n-mK>_D} e^{j2\pi kn/K}, n = 0, 1, \ldots, D-1, m = 0, 1, \ldots, M-1, k = 0, 1, \ldots, K-1$, where $\mathbf{g} \in \mathbb{C}^D$ is called the prototype filter [2]. Let

$$A = \left[\mathbf{g}_{0,0} \cdots \mathbf{g}_{K-1,0} \quad \mathbf{g}_{0,1} \cdots \mathbf{g}_{K-1,1} \quad \mathbf{g}_{0,M-1} \cdots \mathbf{g}_{K-1,M-1}\right] \tag{4.1}$$

be the GFDM transmitter matrix [2] and $\mathbf{x} = \mathbf{Ad}$ be the transmit sample vector, whose nth entry is

$$[\mathbf{x}]_n = \sum_{k=0}^{K-1} \sum_{m=0}^{M-1} [\mathbf{d}]_{k+mK} \left[\mathbf{g}\right]_{<n-mK>_D} e^{j2\pi kn/K}. \tag{4.2}$$

Then, the vector $\mathbf{x}$ passes through parallel-to-serial (P/S) conversion and is further added a CP of length L, generating a GFDM digital baseband transmit signal $x[n]$.

Consider an N-tap wireless channel, which is a causal linear time-invariant (LTI) system with impulse response $h[n]$, where $h[n] = 0, n < 0$ or $n > N-1$. Accordingly, the baseband receive signal is $y[n] = h[n] * x[n] + w[n]$, where $w[n]$ is the complex AWGN with variance N_0. Denote $\mathbf{h} = [h[0]\, h[1] \ldots h[N-1]]^T$ and $\mathbf{w} = [w[0]\, w[1] \cdots : w[D-1]]^T$. Note that the combination of a CP and a linear convolution with a frequency-selective multipath channel can be modeled as a circular convolution [5]. Specifically, after CP removal and serial-to-parallel (S/P) conversion, the receive sample vector is obtained as

$$\mathbf{y} = \mathbf{Hx} + \mathbf{w} = \mathbf{HAd} + \mathbf{w} = \mathbf{W}_D^H diag\left(\mathbf{W}_D \mathbf{Ad}\right) Fh + w, \tag{4.3}$$

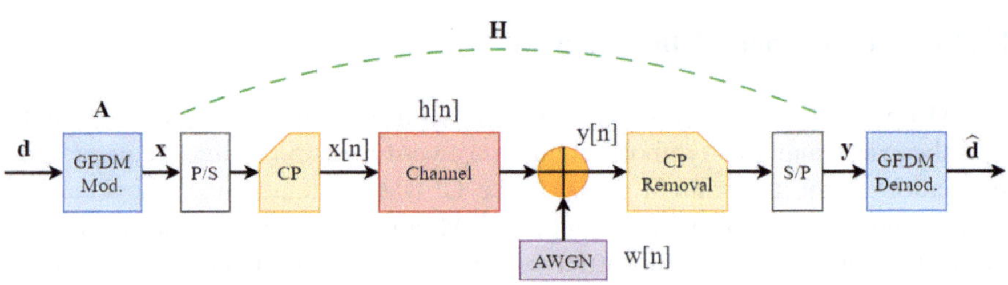

Fig. 4.3 GFDM transceiver model

where $\mathbf{F}_N = \left[\sqrt{D}\mathbf{W}_D\right]_{:,1:N}$, and $\mathbf{H} \in \mathbb{C}^{D\times D}$ is the circulant matrix whose first column is $\left[\mathbf{h}^T \mathbf{0}_{D-N}^T\right]^T$.

Finally, the estimated GFDM block $\hat{\mathbf{d}}$ is obtained after the demodulation process, which varies depending on the target applications.

4.1.3.3 Techniques Used in GFMC

Several key techniques contribute to GFDM's performance and flexibility.

- **Prototype Filter Design:** The choice of prototype filter significantly impacts GFDM's characteristics, influencing out-of-band radiation, inter-symbol interference (ISI), and inter-carrier interference (ICI). Design optimization often considers factors such as the conditional number of the transmitter matrix, out-of-band emission, peak-to-average power ratio (PAPR), and noise enhancement factor (NEF). Various techniques, including those based on frequency shift, FFT, and optimization algorithms, have been proposed for filter design.
- **Tail-biting Cyclic Prefix (CP):** The tail-biting CP differs from OFDM's fixed CP by using the end of the block to create the prefix, reducing overhead and improving spectral efficiency.
- **Low-Complexity Transceiver Implementation:** Efficient algorithms and structures are crucial for practical implementation. Techniques include using sparse matrices, frequency domain processing, and optimized symbol detection algorithms to reduce complexity.
- **Symbol Detection Algorithms:** Beyond conventional linear receivers (Zero-Forcing and Minimum Mean Square Error), advanced symbol detection algorithms, such as those leveraging deep learning, approximate message passing, and parallel interference cancellation, are used to improve performance, particularly in complex scenarios such as MIMO-GFDM.

4.1.3.4 Advantages of GFDM Compare with OFDM

GFDM offers several key advantages over OFDM.

- **Enhanced Spectral Efficiency:** The flexible pulse shaping and tail-biting CP significantly reduce out-of-band emissions compared to OFDM, resulting in improved spectral efficiency and reduced interference with adjacent channels. This is particularly beneficial in crowded spectral environments.
- **Increased Flexibility:** GFDM's adaptability to different pulse shapes and subcarrier allocations makes it suitable for diverse applications with varying bandwidth and latency needs. This adaptability is crucial for handling the heterogeneous traffic demands of future networks.

- **Improved Latency:** By adjusting the number of sub-symbols, GFDM directly influences latency, allowing for faster data transmission when low delay is critical.
- *Robustness to Interference:* The flexible structure and tail-biting CP enhance robustness to interference in time-frequency dispersive channels, making it suitable for environments with high levels of interference.
- **Adaptability to Non-Ideal Conditions:** While GFDM, like OFDM, is susceptible to non-ideal channel conditions such as synchronization errors, I/Q imbalance, and channel estimation challenges, its flexible structure allows for the application of various compensation techniques (e.g., training sequences, blind estimation, and adaptive interference cancellation) to mitigate these effects.

4.1.4 UFMC

Universal Filtered Multi-Carrier (UFMC) modulation represents a sophisticated synthesis of the benefits inherent in both OFDM and filter banks as utilized in FBMC. For UFMC, its entire bandwidth is strategically segmented into sub-bands, each comprising multiple subcarriers. Unlike FBMC, where individual subcarriers undergo filtering, UFMC employs a mechanism in which the subcarriers are filtered collectively in blocks. This approach is grounded in Frequency Division Multiplexing (FDM), wherein the incoming data stream is systematically partitioned into distinct sub-streams. Notably, UFMC offers significant advantages such as the reduction of out-of-band radiation and enhanced time-frequency synchronization—attributes that underscore its efficacy and potential for optimizing communication systems within complex network environments.

4.1.4.1 Transmitter and Receiver

Figure 4.4 illustrates the structure of the UFMC transmitter, which shows that the total available bandwidth is strategically partitioned into sub-bands denoted as B, with each sub-band comprising k subcarriers. This segmentation facilitates efficient data handling, as data bits are allocated to each sub-band and subsequently transformed into parallel streams via a serial-to-parallel converter. The parallel output is then mapped to symbols—a crucial step for effective modulation. Next, these symbols are modulated by the IFFT module, which simplifies design complexity by eliminating the need for individual modulators per subcarrier. Post-IFFT processing involves serialization through a parallel-to-serial converter followed by filtering with a pulse shaping filter of length L. The filtered outputs from all filters are coherently summed and transmitted over the channel. Mathematically, input data X is segmented into B distinct sub-blocks. Each block undergoes transformation via an N-point IFFT represented by matrix 'V', ensuring optimal frequency domain representation before serialization and subsequent filtering through matrix 'F'. This

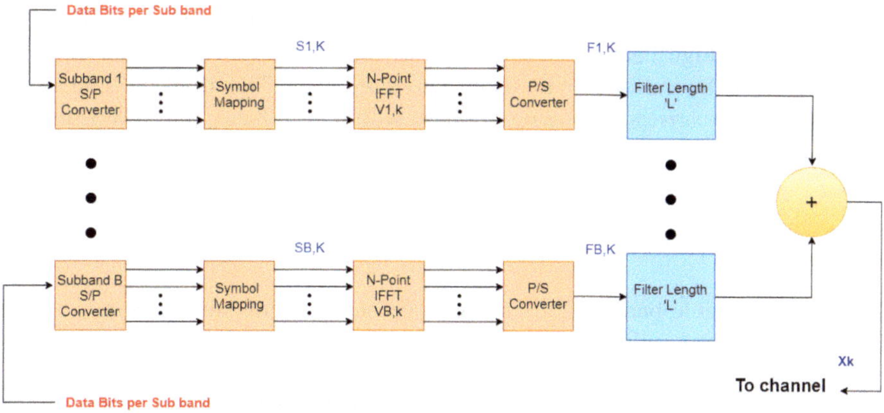

Fig. 4.4 UFMC transmitter diagram

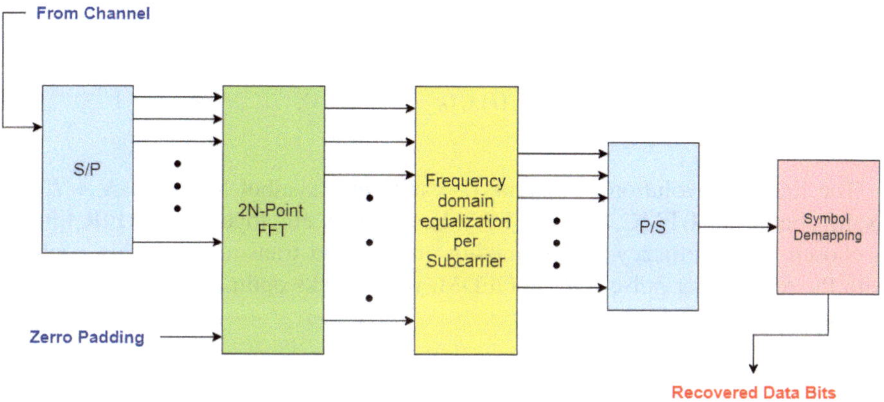

Fig. 4.5 UFMC receiver diagram

structured approach underscores both efficiency and precision in signal transmission within UFMC systems.

Figure 4.5 illustrates the block diagram of the UFMC receiver. Initially, data received from the channel undergoes conversion from serial to parallel format before being processed through an FFT for effective data demodulation. Subsequently, this output is transformed back into a single stream via a parallel-to-serial converter. The symbol decoder then translates these symbols into bits to retrieve the original data.

UFMC offers notable advantages over OFDM, particularly in terms of spectral efficiency. Unlike OFDM, UFMC eliminates the need for CP insertion, thereby utilizing the entire assigned spectrum without redundant bit repetition. In addition, fewer sidelobes are used in UFMC than in OFDM. The reduction in the number of sidelobes significantly minimizes interference with adjacent subcarriers.

In contrast to OFDM's reliance on numerous independently modulated subcarriers that can lead to high Peak-to-Average Power Ratio (PAPR) when added in phase, UFMC divides the available bandwidth into sub-bands. This division reduces the likelihood of subcarriers adding up in phase within each band, thus lowering peak power and consequently reducing PAPR better than OFDM systems.

4.1.4.2 Techniques in UFMC

The whole available bandwidth is decomposed into several sub-band B, which in turn is divided into multiple sub-groups. Let N denote the number of subcarriers. In the context of signal processing, an N-point Discrete Fourier Transform (DFT) is executed to facilitate the transformation from the frequency domain back into the time domain. Subsequently, this output signal undergoes filtering through a length-L FIR filter. The mathematical representation of this filtered output signal can be expressed as

$$y_i(k) = x_i * f_i = \sum_{l=0}^{L-1} f_i(l)x_i(k-l);\ k = 0, \ldots\ldots, N+L-1 \tag{4.4}$$

The linear convolution of x_i and f_i results in a symbol length of $N + L - 1$. In the context of UFMC, Dolph-Chebyshev filter is employed as the FIR filter. It is chosen for its efficacy in minimizing out-of-band transmission. This contrasts with the rectangular pulse filter in OFDM, which lacks optimality in both time and frequency domains.

4.1.4.3 Advantages and Disadvantages of UFMC

UFMC shares many similarities with OFDM in terms of properties. OFDM has been used in LTE and LTE-A systems and is also considered for future 5G networks. However, UFMC has several advantages over OFDM. The most significant and apparent difference between the two techniques is the absence of the CP. UFMC adds a filter to each sub-band, which reduces the sidelobes outside the sub-band. This enhances resistance to interference from any carrier noise sources. OFDM effectively uses FFT/IFFT and has simple equalization schemes, but spectral efficiency is compromised due to CPs.

When examining UFMC's advantages and disadvantages, it is essential to note its similarities with OFDM regarding properties. While OFDM has been integral to LTE and LTE-A systems and is considered for 5G networks, UFMC presents notable advantages. A key distinction between these techniques is that a CP is not employed in UFMC. By incorporating a filter into each sub-band, UFMC effectively reduces sidelobes outside these bands, thereby enhancing resistance to interference from carrier noise sources.

Although OFDM benefits from efficient FFT/IFFT operations and straightforward equalization schemes, its spectral efficiency degrades because of CPs. The primary advantage of UFMC over OFDM is attributed to the Dolph-Chebyshev filter's characteristics that significantly mitigate sidelobe impact on adjacent subcarriers. Furthermore, UFMC provides substantial flexibility when implementing MIMO techniques (MIMO-UFMC). However, it should be acknowledged that this comes at the cost of increased computational complexity associated with UFMC implementations.

During the standardization process for 5G, UFMC was evaluated but ultimately not selected as the primary waveform. Instead, Cyclic Prefix Orthogonal Frequency Division Multiplexing (CP-OFDM) has been established as the predominant modulation scheme in 5G New Radio (NR). Nevertheless, ongoing research into UFMC and alternative modulation techniques is crucial. Such exploration contributes significantly to advancing wireless communication technologies by potentially addressing limitations of current systems and enhancing future network performance.

4.1.5 Weighted Overlap and Add (WOLA) Based OFDM

4.1.5.1 Introduction

The results prove WOLA-OFDM to be a potential multi-carrier waveform for future wireless standards due to the simple implementation using time domain windowing, efficient OOB suppressions, better robustness to asynchronism between users and compatibility with MIMO.

In the seminal work by Qualcomm (2016) [6], several waveform candidates from existing literature were meticulously evaluated for their applicability in diverse 5G NR RAN service and deployment scenarios. Their study group conducted a comprehensive analysis of the spectrum and PAPR across several multi-carrier waveforms, including CP-OFDM, WOLA-OFDM, UFMC, F-OFDM, and FBMC. A critical issue addressed was the reduction of out-of-band (OOB) components inherent from OFDM systems. To mitigate this challenge, the employment of filters at the transmitter such as the Root Raised Cosine (RRC) pulse were proposed.

Further exploration by Zayani et al. in 2016 [7] delved into WOLA-OFDM's performance under the relaxed synchronization scenario. This investigation compared OOB emission levels, robustness against asynchronism, and channel delay spread of WOLA-OFDM with CP-OFDM and UFMC counterparts. The findings underscored WOLA-OFDM's potential as a formidable multi-carrier waveform for next-generation wireless standards due to its straightforward implementation via time-domain windowing technique, effective OOB suppression capability, enhanced resilience to user asynchronism issues, and compatibility with MIMO systems.

WOLA, initially introduced by Qualcomm Incorporated in 2016 [6], represents an innovative variant of OFDM poised for integration into 5G frameworks. Subsequent studies [7, 8] have further examined its efficacy within 5G environments alongside traditional OFDM approaches. Fundamentally aimed at achieving superior frequency response relative to standard OFDM methods, WOLA replaces the conventional rectangular prototype filter with pulses featuring softer edges on both sides—thereby facilitating significant reduction in OOB transmission while ensuring well-localized frequency responses through time-domain windowing operation [9].

4.1.5.2 Block Diagram

Figure 4.6 illustrates the entire structure of a typical WOLA-OFDM system, which is a sophisticated evolution of the traditional OFDM framework. This system introduces several key modifications, most notably replacing the conventional CP block with a CP+CS block, followed by time-domain windowing as previously discussed. Specifically, besides adding a CP at the beginning of each transmited symbol, some initial segment of the original symbol is also appended at the end of that symbol. This extension is termed as "cyclic suffix" (CS), whose length is denoted by W_{Tx}. Then, the total length of a transmit data symbol in time-domain can be calculated as

$$L = N_S + CP + W_{Tx}, \tag{4.5}$$

where N_S denotes the length of the data symbol.

To further refine signal integrity at transitions between symbols, the edges are softened using a time-domain windowing function with length L. The Meyer's RRC pulse is typically employed for this purpose because of its ability to create smooth transitions between consecutive symbols. Consequently, ovelapping still exists in the transition regions between consecutive symbols, but this overlap does not increase overhead compared to standard OFDM systems although the CS is added.

On reception within WOLA systems—as depicted in Fig. 4.6—a pre-processing step involves applying another windowing function across received symbols with length extending over $N_S + 2W_{Tx}$. To mitigate any adverse effect from windowing process on these signals' integrity, we overlap-and-add the edge transition samples

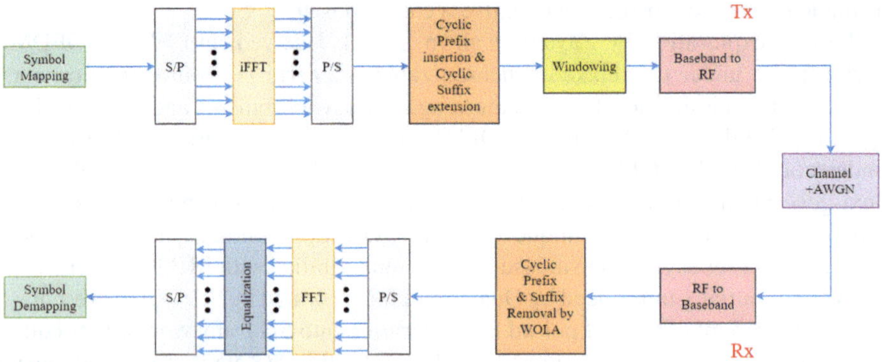

Fig. 4.6 WOLA-OFDM block diagram

from the right to the left and vice versa. This effectively neutralizes potential distortion before removing the extension altogether; thus forming final processed data symbol with the length of N_S, which is ready for subsequent FFT operation while maintaining desired precision level throughout all stages involved therein.

4.1.5.3 Advantages of WOLA Compared with OFDM

As described above, WOLA-OFDM can have some advantages compared with OFDM. First, due to the soften edges of time-domain symbols, WOLA transmission spectrum has significant suppression of the OOB emission which is a main problem in OFDM. WOLA also has better robustness to asynchronous frequency environment as well as to channel delay spread. Finally, WOLA has a possibility to obtain higher data rate transmission with the trade-off in spectrum.

As previously outlined, WOLA-OFDM presents several advantages over traditional OFDM systems. Firstly, the softened edges of time-domain symbols in WOLA result in a notable reduction of out-of-band (OOB) emission, which can resolve a significant challenge faced by OFDM. Additionally, WOLA exhibits enhanced robustness in asynchronous frequency environment and demonstrates improved resilience to channel delay spread. Furthermore, there exists the potential for achieving higher data transmission rate with WOLA by strategically managing the spectrum trade-off. These attributes underscore the efficacy and adaptability of WOLA-OFDM in modern communication systems.

4.2 General Setup

4.2.1 Simulation Software

MATLAB is a software that provides a numerical computing and programming environment, designed by MathWorks. MATLAB enables numerical computations with matrices, plotting functions and data charts, executing algorithms, creating user interfaces, and interfacing with programs written in various programming languages. In this project, MATLAB will be utilized to simulate the signals of modulation techniques (Table 4.3).

4.2.2 Simulation of F-OFDM Technique

The simulation of F-OFDM technique in this section includes the simulation of the signal spectrum and the bit error rate (BER).

Table 4.3 Description of the simulation parameters for 5G modulation waveforms module

Symbol	Description
bitsIn	Input bit sequence
G	Guard interval
NFFT	Number of FFT points
bsc	Number of bits per sub-carrier (2: QPSK, 4: 16QAM, 6: 64QAM, 8: 256QAM)
cpLen	Length of the CP
L	Filter length
K	Filter coefficient in FBMC
rbSize	Number of sub-carriers per resource block
numRBs	Number of resource blocks
rxSig	Received signal
txSig	Transmitted signal
psd	Power spectral density per sub-band
snrdB	Signal-to-noise ratio (SNR) in dB
BER	Bit error rate (BER)

4.2.2.1 Signal Spectrum Simulation

Figure 4.7 illustrates the flowchart of the transmission part of f-OFDM simulation module.

1. *Step 1:* First, the system input parameters are determined from the user interface. Then, the program creates a random bit sequence using the randi function. The number of Resource Blocks is set to 50, the CP length is 72, the number of sub-carriers per Resource Block is 12, and the number of bits per sub-carrier is set to $2n$ (2 for QPSK, 4 for 16QAM, 6 for 64QAM, 8 for 256QAM). The number of sub-carriers in the sub-band is calculated by multiplying the number of Resource Blocks with the number of sub-carriers per Resource Block.
2. *Step 2:* QAM modulation is implemented, then the Inverse Fast Fourier Transform (IFFT) is calculated by using the MATLAB function `y = ifft(X)`.
3. *Step 3:* The CP is added into the OFDM signal.
4. *Step 4:* In this step, we design a low-pass filter using the sinc function to determine the signals 0 and 1.

$$sinc(t) = \begin{cases} \frac{sin(\pi t)}{\pi t} \text{ for } t \neq 0 \\ 1 \text{ for } t = 0 \end{cases} \tag{4.6}$$

This signal traverses through the wireless media to reach the receiver, where it undergoes a reverse processing sequence akin to that at the transmitter. Initially,

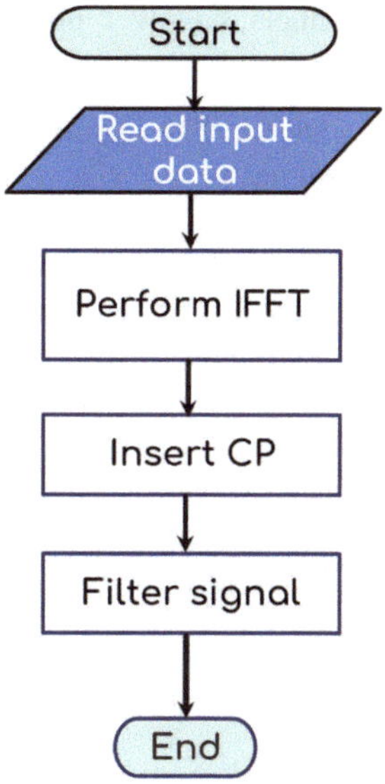

Fig. 4.7 Flowchart of F-OFDM signal transmission simulation

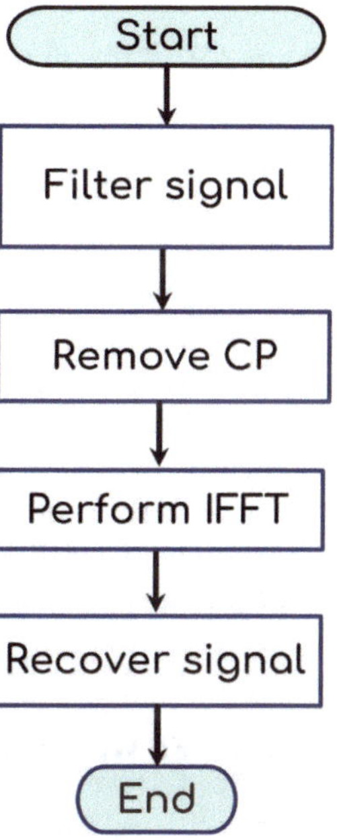

Fig. 4.8 Flowchart of F-OFDM signal reception simulation

filtering is applied, followed by CP removal and FFT demodulation to reconstruct the bit stream. The concluding phase involves data recording.

Figure 4.8 delineates a flowchart representing the simulation algorithm on the receiver side within an F-OFDM system framework. This algorithm encompasses three methodical steps for executing signal reception without considering fading channels; however, noise is introduced into the received signal to attain the specified SNR.

1. *Step 1:* The signal of interest is filtered using a low-pass filter; then the resulted signal is fed to a matched filter. Then, the signal goes through the CP-OFDM receiver.
2. *Step 2:* CP is removed by using the standard CP-OFDM receiver.
3. *Step 3:* FFT is performed by using the function `Y = fft(X)`. This function computes the discrete Fourier transform (DFT) of X using the Fast Fourier Transform (FFT) algorithm. For vector **X** inputs, `fft(X)` yields its Fourier transform; for matrix **X** inputs, it returns a matrix **Y**, where each column of **Y** is the FFT of the corresponding column of **X**. Finally, for multi-dimensional arrays, transformations occur along vectors in the first non-singleton dimension. Post-FFT execution facilitates demodulation based on the bit count on each sub-carrier $2n$—such as 2 for QPSK or up to 8 for 256QAM—ensuring that received signal closely mirror their original counterpart.

4.2.2.2 BER Simulation

Figure 4.9 is the flowchart of the BER calculation algorithm simulation, which includes the following steps:

1. *Step 1:* We generate the input signal including SNR in dB as an array with n elements. BER is also an array with values set to 0, and it has n elements such that the lengths of the two arrays are equal.
2. *Step 2:* We create a For loop with K running from 1 to the length of the `snrdB` array. Inside the For loop, Additive White Gaussian Noise (AWGN) will be added to the signal using the `awgn(x, snr, 'measured')` function in MATLAB. This simulation will use the `comm.RectangularQAMDemodulator(M, Name, Value)` function to demodulate using the rectangular QAM constellation and create an error rate calculation object using the comm.ErrorRate function to compute the BER values.
3. *Step 3:* BER values are output and the resulted graph is plot.

4.2.2.3 FBMC Technique Simulation

In FBMC, at frequencies that are an integer multiples of $1/M$, only one filter frequency response is non-zero. The first filter in the bandwidth is the filter associated with the 0 frequency carrier, known as the prototype filter, and the states of other filters are derived from it by frequency shifting.

In the frequency domain, additional coefficients are inserted between existing coefficients, which allows for better control of the filter's frequency response. Furthermore, prototype filters are characterized by the overlapping factor K, which is

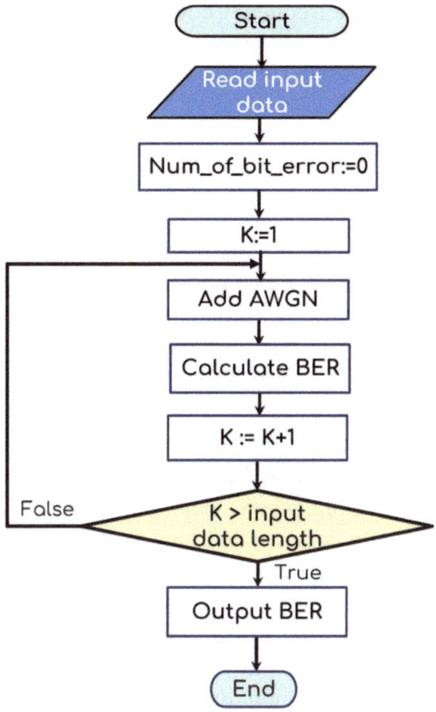

Fig. 4.9 Flowchart of BER calculation algorithm simulation

Table 4.4 Values of the prototype filter coefficients for each K

Value of K	h_0	h_1	h_2	h_3
2	1	$\frac{\sqrt{2}}{2}$	–	–
3	1	0.911438	0.411428	–
4	1	0.971960	$\frac{\sqrt{2}}{2}$	0.235147

the ratio of the filter's impulse response time to the symbol period:

$$K = \frac{\theta}{T}$$

The Table 4.4 shows the frequency coefficients of the Half Nyquist filter for K = 2, 3, and 4. Thus, the band-pass filter is obtained by shifting the frequency by k/M. Similar to the FFT case mentioned above, the filter with index k is obtained by multiplying the prototype filter coefficients by $e^{\frac{j2\pi K}{M}}$, where M is the number of samples.

In the time domain, the impulse response of the interference filter is calculated by the formula:

$$g(t) = [G_2 + 2G_1 cos(2\pi \times \frac{t}{2K})]e^{j2\pi \times \frac{t}{2T}}, \tag{4.7}$$

where the frequency coefficients in the filter $G_K = H_k H_{K-k}$ with $k = 1, ..., K - 1$. The set of coefficients is symmetric, and for $K = 4$:

$$G_1 = 0,228553$$
$$G_2 = 0,228553$$
$$G_3 = 0,5$$

In the subsequent section, we rigorously define the modulation scheme employed to mitigate interference. Specifically, the expression $e^{j2\pi \times \frac{t}{2T}} = \cos\left(\frac{\pi t}{T}\right) + j \sin\left(\frac{\pi t}{T}\right)$ exemplifies the inherent symmetry of frequency coefficients. This characteristic ensures that the imaginary component of $g(t)$ intersects the zero axis at $nT, n \in \mathbb{Z}$, where T is the symbol period, whereas its real component does so at $(2k+1)\frac{T}{2}, k \in \mathbb{Z}$.

Frequency spreading is executed by the appropriate filter. The term "spreading" aptly named as it disperses frequency samples across various sub-channels. In principle, when a band-pass filter characterized by an overlap factor K is applied, a single data symbol can modulate $2K - 1$ carriers. For the transmitter side, the band-pass filter can be implemented by an IFFT of size KM, which generates all necessary carriers. To implement frequency spreading, the IFFT inputs are multiplied by the frequency coefficients, then these $2K - 1$ results are indexed as $(i-1)K + 1, ..., (i+1)K - 1$.

For every input dataset processed, the IFFT yields a block comprising KM samples. Consequently, this results in a symbol rate of $\frac{1}{M}$, with K continuous overlapping outputs in time domain by IFFT.

4.2.2.4 Signal Spectrum Simulation

The flowchart in Fig. 4.10 simulates FBMC signal transmission and consists of the following 5 steps:

1. *Step 1:* First, The system input parameters are determined, in particular, guard interval, filter overlap factor K, and the number of bits per sub-carrier 2^n (for examples, 2 bits for QPSK, 4 bits for 16QAM, or 8 bits for 256QAM).
2. *Step 2:* Next, Table 4.4 is utilized to calculate the frequency coefficients of the half-Nyquist filter for various overlap factors, such as $K = 2, 3,$ and 4. This is the prototype filter for the zero frequency carrier and also the basis for other sub-carriers. The overlap factor (K) is essential since it dictates how many multi-carrier symbols overlap in time. The order of this prototype filter can be optimally selected as $2K - 1$. Subsequently, in FBMC simulation processes, it is necessary to construct a symmetric filter configuration while also computing how many complex symbols correspond to each OFDM symbol.
3. *Step 3:* Then, we repetitively generate the random input data sequence in a `For` loop that runs from 1 to the number of symbols n, for example, 100, by using the `randi` function.

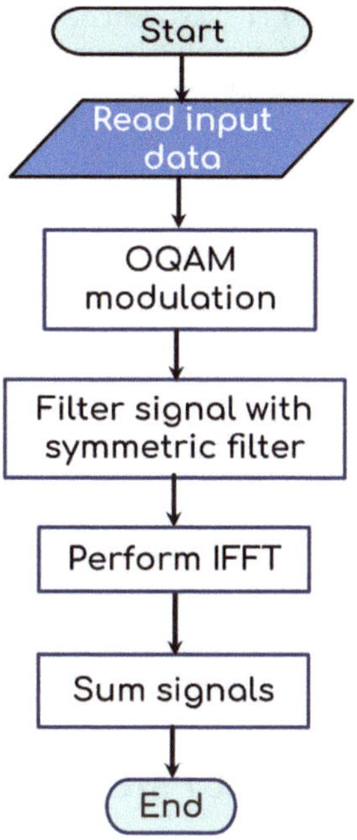

Fig. 4.10 Flowchart of FBMC signal transmission simulation

4. *Step 4:* OQAM modulation is performed and sampled by k. The guard interval is added, and the signal is filtered with the prototype filter. The transmitted signal will be the sum of the real and imaginary parts.
5. *Step 5:* After completing the `For` loop, the signal is simulated using the `plot` function.

4.2.2.5 Bit Error Rate (BER) Simulation

After passing through the transmission channel, the signal is obtained at the receiver and the reverse process of the transmitter is performed. Figure 4.11 depicts the signal reception algorithm flowchart, which includes three steps:

1. *Step 1:* AWGN is added to the received signal using the function `awgn(x, snr)` and then FFT is calculated.
2. *Step 2:* The signal is filtered using the function `filter(b, a, x)`. This function allows filtering the input data using a rational transfer function determined by the corresponding numerator and denominator coefficients, `b` and `a`. Subsequently, the guard interval is removed.
3. *Step 3:* After OQAM demodulation, the original signal is recovered, and the bit error rate is calculated using an algorithm similar to F-OFDM as shown in Fig. 4.9.

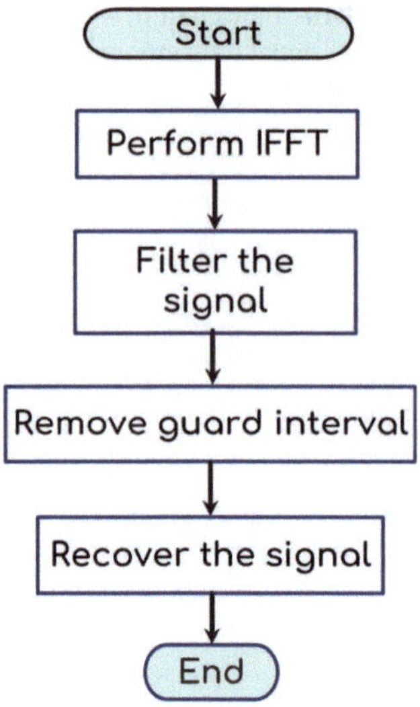

Fig. 4.11 Flowchart of FBMC Signal Reception Simulation

4.2.3 UFMC Technique Simulation

4.2.3.1 Spectrum Signal Simulation

1. *Step 1:* First, the input parameters of the system need to be defined. These parameters can be modified to explore their impact on the system.
2. *Step 2:* A random input data sequence is generated using the randi function. The filter is designed using the function w = `chebwin(L, r)`, which returns a column vector **w** containing the Chebyshev window of length L.
3. *Step 3:* We perform IFFT using the `ifft` function in MATLAB, and then filter the signal for each sub-band using the Chebyshev filter designed in Step 2 (Fig. 4.12).

4.2.3.2 Bit Error Rate (BER) Simulation

After passing through the transmission channel, the receiver obtains the signal and performs the reverse process of the transmitter. Figure 4.13 depicts the signal reception algorithm flowchart, which includes three steps:

1. *Step 1:* AWGN is added to the received signal using the function `awgn(x, snr)`, followed by performing FFT.
2. *Step 2:* The signal is filtered using the Chebyshev filter, similar to the transmitter. Subsequently, the BER analysis and calculation are performed according to the BER calculation scheme shown in Fig. 4.9. The BER signal graph is plotted using the plot function, as with other modulation techniques.

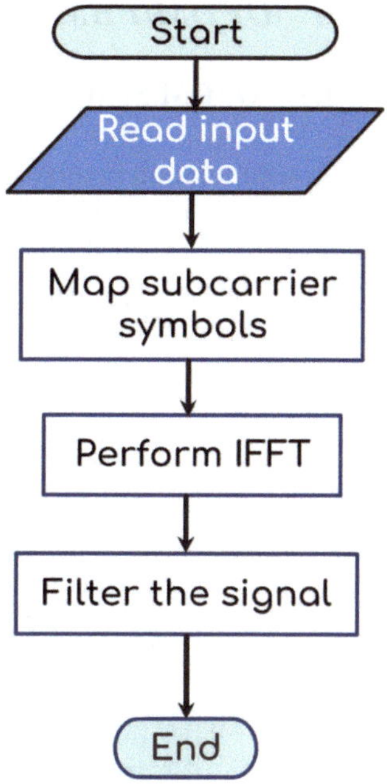

Fig. 4.12 Flowchart of UFMC signal transmission simulation

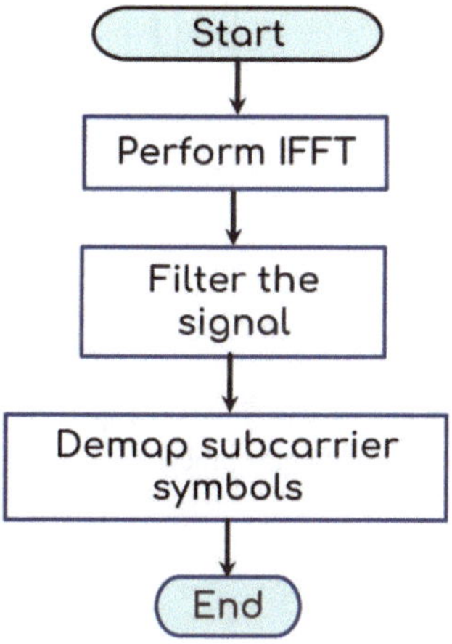

Fig. 4.13 Flowchart of UFMC signal reception simulation

4.3 Simulation Results

This section presents the simulation results related to the modulation waveforms that are candidates for 5G wireless communication standards, including FBMC, f-OFDM, UFMC, GFDM, WOLA, compared with the benchmark system, i.e., the OFDM waveforms.

4.3.1 OFDM/FBMC/WOLA/f-OFDM/UFMC

4.3.1.1 OFDM Signal Simulation

The first simulation involves Orthogonal Frequency-Division Multiplexing (OFDM) with 600 subcarriers (Fig. 4.14). Here, attention is given to the CP. The larger the CP, the poorer the spectral efficiency and the higher the Peak-to-Average Power Ratio (PAPR). The total spectrum of the OFDM signal is the sum of all the spectra of each subcarrier (Fig. 4.15).

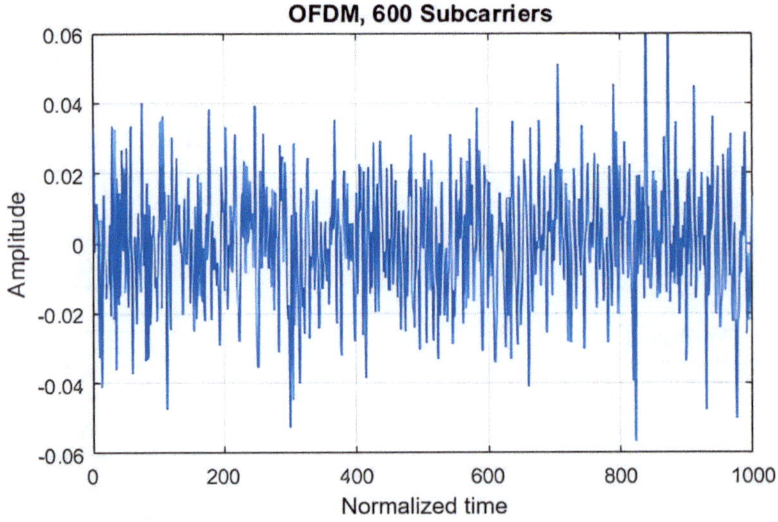

Fig. 4.14 OFDM signal in the time domain

Fig. 4.15 OFDM spectrum signal with 600 subcarriers

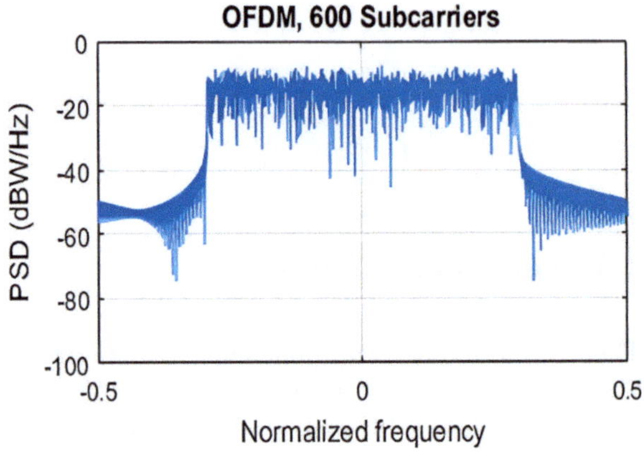

4.3.1.2 F-OFDM Signal Simulation

The F-OFDM simulation involves 50 Resource Blocks and 12 Subcarriers. A well-designed filter is applied to the OFDM symbols in the time domain to improve the out-of-band radiation of the subband signal while maintaining the orthogonality of the complex-domain OFDM symbols. In F-OFDM, the subband OFDM signal is transmitted through a designed filter. Since the filter bandwidth matches the signal bandwidth, only a few subcarriers near the edge are affected. An important consideration is that the filter length can be allowed to exceed the CP length for F-OFDM. The inter-symbol interference that arises is minimized by using the filter. Figures 4.16 and 4.17 show examples of an f-OFDM signal in time domain and its spectrum, respectively.

4.3.1.3 FBMC Signal Simulation

In Fig. 4.18, since FBMC does not use a CP, it avoids reducing transmission throughput and wasting power as in OFDM. FBMC utilizes the available channel capacity much better and can provide higher data rates within a given radio spectrum bandwidth, resulting in higher spectral efficiency (Fig. 4.19).

The power spectral density of the transmitted FBMC signal is plotted to highlight the low out-of-band leakage. Comparing the power spectral density of OFDM and

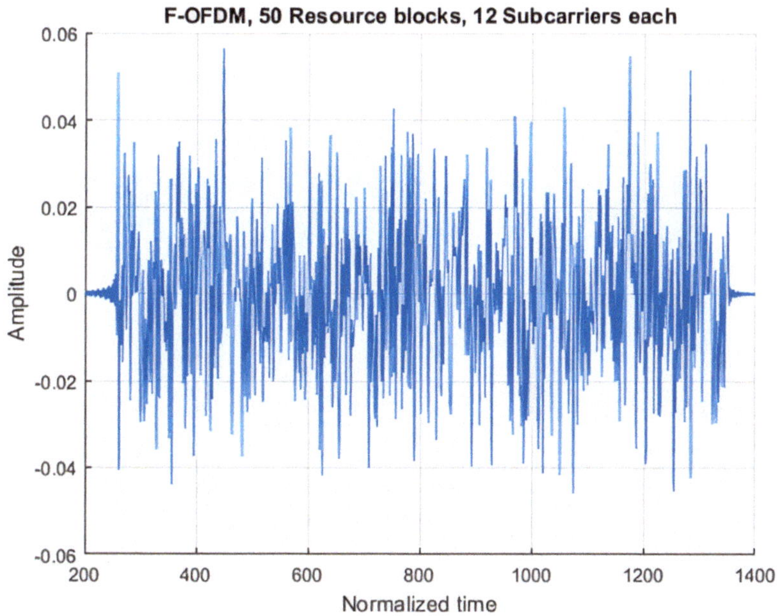

Fig. 4.16 F-OFDM signal in the time domain

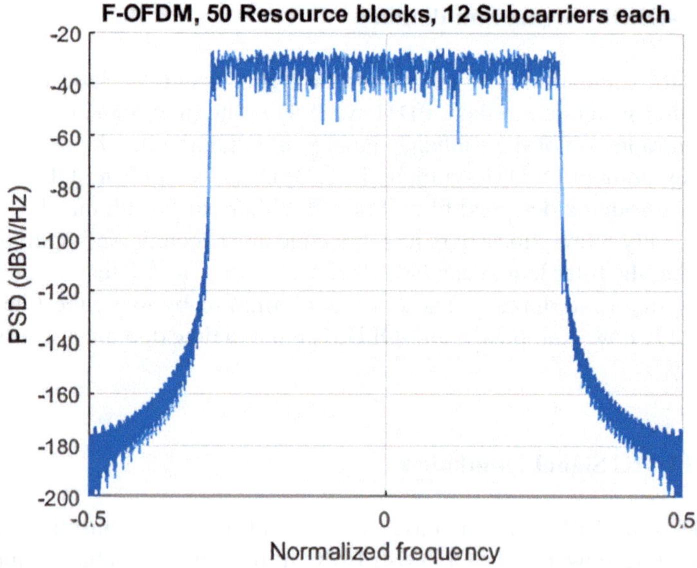

Fig. 4.17 F-OFDM spectrum signal

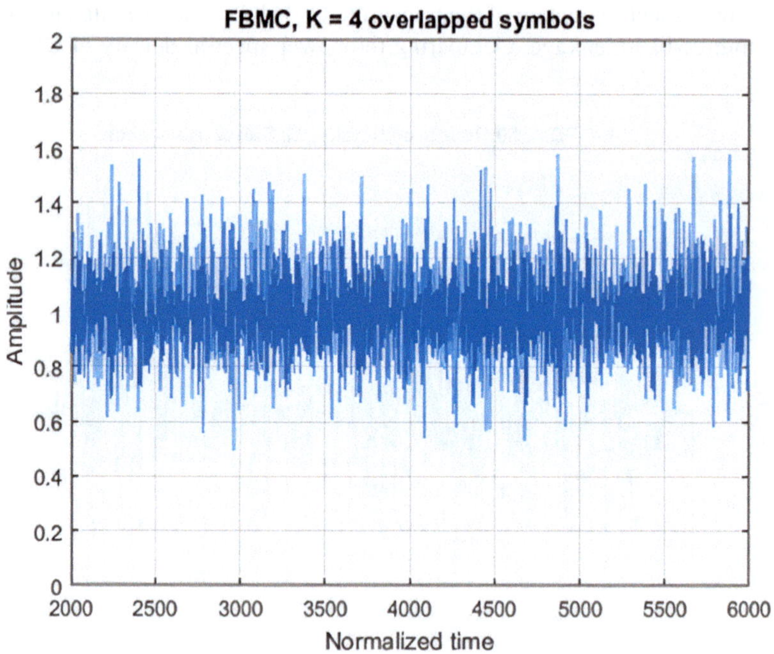

Fig. 4.18 FBMC signal in the time domain

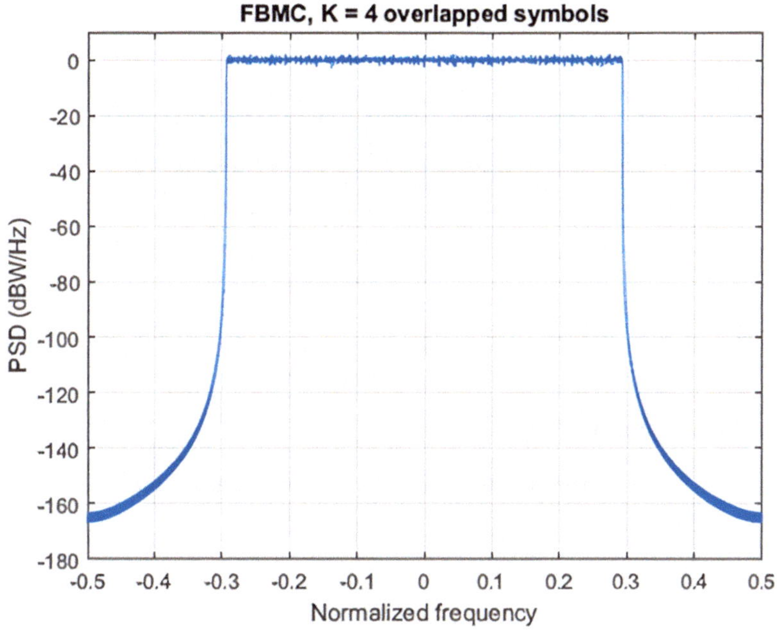

Fig. 4.19 FBMC spectrum signal with k = 4

FBMC schemes, FBMC exhibits lower sidelobes. This allows for higher utilization of the allocated spectrum, leading to increased spectral efficiency. When subcarriers are modulated in an OFDM system, sidelobes spread out on both sides. With bandpass filtering systems, the filters are employed, resulting in much cleaner transmission.

4.3.1.4 UFMC Signal Simulation

In the simulation shown in Figs. 4.20 (time domain signal) and 4.21 (spectrum), the entire carrier band (N) is divided into sub-bands. Each sub-band has a fixed number of subcarriers, and not all sub-bands need to be used for a given transmission (Fig. 4.21).

4.3.1.5 BER Simulation of F-OFDM, FBMC, and UFMC Modulation Techniques

Figure 4.22 shows the BER of F-OFDM. The simulation indicates that the bit error rate is lowest when the filter length is 513. The BER plot for FBMC shows that the red line represents the lowest bit error rate. Therefore, when K = 4, the performance of the FBMC system is optimal (Fig. 4.23).

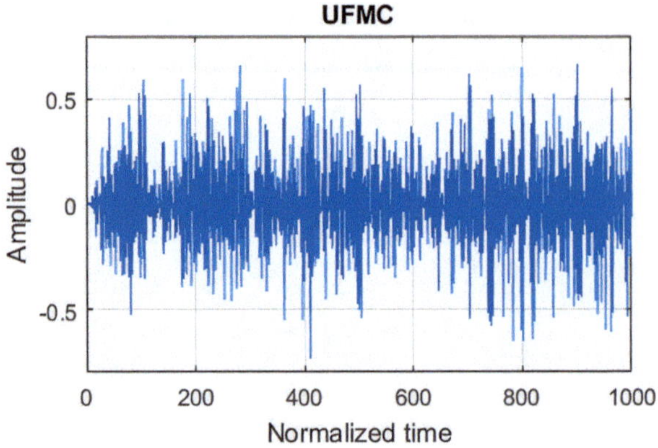

Fig. 4.20 UFMC signal in the time domain

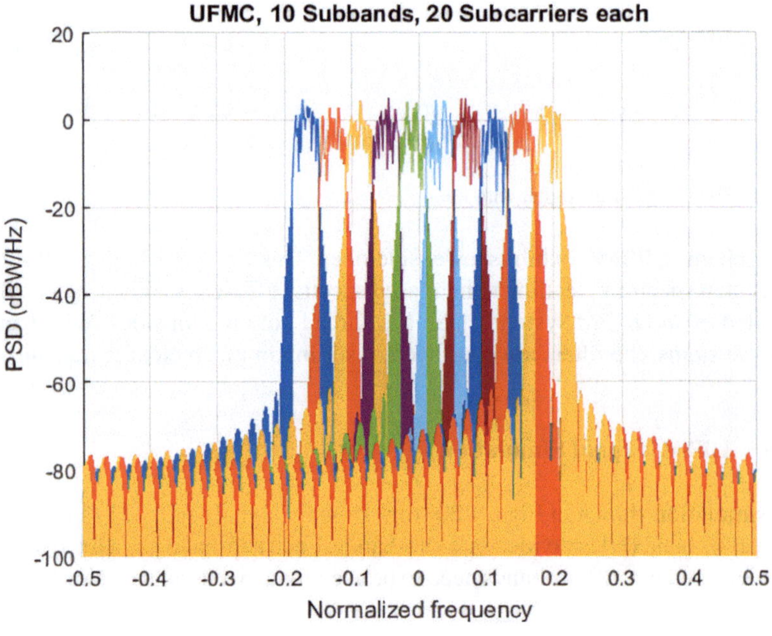

Fig. 4.21 UFMC spectrum signal

Figure 4.24 illustrates the BER of UFMC, showing that the bit error rate is lowest when the number of sub-bands is 16. Therefore, the number of sub-bands is inversely proportional to the bit error rate.

The simulation in Fig. 4.25 demonstrates that the BER of UFMC is the lowest. This is because UFMC combines the advantages of OFDM and the filter bank in FBMC. In UFMC, the total bandwidth is divided into sub-bands. Each sub-band contains a number of subcarriers. Instead of filtering each subcarrier individually as

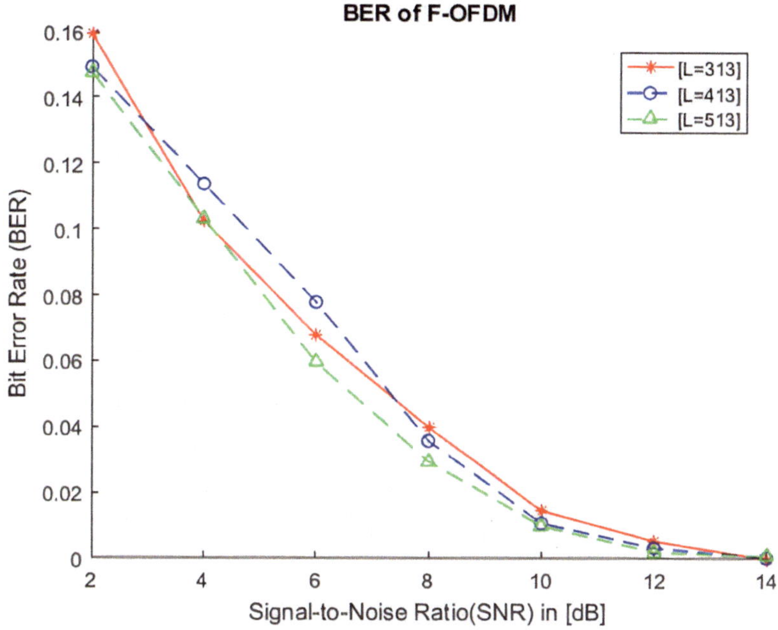

Fig. 4.22 BER in F-OFDM with varying filter lengths

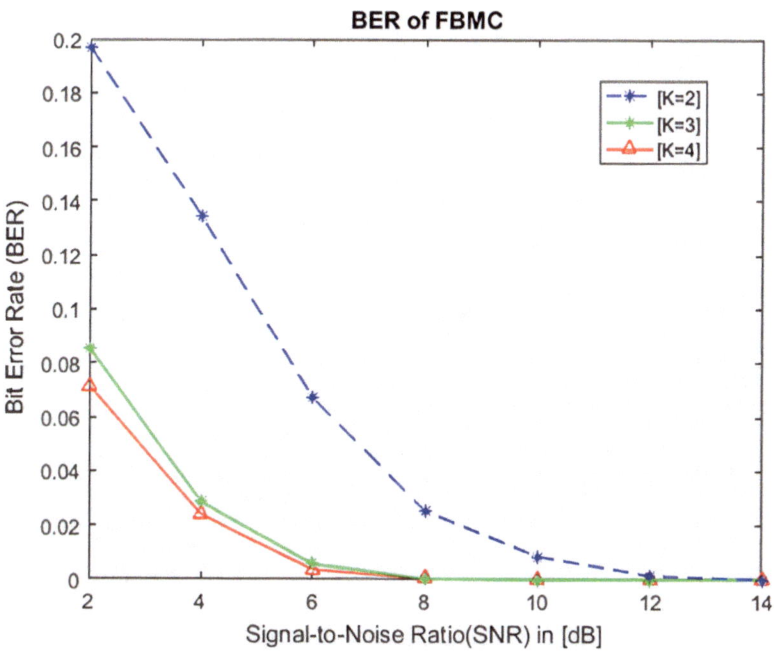

Fig. 4.23 BER in FBMC with different K factors

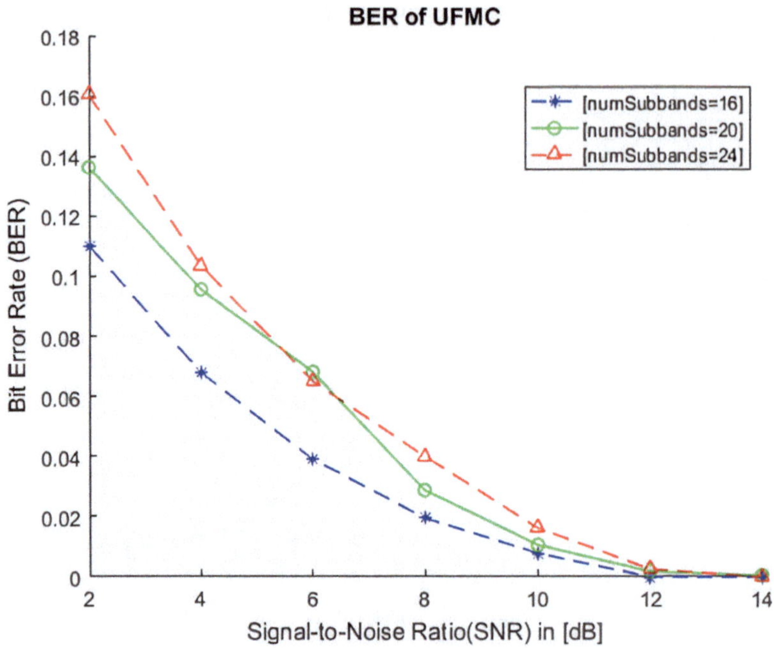

Fig. 4.24 BER in UFMC with varying number of sub-bands

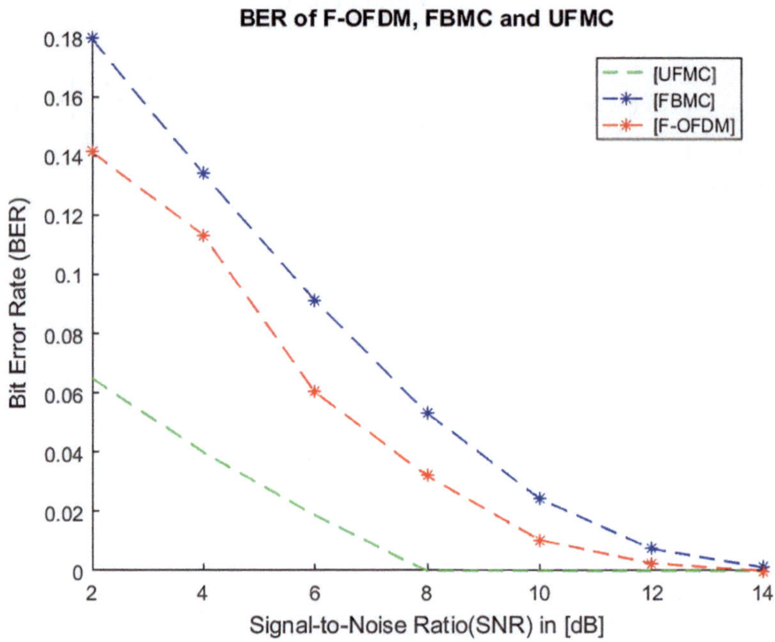

Fig. 4.25 BER in UFMC, FBMC, and F-OFDM

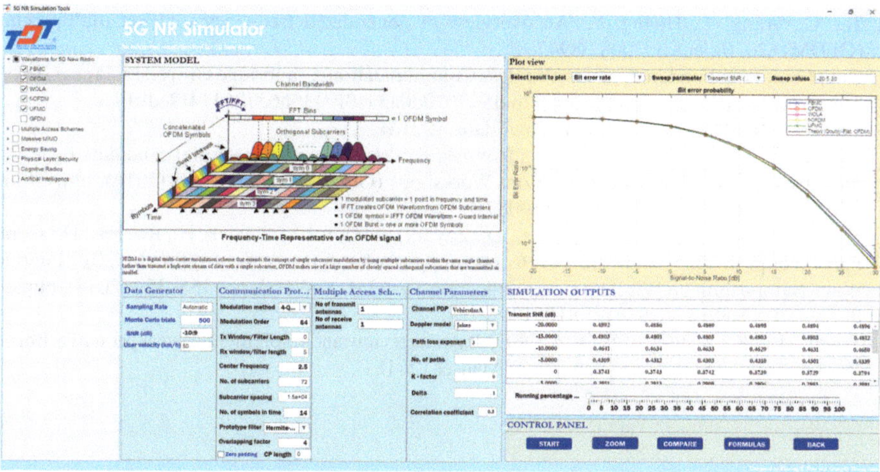

Fig. 4.26 User interface for UFMC

in FBMC, UFMC filters a block of subcarriers, resulting in a lower BER compared with FBMC and F-OFDM.

4.3.1.6 User Interface Design

The interface is designed in four parts, as shown in Fig. 4.26. The left-hand side is a check-node tree to select the considered technology for simulation. In the middle, on the top we have a block diagram of the system model and a textbox to show the brief explanation of the model, while at the bottom we have 4 input panels, where the users can input their selected parameters. The right-hand side is the output section, including an axes on the top part for plotting the output curves such as PSD, time-domain signals, or BER curves, and a table to show the resulted numerical results. The bottom right of the user interface shows the controlled buttons for the users.

References

1. Fettweis, G., Krondorf, M., Bittner, S.: Gfdm - generalized frequency division multiplexing. In: VTC Spring 2009 - IEEE 69th Vehicular Technology Conference, pp. 1–4 (2009). https://doi.org/10.1109/VETECS.2009.5073571
2. Michailow, N., Matthe, M., Gaspar, I. et al.: Generalized frequency division multiplexing for 5th generation cellular networks. IEEE Trans. Commun. **62**(9), 3045–3061 (2014)
3. Farhang, A., Marchetti, N., Doyle, L.E.: Low complexity gfdm receiver design: A new approach. In: 2015 IEEE International Conference on Communications (ICC), pp. 4775–4780 (2015). https://doi.org/10.1109/ICC.2015.7249078

4. Tai, C., Wang, T., Huang, Y.: An overview of generalized frequency division multiplexing (GFDM) (2020). arXiv:2008.08947
5. David Tse, P.V.: Fundamentals of Wireless Communication. Cambridge University Press (2005). http://gen.lib.rus.ec/book/index.php?md5=7010d0a3a40f9335b63735114f2ed017
6. Qualcom: R1-162199 - waveform candidates (2016)
7. Zayani, R., Medjahdi, Y., Shaiek, H., Roviras, D.: Wola-ofdm: A potential candidate for asynchronous 5g. In: 2016 IEEE Globecom Workshops (GC Wkshps), pp. 1–5 (2016). https://doi.org/10.1109/GLOCOMW.2016.7849087
8. Gerzaguet, R., Medjahdi, Y., Demmer, D., Zayani, R., Dor, J.B., Shaiek, H., Roviras, D.: Comparison of promising candidate waveforms for 5g: Wola-ofdm versus bf-ofdm. In: 2017 International Symposium on Wireless Communication Systems (ISWCS), pp. 355–359 (2017). https://doi.org/10.1109/ISWCS.2017.8108139
9. Singh, M.G.K.: A study of WOLA-OFDM: An overview and implementation aspects. Technical Report, Indian Institute of Technology Madras (2018)

Chapter 5
Multiple Access Schemes for 5G/6G Networks

5.1 Non-Orthogonal Multiple Access (NOMA)

5.1.1 Necessity of the Topic

Wireless technology has fundamentally transformed communication paradigms, evolving from the inaugural first-generation (1G) systems of the 1980s to the extensively implemented fifth-generation (5G) networks of the 2020s. This progression not only signifies a technological advancement but also heralds a transformation in related applications, contributing significantly to enhanced living standards. The anticipated proliferation of smart objects within homes, offices, streets, and cities is expected to culminate in an interconnected smart world [1]. Such an environment will inevitably lead to a substantial increase in active connections, heightened data traffic demands, and a broader spectrum of supported services.

Existing communication architecture cannot fulfill 5G's needs. For example, 5G requires connection speeds up to 1,000 times faster than current technology can provide. Also, from transmitter side to receiver side, 5G delays should be less than 1 ms , while 4G targets a 5 ms delay speed. To meet these requirements, 5G will apply several disruptive techniques, including new radio. The current radio resources represented by the wireless spectrum and transmission power are scarce which make it not easy to meet such challenges with orthogonal multiple access (OMA) as a backbone air interface, which includes methods such as TDMA, CDMA, FDMA, and OFDMA.

However, current communication architectures are inadequate for meeting 5G's stringent requirements. For instance, 5G necessitates connection speeds up to 1,000 times faster than those achievable with existing technologies. Additionally, it mandates end-to-end transmission delays below one millisecond—an ambitious target compared to the five-millisecond delay objective of 4G systems. To address these challenges effectively and meet such rigorous specifications, several disruptive techniques will be employed by 5G networks; among them is the implementation of new

P. T. Tran et al., *Developing MATLAB Comprehensive Simulation for 5G/6G Wireless Communications*, Signals and Communication Technology,
https://doi.org/10.1007/978-981-95-1117-4_5

radio technologies. The scarcity inherent in current radio resources—specifically wireless spectrum and transmission power—presents significant challenges when utilizing orthogonal multiple access (OMA) as a backbone air interface method (including TDMA, CDMA, FDMA, and OFDMA), necessitating innovative approaches for optimal resource allocation within this advanced framework.

The exploration of NOMA as a promising solution in communication systems has been extensively documented in the literature [2, 3]. NOMA is distinguished by its ability to enhance system capacity and SE beyond what orthogonal counterparts offer. The core concept of NOMA involves serving multiple users within a single resource slot, e.g. time, frequency, or code [4], by employing successive interference cancellation (SIC) [5, 6]. This approach marks a significant improvement from traditional OMA technologies [7], addressing critical challenges such as energy efficiency (EE), latency reduction, and fairness among user devices [4–6].

At the transmitter, NOMA operates by superimposing multiple signals with varying power levels into a single superposition coded (SC) signal [6]. In this configuration, stronger user devices with superior channel conditions are assigned lower power compared to their weaker counterparts (with poorer channel conditions) [5, 8–10]. On the receiver side, SIC plays a pivotal role; it enables users to decode the strongest signal first while treating others as interference. If this decoded signal corresponds to its own data stream, the operation is ceased. Otherwise, the receiver continuously subtracts these decoded signals from the SC signal and proceeds to decode subsequent signals in order of strength. This methodology not only optimizes resource utilization but also enhances overall network performance through intelligent power distribution and interference management strategies.

In the context of NOMA, a notable challenge arises from the coexistence of near and far users, which can lead to performance degradation for those located farther from the BTS. To address this issue, an efficient variation has been proposed that leverages cooperative transmission within NOMA systems, as detailed in recent studies [2, 3]. The fundamental concept behind this cooperative NOMA (C-NOMA) scheme is to utilize users situated closer to the BTS as relays to assist the communication between the BTS and the far users, which can enhance both coverage and overall system performance.

For C-NOMA frameworks, two primary relaying strategies are identified: decode-and-forward (DF) and amplify-and-forward (AF) [5]. A significant advantage of integrating cooperative communication into NOMA systems lies in the application of SIC at near-user locations. This allows these proximate users to possess knowledge of far-user information, thereby naturally be assigned as DF relays for transmitting data to distant counterparts.

Furthermore, in C-NOMA, distant mobile users devices typically obtain the signals transmitted from the BTS and relayed through cooperative relay stations [11, 12], due to obstacles in the transmission path [8, 9, 13]. Conversely, in scenarios devoid of such obstructions, these remote users can simultaneously acquire signals both from relay stations and directly from the BTS itself, forming a direct-link C-NOMA relaying configuration [14–16].

Currently, a multitude of research groups are rigorously investigating NOMA for current 5G and future network generations. The team at Middlesex University [17, 18] has been concentrating on energy-related challenges within NOMA systems in Heterogeneous Cloud Radio Access Networks (HCRAN), alongside spectrum sensing in cognitive radio wireless networks. Meanwhile, a Manchester University research group [19] has been researching cooperative NOMA, relay selection algorithms combined with energy harvesting (EH) in NOMA-based networks. The research group in Ton Duc Thang University has been working on the combination of NOMA and RF energy harvesting in full-duplex relaying networks [20, 21].

In Zhou et al.'s study [15], a dynamic DF relaying scheme for C-NOMA downlink communication was introduced, in which the system OP expression was obtained through stochastic geometry theory. Liu's 2018 work [22] proposed three novel relaying variations to DF in C-NOMA systems that demonstrated superior performance compared to traditional DF relaying in the scenarios of no direct link and with non-relayed signal superposition across multiple users. Yue's 2017 analysis [14] explored a DF-based C-NOMA network featuring a direct link between the transmitter and weaker user device. In 2018, Duan et al. [23] investigation into a cooperative device-to-device (D2D) NOMA network revealed that simultaneous communication between the BTS and all mobile users could be achieved using either single-signal decoding or maximum ratio combining (MRC). The numerical results from simulation indicated an enhanced total average rate and OP over basic NOMA schemes. Li et al.'s protocol [3] allowed flexible switching between two modes (direct or indirect) in a C-NOMA network involving two user devices, analytically demonstrating the superiority over traditional protocols. In the same year, Ye et al. [24] presented the OP analysis of a dual-DF NOMA network with SWIPT and the availability of the direct link.

Regarding the simulation of NOMA techniques in this module, we introduce the performance evaluation of basic NOMA-based cellular networks as well as some of their advanced variations, especially the combination of NOMA and wireless power transfer. The main contributions of this simulation module are as follows:

1. Performance Metrics Analysis: Performance metrics such as OP or throughput, BER, and average achievable rate are analyzed to compare the performance of NOMA and OMA in 5G cellular network. This basic simulation is developed based on the tools provided by TU Vienna [25].
2. Evaluation of advanced protocols for NOMA: The study explores the application of other advanced 5G/6G technologies such as full-duplex communication, massive MIMO, and SWIPT mechanisms within cooperative relay-assisted NOMA systems.
3. Mathematical Expressions for System Performance: Explicit mathematical expressions for system performance metrics, including OP, throughput, average rate, and EE, are derived and compared with Monte Carlo simulation results for various NOMA schemes, with different cooperative relaying strategies, with or without direct link.

4. Impact of system parameters: A thorough investigation into the impact of several critical parameters, including power allocation factor, data transmission rate, BTS-relay distance, number of users, EH parameters (time-switching or power-splitting factor), on the key performance indicators (OP, throughput, BER, and EE), has been conducted. The impact of these parameters on the system performance is assessed based on numerical simulation results to identify performance variations. These analytical insights are instrumental in determining optimal values for the system parameters. This approach ensures an equilibrium is achieved between key performance metrics and user equipment capability while maintaining system integrity.

5.1.2 *Introduction to Non-Orthogonal Multiple Access (NOMA)*

NOMA has recently been demonstrated as one of the most promising techniques for deployment in 5G and next-generation networks. It effectively addresses the limitations of current technologies, such as spectral efficiency, latency, and fairness among user devices [26–29]. The advantages of NOMA include the following:

- NOMA enhances spectral efficiency by allowing multiple users to simultaneously share the same resources. In OMA, resources are divided into orthogonal sub-channels, with each user occupying a separate sub-channel. In contrast, NOMA enables multiple users to share the same sub-channel, leading to more efficient resource utilization.
- NOMA employs a technique known as superposition coding, where users with better channel conditions are assigned lower power levels, while users with poorer channel conditions are assigned higher power levels. This allows the base station to serve multiple users simultaneously using the same frequency-time resources.
- NOMA uses SIC at the receiver to separate and decode signals from different users. The user with the stronger received signal is decoded first, and its signal is removed from the combined received signal. This reduces interference for the subsequent users, enhancing overall system capacity.
- NOMA provides flexible resource allocation and prioritization among users. By assigning different power levels or codebooks, NOMA can offer varying levels of fairness in terms of quality of service (QoS) and data rates to different users, based on their channel conditions and service requirements.
- These advantages make NOMA a strong candidate for improving spectral efficiency and overall system capacity compared to conventional OMA techniques. However, it is important to note that NOMA introduces additional complexity in signal processing and receiver design, requiring careful optimization and consideration of practical implementation challenges.

In C-NOMA, powering relay user devices is crucial to maintaining their operation. Energy harvested from radio frequency (RF) signals provided by the source is a viable energy source within cooperative NOMA systems. Furthermore, energy constraints are a persistent challenge in mobile and wireless networks. Energy for these networks can be sourced from batteries or externally harvested energy [27].

5.1.3 Classification of Non-Orthogonal Multiple Access (NOMA)

There are various methods to classify NOMA. Figure 5.1 illustrates the classification of NOMA based on its operational mechanism, whether cooperative or non-cooperative. Cooperative NOMA is further categorized into single-relay NOMA transmission (C-NOMA) and multi-relay NOMA transmission. Conversely, non-cooperative NOMA can be categorized by resource domains, including power domain, such as Power-domain NOMA (PD-NOMA), code domain such as Sparse Code Multiple Access (SCMA), Multi-User Shared Access (MUSA), Low-Density Spreading (LDS), and multiple domains suchas Outer Loop Link Adaptation NOMA (OLLA-NOMA), Pattern Division Multiple Access (PDMA), Topology Interference Management in NOMA (TIM-NOMA).

However, this section focuses on detailing power-domain NOMA and cooperative NOMA (C-NOMA) as the theoretical foundation for studies related to NOMA in general. SCMA will also be presented in the next section.

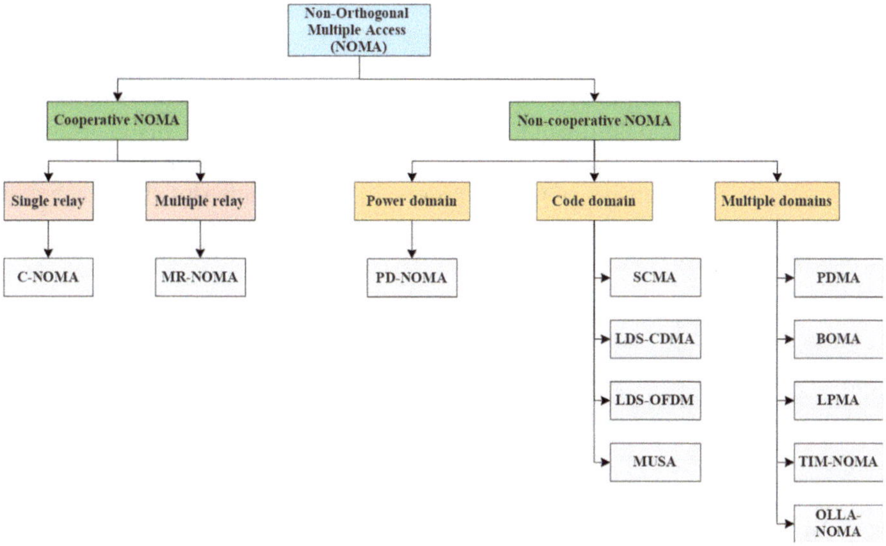

Fig. 5.1 NOMA Classification

5.1.4 Power-Domain NOMA

The power-domain NOMA technique [26] can serve multiple user devices within the same time slot. User devices with different channel conditions are assigned different power levels. The power allocation principle is as follows: user devices with weaker channel conditions are allocated more power than those with better channel conditions to ensure fairness among the devices [28–30]. In power-domain NOMA downlink, information is transmitted to the user devices using SC at the transmitter side and decoded using successive interference cancellation (SIC) at the receiver side, as illustrated in Fig. 5.2.

In Fig. 5.2, the base station transmits an SC signal to UE1, where UE1 uses the SIC technique to decode UE2's signal first (since UE2 has worse channel quality than UE1 and cannot receive the broadcast signal from the base station). Afterward, UE1 decodes its own signal. A common illustration of power-domain NOMA is the MUST model, shown in Fig. 5.3. This is a two-user power-domain NOMA downlink model, where the base station (BTS) simultaneously serves two single-antenna user devices on the same OFDMA subcarrier. The power is allocated to the user device UE2 higher than that of UE1, as UE2 is farther from the transmitter than UE1. The SIC process takes place at the user device UE1. Meanwhile, user device UE2 decodes its own signal and treats the signal from UE1 as interference.

5.1.5 Cooperative Non-Orthogonal Multiple Access (Cooperative NOMA)

The fundamental principle of cooperative NOMA lies in the transmission process between the source and destination, facilitated by one or more NOMA-enabled user devices acting as relay stations [31, 32]. Specifically, the transmission process in cooperative NOMA consists of two time slots. In the first time slot, the BTS broad-

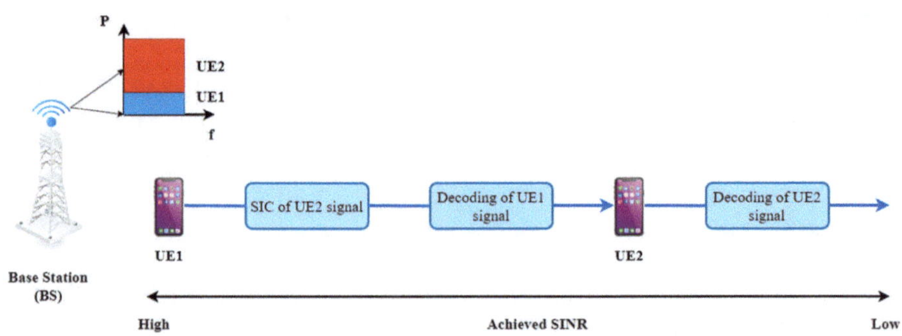

Fig. 5.2 NOMA applies SIC at the receiver side in the downlink

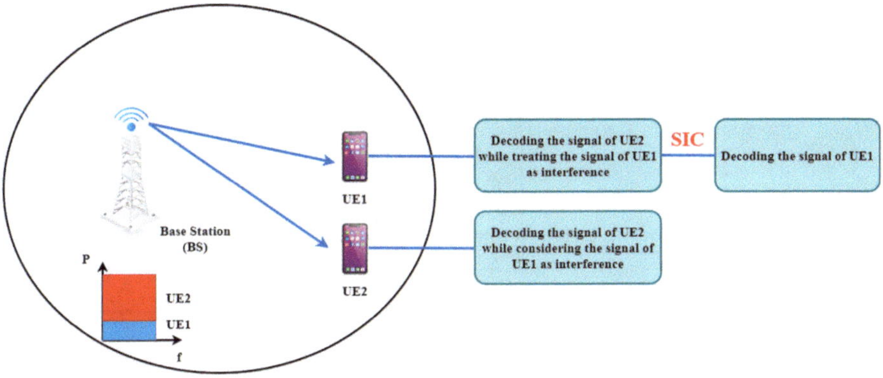

Fig. 5.3 Example of NOMA downlink mechanism with two user devices UE1 and UE2 and one subcarrier

casts superposed information to the user devices. In the second time slot, the user device with a better channel condition acts as a relay station to forward the decoded information to the user device with a poorer channel condition [33]. Similar to Fig. 5.3, UE1 uses the SIC technique to decode UE2's signal first, then decodes its own signal. Here, the difference is that UE1 then sends the decoded UE2 signal to UE2, where UE2 receives and decodes its own signal. As a result, the reliability of the user device with poorer channel conditions is significantly improved.

Compared with traditional NOMA, the core advantage of cooperative NOMA is its ability to achieve low system redundancy, enhance fairness, and improve overall performance gains [29]. In cooperative communication, the coverage area is expanded, and channel capacity is increased while mitigating adverse effects caused by multipath fading [34, 35].

In cooperative communication systems, relay stations forward received information signals to relevant destination nodes by employing forwarding protocols such as amplify-and-forward (AF) [36, 37], decode-and-forward (DF) [37]-[14], and quantize-and-map (QMF) [38, 39]. Based on the relay's operation mechanism, relaying can be classified into half-duplex (HD) and full-duplex (FD) modes [40, 41]. Unlike HD relaying, FD relaying enables simultaneous transmission and reception within the same frequency band and time slot [42]. Consequently, FD relaying offers higher spectral efficiency compared to HD relaying [43]. Therefore, the integration of cooperative communication with NOMA represents a crucial solution to enhance the system performance of NOMA. Figure 5.4 illustrates a cooperative NOMA system.

- Single-Relay Channel [44]: In single-relay channel systems, a user device with favorable channel conditions typically serves as a relay to assist in forwarding data information from the source station to the destination node. Additionally, the relay station may also be a dedicated relay device.
- Multi-Relay Channel [45]: In contrast to single-relay systems, multi-relay channel systems consist of multiple relays, as illustrated in Fig. 5.5. To forward informa-

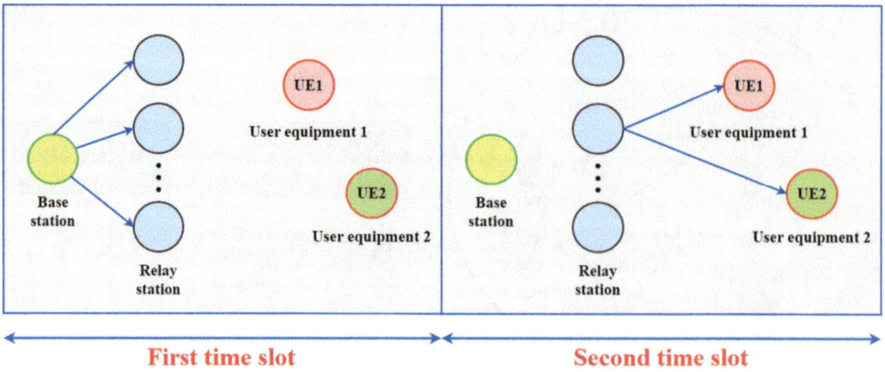

Fig. 5.4 Illustration of the system model of cooperative NOMA network

tion from the source station to the destination node, the relay with the best channel condition is usually selected to transmit the information. Furthermore, in forwarding data from the source station to the destination node, the information may pass through multiple relay stations.

5.1.6 Downlink NOMA

The NOMA technique can be implemented for both downlink and uplink communication in 5G networks. This section presents downlink NOMA as the foundational focus of the research.

In downlink NOMA, the transmitter, i.e., the BTS, transmits a combined signal, which is the superposition of the desired signals for multiple user devices, each assigned different power allocation coefficients [32]. At the receiver side of each user device, SIC is assumed to be successfully performed until the signal intended for the user device is decoded.

User devices are assigned power allocation coefficients inversely proportional to their channel conditions. Specifically, the user device with the highest power allocation will treat the signals intended for other user devices as interference and decode its own signal directly without any SIC process. However, other user devices must perform SIC.

During the SIC process, at the receiver of each user device, the stronger interfering signals relative to the desired signal of that user device are first detected. These signals are then subtracted from the received signal, and this process is repeated until the signal intended for the specific user device is isolated. Finally, each user device decodes its signal by treating the signals of other user devices with lower power coefficients as interference.

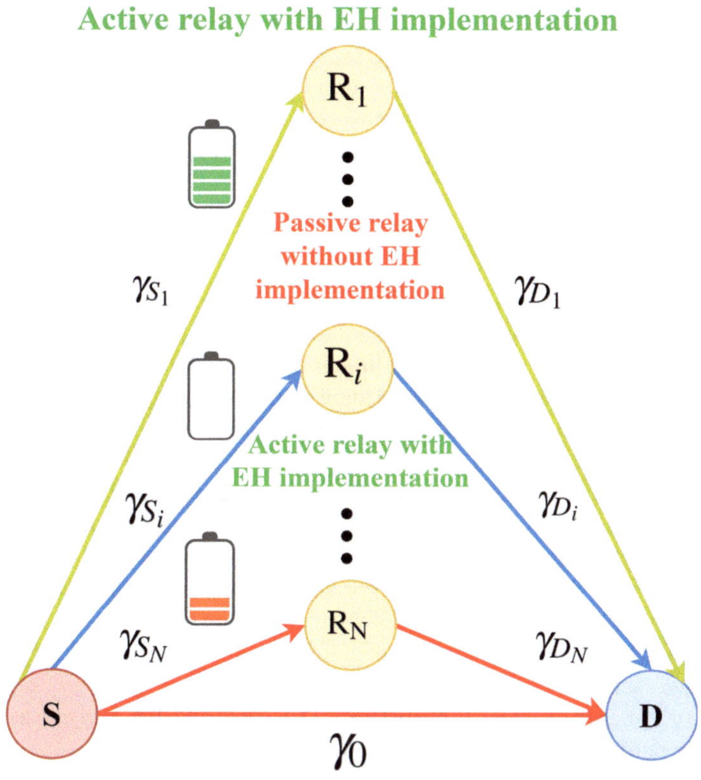

Fig. 5.5 Relay stations help forward EH execution information and store it in battery to help cooperative communication between the base station and the destination node

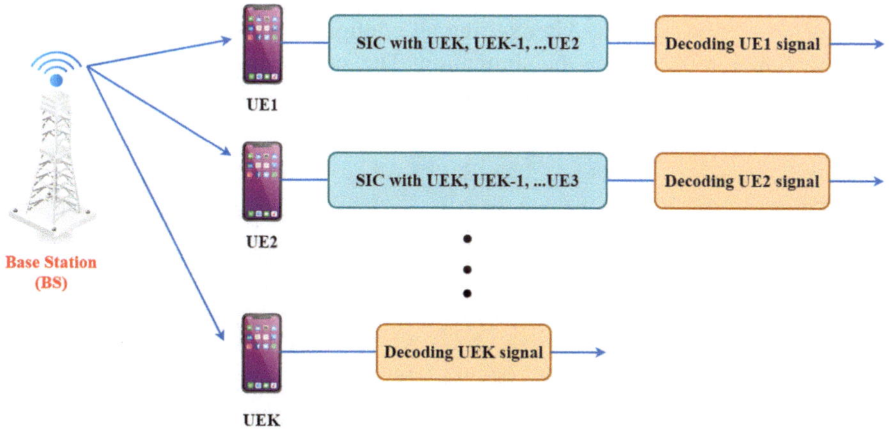

Fig. 5.6 Downlink NOMA for K user devices

Figure 5.6 illustrates a downlink NOMA system consisting of a BS and K user devices equipped with SIC receivers.

5.1.7 Simulation of NOMA

5.1.7.1 Downlink NOMA Examples

In this part, we explain how to set up an example of simulation scenario for downlink NOMA-based cellular network. This simulation is taken over from [25]. The objective of this scenario is to elucidate the functionality of NOMA) within a simulation environment and to illustrate the performance enhancement provided by the 3GPP Multi-User Superposition Transmission (MUST). We configure two distinct cells: one utilizing OMA and the other employing NOMA. In both cells, the base station allocates bandwidth equally between two user equipments (UEs) with favorable channel conditions, referred to as strong users. However, due to the capabilities of MUST in the second base station, it is feasible to superimpose these strong UEs with an additional pair of cell-edge UEs, categorized as weak users. This setup exemplifies how MUST can effectively optimize resource allocation by leveraging superposition coding techniques. The simulation algorithms should go over the following steps:

First, we set up the topology by inputing the `topology.nodes` and `topology.primaryLinks` properties. The BTS and the UE are assigned by the variable names like `BS1, BS2` (for the BTS) and `UE1, UE2, ...`. Secondly, the strong users and the weak users are configured by setting the `simulation.pathloss` property. Thirdly, the multiple acess scheme (OMA or NOMA) is selected by declaring the multiuser transmission mode, allocating the power factor for each user, and assign the scheduling by adjusting the `sschedule.fixed ScchedulеDL` properties. After setting all configuration parameters, we can call the core function to execute the simulation. The sample codes for this simulation scenarios is shown in the following listing.

```
% This script is taken from the 5G Link Level Simulator:
% Pratschner, S., Tahir, B., Nissel, R., Marijanovic, L., Mussbah
    , M., Kirev, K., Schwarz,
% S., Rupp, M.: The vienna 5g link level simulator v1.2. nt.
% tuwien. ac. at/wpcontent/uploads/2020/05/5gll usermanual. pdf,
    Seen In 19, 25 (2021)

%% Topology
% Specifiy all the nodes in ascending order with starting
% index of 1 (BS0 or UE0 is not allowed).
```

```
scStr.topology.nodes                    = ['BS1,BS2,UE1,UE2,UE3,UE4,UE5,
    UE6'];

% Primary (desired) links
scStr.topology.primaryLinks             = [ 'BS1:UE1,' ...
                                            'BS1:UE2,' ...
                                            'BS2:UE3,' ...
                                            'BS2:UE4,' ...
                                            'BS2:UE5,' ...
                                            'BS2:UE6'
                                            ];

% Interference Links
scStr.topology.interferenceGeneration          = 'Automatic';
scStr.topology.attenuation                     = 300;
                               % in dB, set a very high attenuation
    level to virtually decouple the two BS
scStr.topology.interferingLinks                = [];

%% General Simulation Parameters
% set link types to simulate
scStr.simulation.simulateDownlink              = true;
                               % downlink
scStr.simulation.simulateUplink                = false;
                               % uplink
scStr.simulation.simulateD2D                   = false;
                               % device to device links

% Number of simulation frames
scStr.simulation.nFrames                       = 100;
                               % Number of frames to simulate per sweep
     value, adjust to obtain sufficiently small confidence
    intervals.

%% Physical Transmission Parameters
scStr.simulation.centerFrequency               = 2.5e9;
                               % center frequency
scStr.simulation.txPowerBaseStation            = [30];
                               % per BS; base station total transmit
    power in dBm
scStr.simulation.txPowerUser                   = [30];
                               % per UE; user total transmit power in
    dBm
```

```
scStr.simulation.nAntennasBaseStation           = [2];
                        % per BS; number of antennas at the base station
scStr.simulation.nAntennasUser                  = [2];
                        % per UE; number of antennas at the user
scStr.simulation.userVelocity                   = [0];
                        % per UE; velocity in m/s

% Links to UE1 and UE3 have pathloss of 80, UE2 and UE4 of 90, and UE5 and UE6 of 110 and 115, respectively (cell edge).
scStr.simulation.pathloss                       = [80,90,80,90,110,115];    % per Link, channel pathloss in dB, this is most likely swept over

%% Channel Parameters
scStr.channel.dopplerModel                      = 'Discrete-Jakes';
scStr.channel.timeCorrelation                   = false;
scStr.channel.spatialCorrelation                = 'none';
scStr.channel.nPaths                            = 50;
scStr.channel.powerDelayProfile                 = 'PedestrianA';
scStr.channel.K                                 = 0;
scStr.channel.delta                             = 1;

%% Channel Estimation and Equalization Parameters
scStr.simulation.channelEstimationMethod        = 'Approximate-Perfect';
scStr.simulation.noisePowerEstimation           = false;
scStr.simulation.pilotPattern                   = 'LTE Downlink';
scStr.simulation.equalizerType                  = 'One-Tap';
scStr.simulation.receiverTypeMIMO               = 'MMSE';

%% MIMO Parameters
% Layer mapping
scStr.layerMapping.mode                         = '5G';
scStr.layerMapping.table.Uplink                 = {1;2;[1,2]};
scStr.layerMapping.table.Downlink               = {1;2;[1,2]};

% MIMO mode
scStr.modulation.transmissionMode               = 'CLSM';
scStr.modulation.delayDiversity                 = 1;
```

```
scStr.modulation.nStreams                    = [2];                    % per Link; number of active spatial streams
% Per link, precoder selection (used when feedback is disabled)
scStr.modulation.precodingMatrix{1}          = 1/sqrt(2)*ones(2,2);   % per Link; employed precoding matrix

%% Modulation Parameters
% waveform
scStr.modulation.waveform                    = {'OFDM'};

% time and bandwidth setup (number of subcarriers, frame duration, CP
% length, sampling rate)
scStr.modulation.numerOfSubcarriers          = [72];                   % per BS; number of used subcarriers
scStr.modulation.subcarrierSpacing           = [15e3];                 % per BS; per base station in Hz
scStr.modulation.nSymbolsTotal               = [15];                   % per BS; total number of time-symbols per frame, the frame duration will be nSymbolsTotal/subcarrierSpacing
scStr.modulation.nGuardSymbols               = [1];                    % per BS; select how many of the total time-symbols will be used as guard symbols (cyclic prefix in OFDM)
scStr.modulation.samplingRate                = 'Automatic';            % sampling rate has to be the same for all nodes (across all base stations) either numeric value for manual setting or 'Automatic'

%% Channel Coding Parameters
% All links are operating with the same coding parameters, enter it only once.
scStr.coding.code                            = {'Turbo'};
scStr.coding.decoding                        = {'MAX-Log-MAP'};
scStr.coding.decodingIterations              = [8];

%% Schedule
% static schedule per base station

% BS1 does Orthogonal Multiple Access
```

```
scStr.schedule.fixedScheduleDL{1}            = ['UE1:36,UE2:36'];    % schedule for BS1 Downlink
scStr.schedule.fixedScheduleUL{1}            = [];                   % No uplink for BS1.

% BS2 does MUST operation
scStr.schedule.fixedScheduleDL{2}            = ['UE3:36,UE4:36,UE5:UE3,UE6:UE4'];    % schedule for BS2 Downlink
scStr.schedule.fixedScheduleUL{2}            = [];                   % No uplink for BS2.

% Other setups
% Due to page limitation, other common setups are skipped

% Downlink simulation
    for iBS = 1:nBS
            BS{iBS}.generateTransmitSignal(Links);
        end

        for iUE = 1:nUE
            UEID = UE{iUE}.ID;
            primaryLink = Links{UE{iUE}.TransmitBS(1), UEID};
            if primaryLink.isScheduled
                primaryLink.generateReceiveSignal();
                UETotalSignal = primaryLink.ReceiveSignal;

                % Collect signals from all other BSs
                for iBS = 2:length(UE{iUE}.TransmitBS)
                    currentLink = Links{UE{iUE}.TransmitBS(iBS), UEID};
                    currentLink.generateReceiveSignal();
                    UETotalSignal = Channel.addSignals(UETotalSignal, currentLink.ReceiveSignal);
                end
                % correct signal length
                UETotalSignal = Channel.correctSignalLength(UETotalSignal, primaryLink.Modulator.WaveformObject.Nr.SamplesTotal);

                % add noise
```

```
                    UETotalSignal = UETotalSignal + Channel.AWGN(
                        simParams.phy.noisePower, length(
                        UETotalSignal), UE{iUE}.nAntennas );

                    % process received signal
                    UE{iUE}.processReceiveSignal(UETotalSignal,
                        Links, simParams);

                    % Collect the results which is now stored in
                        the primary link
                    primaryLink.calculateSNR(simParams.constants.
                        BOLTZMANN, simParams.phy.temperature);
                    perSweepResults{UE{iUE}.TransmitBS(1), UEID,
                        iFrame} = primaryLink.getResults(
                        simParams.simulation.saveData);
                end
            end

% Output results: please refers
```

5.1.8 SWIPT in C-NOMA Simulation

Another simulation example in this module is a comprehensive system model that integrates three advanced techniques: EH, FD, and NOMA, as depicted in Fig. 5.7. This scenario is generated according to [20], in which a source node (S) want to transmit wo distinct messages to two mobile users D1 (far user) and D2 (near user) by facilitating a FD relay node (R) and a power-domain NOMA mechanism. Both S and R are required to harvest energy from RF signals generated effectively by a a power beacon (PB). The relay node (R) features dual antennas with self-interference (SI) cancellation capability. All other nodes are single-antenna devices.

Following the EH phase with the length αT, nodes S and R convert the harvested RF energy into usable power for signal transmission. Then, the source node S transmits the superimposed message to the relay R, while R simultaneously decodes the messages from previous blocks and forwards them to D_1 and D_2. Figure 5.8 illustrates the scheduling protocol, where T denotes the duration of one transmission block and α is the time-switching factor of EH.

For this simulation, we apply the theoretical formulas developed in [20] to rigorously assess the system performance of the considered model. To ensure robustness and accuracy, Monte-Carlo simulations are employed as a verification tool for these analytical results. This dual approach not only strengthens the validity of our findings but also provides a comprehensive understanding of the system's operational dynamics.

5.1.9 Simulation Results

The BER performance result at the mobile users in the downlink NOMA-based cellular network described in the previous subsection is illustrated in Fig. 5.9. It can be observed from this figure that the BER values decreases with the transmit SNR, while the users that exploit the NOMA scheme can achieve better BER compared to the OMA users. Figure 5.9 also shows that the fairness can be achieved between users within a NOMA cell, regardless the distance from the users to the BTS.

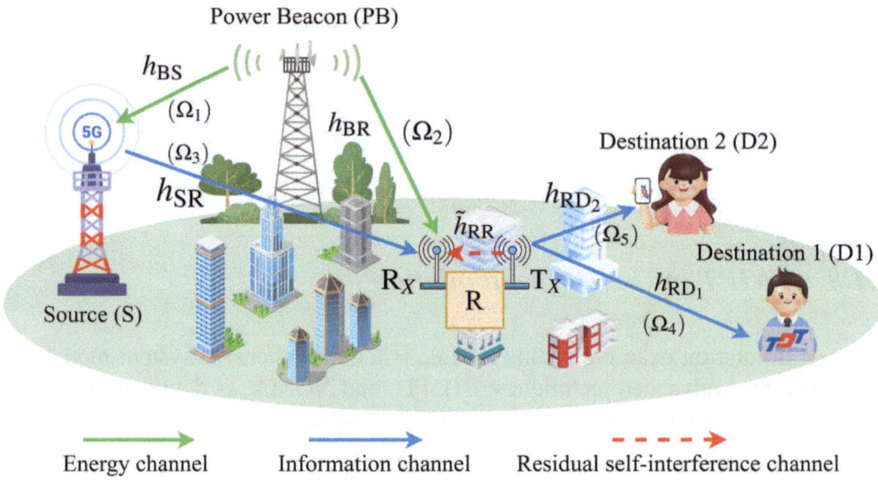

Fig. 5.7 Simulation model for SWIPT C-NOMA network

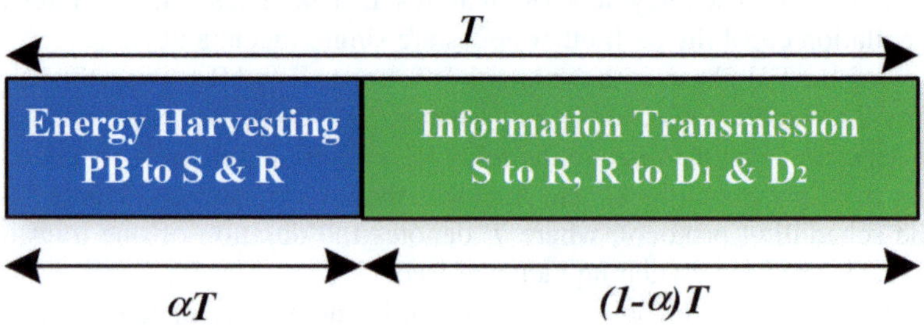

Fig. 5.8 Scheduling protocol for the SWIPT C-NOMA network in simulation

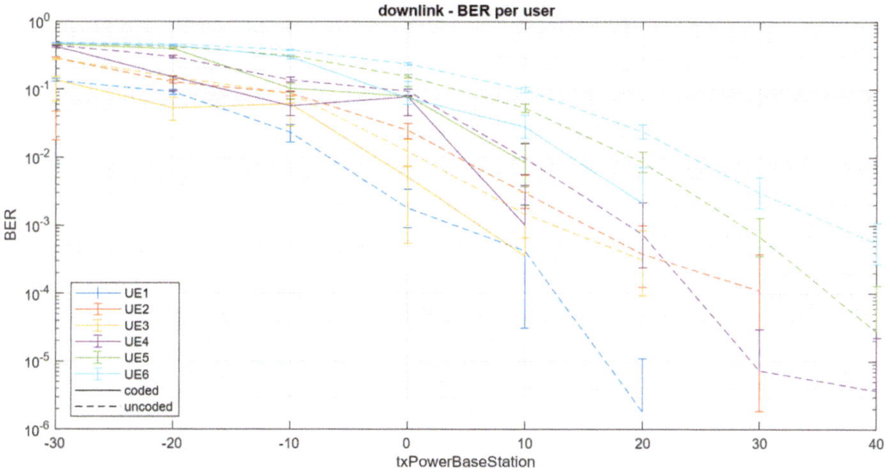

Fig. 5.9 BER performance of different users in NOMA and OMA cellular networks

In the following paragraphs, the simulation results for the SWIPT C-NOMA network are illustrated. The system performance is rigorously assessed across various average SNR values, which is defined as the quotient of the transmit power of PB and the variance of AWGN, i.e. SNR $= P/\sigma^2$. For our simulations, we have meticulously selected parameters to evaluate system performance: power allocation coefficients are set at $a_1 = 0.65$ and $a_2 = 0.35$, with an EH efficiency η of 0.85 at each node. The average channel gains are assigned as $\Omega_1 = \Omega_2 = \Omega_3 = \Omega_5 = 1$, $\Omega_4 = 0.7$. These simulation results were derived from an extensive dataset comprising 10^6 channel realizations, ensuring robust statistical significance and reliability in our findings for academic scrutiny and further research implications.

Figure 5.10 illustrates the throughput of the EH-FD-NOMA system versus average SNR with various data transmission rate $\mathcal{R} = 0.3$; 0.5; 0.7 bit/s/Hz. It shows that at low SNR regime (i.e. SNR < 10 dB), the highest throughput is obtained with the lowest data rate ($\mathcal{R} = 0.3$). When SNR increases, the optimal throughput can be obtained at higher data rate, such as $\mathcal{R} = 0.7$ when SNR > 26 dB. This suggests that by strategically selecting the data transmission rate in accordance with the available transmit power from the PB, we can achieve optimal throughput and thereby enhance overall system performance effectively.

Figure 5.11 illustrates the influence of EH time-switching ratio α on the OP of the system of interest with a data rate $\mathcal{R} = 0.3$ bit/s/Hz and an interference level $\tilde{\Omega} = -30$ dB. The figure clearly demonstrates that there exists an optimal value of α that aligns with a specific transmit power of the PB. Thus, by precise parameter tuning for the EH, we can effectively enhance the system performance.

In Fig. 5.12, we observe the ergodic capacity of the considered network (including separate capacity of each user and the sum-capacity) across varying time-switching factor α, with the same interference level as Fig. 5.11. The data illustrates a positive

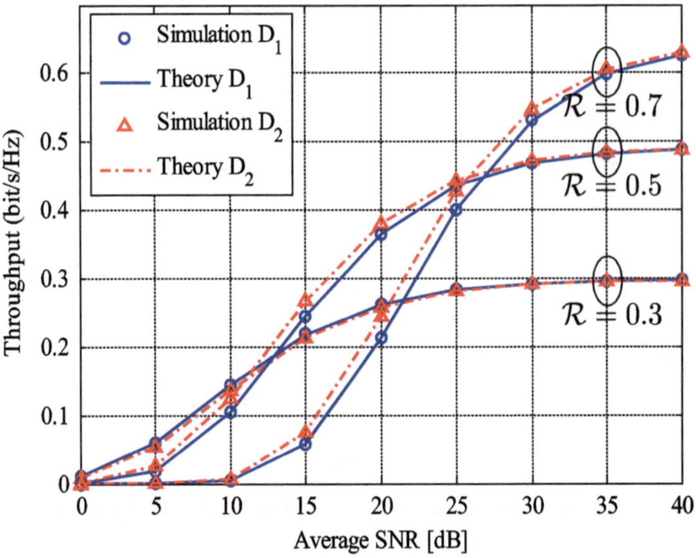

Fig. 5.10 Throughput of the SWIPT C-NOMA system with various data rate $\mathcal{R} = 0.3; 0.5; 0.7$ bit/s/Hz

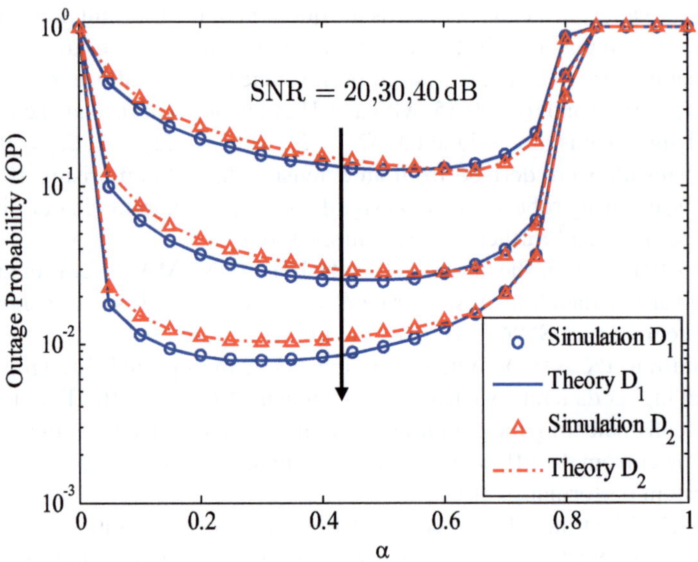

Fig. 5.11 Effect of time-switching factor on the OP with SNR = 20, 30, 40 dB, $\mathcal{R} = 0.3$ bit/s/Hz and $\tilde{\Omega} = -30$ dB

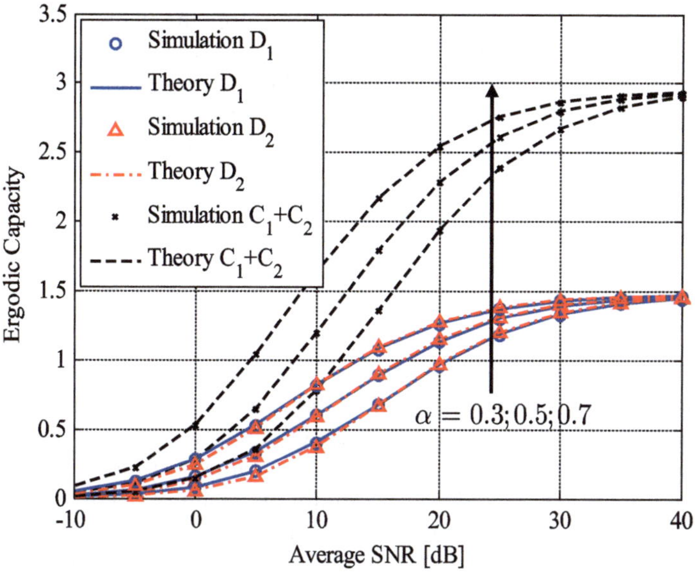

Fig. 5.12 Effect of time-switching factor and transmit SNR on the ergodic capacity with $\tilde{\Omega} = -30$ dB

correlation between increasing values of α and enhanced ergodic capacities for users D1 and D2, but as high SNR levels, such as more than 40 dB, the ergodic capacities converge to the same value for three values of α.

5.2 Sparse Code Multiple Access (SCMA)

In the realm of future 5G wireless networks, accommodating diverse traffic types with stringent requirements is a critical challenge [46]. These requirements encompass massive connectivity, enhanced QoS, increased throughput, reduced latency, and minimized control signaling overhead [47]. Traditional OMA-based systems may fall short in supporting the connectivity demands of tremendous number of devices for futuristic applications. Recently, researchers have pivoted toward NOMA systems to achieve superior spectral efficiency and reduced latency in 5G/6G networks.

Among the various forms of NOMA being explored, including power domain, code domain, and pattern division multiple access, Sparse Code Multiple Access (SCMA) emerges as a promising code domain technique. In SCMA, each user can utilize multiple orthogonal resources (including time slots, frequency bands, and codes) for communication [48]. This approach employs multi-dimensional codebooks to enhance shaping gain and can be viewed as an extension of the low-density spreading (LDS) multiple access technique. The recent surge in interest from researchers underscores SCMA's potential for 5G and beyond wireless networks.

This section delves into the simulation of SCMA as a code-domain-based NOMA scheme aimed at boosting spectral efficiency in wireless networks [48]. We conduct a thorough analysis of SCMA system design and detection, while emphasizing the significance of codebook design on system performance.

5.2.1 Introduction

Multiple access serves as a fundamental technique in wireless communication, designed to facilitate simultaneous access for multiple users to limited resources in an efficient manner. Traditional multiuser communication systems predominantly employ OMA schemes, where users are orthogonal concerning specific resources such as time, frequency, or codes [49]. Within the realm of OMA schemes, multi-carrier CDMA (MC-CDMA) exemplifies a multiple access strategy by distributing modulated QAM symbols across OFDMA tones using generally complex spreading sequences. This method effectively replicates a QAM symbol over several tones. A noteworthy variant of CDMA is Low-Density Signature (LDS), which utilizes low-density spreading sequences and supports the implementation of a message passing algorithm (MPA) receiver, which requires low computational complexity but delivers performance comparable to the optimal maximum likelihood (ML) receiver [50].

In recent years, the concept of NOMA has garnered significant research interest, primarily due to its potential to support massive connectivity, reduce network latency, and improve spectral efficiency beyond what OMA can achieve. In a NOMA system, multiple users are superimposed on the same physical resource, such as power, frequency, time, or code. In 2013, Nikopour and Baligh [51] introduced SCMA, a significant advancement in the field of NOMA within the code domain..

SCMA presents a sophisticated multi-dimensional codebook-based non-orthogonal spreading mechanism, which can be viewed as an extension of LDS. Different from LDS, SCMA innovatively combines bit-to-QAM mapping and sparse sequence spreading into a singular, cohesive process. This integration allows for the direct mapping of a set of bits to a sparse complex vector, termed a codeword. Each layer utilizes a unique SCMA codebook set, with the non-orthogonal layers superimposed on one another.

The primary performance improvement of SCMA is obtained from the shaping and coding gains [52] achieved through its multi-dimensional constellations, which surpass the simple repetition-based spreading of LDS. Additionally, SCMA maintains low-complexity receiver design by utilizing the MPA detection technique, thanks to the inherent sparsity of its codewords. Furthermore, SCMA facilitates system overloading, allowing for a number of multiplexed layers that exceed the spreading factor or codeword length.

Since Nikopour's seminal work in 2013 [51], three principal research trajectories have emerged within SCMA schemes: codebook design, multiuser detection (MD), and resource allocation. The first area focuses on designing and optimizing methods to develop effective multidimensional constellations tailored to specific channel conditions. This aims to enhance key performance metrics such as BER, SE, and PAPR [47, 53, 54]. The second research direction involves creating low-complexity receiver algorithms for user signal separation, which are more efficient or less complexity compared to the original MPA without compromising BER performance [55, 56]. Lastly, resource allocation seeks to optimize both user-specific spreading sequences and power distribution among SCMA users [57, 58].

This section offers a comprehensive introduction to SCMA alongside a systematic methodology for designing SCMA codebooks by leveraging lattice constellation design principles. Simulation methods and results are presented as well, illustrating SCMA's superior reliability and SE compared to LDS and OFDMA.

5.2.2 System Model

In the context of an uplink SCMA scheme, we examine a system where J users transmit their information to a BTS by distributing their signals across K orthogonal resource elements, e.g. time slots or frequency bands, within the wireless medium.[1] Importantly, our analysis predominantly focuses on overloaded scenarios characterized by the condition $J > K$.

An SCMA encoder has J layers and we assume one layer per user. In SCMA, each user's data is directly encoded into a multi-dimensional complex vector for transmission purposes. Each user possesses a codebook denoted as $\mathcal{X}_j$, where $j = 1, 2, \ldots, J$. This codebook comprises M_j complex columns, each with a dimensionality of K. For simplicity and without loss of generality, we assume uniform constellation size and length across all layers; thus, $M_j = M$ and $N_j = N$ for every user index j. Consequently, the structure of each user's codebook $\mathcal{X}_j$ is defined by dimensions $K \times M$, consisting of complex elements. It can be expressed as $\mathcal{X}_j = \left[\mathbf{x}_{j,1}, \mathbf{x}_{j,2}, \ldots, \mathbf{x}_{j,M}\right]$, belonging to the space $\mathbb{C}^{K \times M}$. For instance, if we set $M = 4$, then each data symbol encodes $\log_2(M) = 2$ bits. Each column within the codebook $\mathcal{X}_j$, referred to as a codeword, corresponds uniquely to one data symbol.

Let an intermediate codebook denoted by $C_j \subseteq \mathbb{C}^{N_j}$ represents an M-ary modulation scheme in an N_j-dimensional space associated with user j. To facilitate symbol spreading of $\mathbf{c}_j \in C_j$ from this set, a binary matrix denoted as $\mathbf{V}_j \in {\mathbb{F}_2}^{K \times N_j}$ executes a linear spreading operation. Analogous to optical orthogonal codes (OOC) [59], this mapping process can be depicted using a binary vector $\mathbf{f}$ with length K, which specifies the positions corresponding to nonzero entries within the given codebook structure.

Figure 5.13 provides a detailed representation of the SCMA encoder process for user j. The SCMA encoder is composed of J distinct layers, each operating independently. Within each layer, the constellation function μ_j generates a set C_j consisting

[1] This configuration is typically denoted as a $J \times K$ SCMA system.

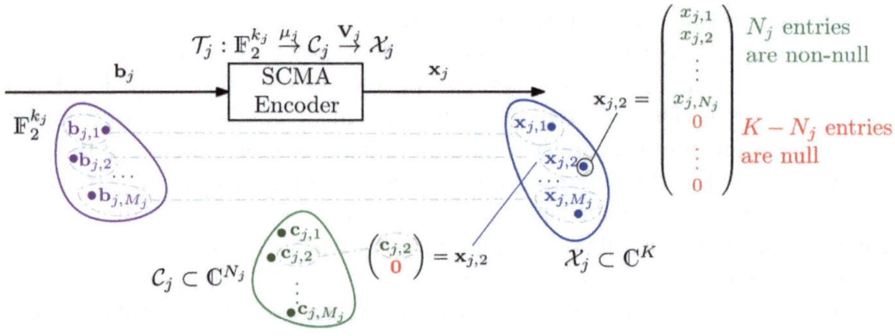

Fig. 5.13 Illustration of the SCMA encoder $\mathcal{T}_j$ of user j. *Source* [55]

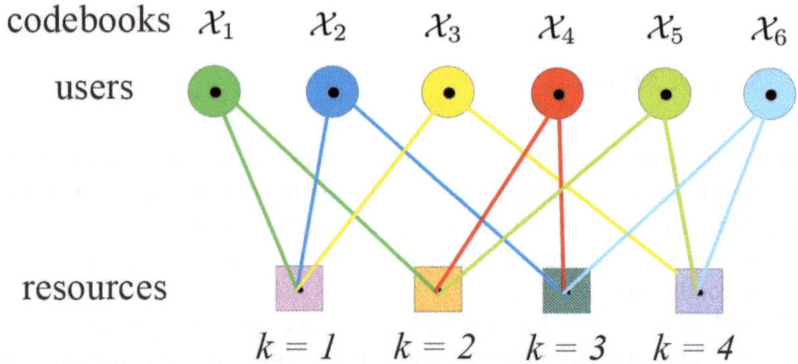

Fig. 5.14 Factor graph representation of an SCMA system with $J = 6$, $K = 4$, $N = 2$, and $d_f = 3$

of M_j symbols, where each symbol has a length of N_j. Subsequently, the mapping matrix $\mathbf{V}_j$ projects these N_j-dimensional constellation points into SCMA codewords, thereby forming the codeword set $\mathcal{X}_j$, where each codeword is organized as one column. Users then select their respective codewords from their designated codebooks; these selected codewords are aggregated as illustrated in Fig. 5.15. It is crucial to note that $\mathcal{T}_j$ is determined by K, M_j, μ_j, N_j, C_j, and $\mathbf{V}_j$. The sparse codebook $\mathcal{X}_j$ is entirely determined by C_j and matrix $\mathbf{V}_j$. The overall structure of SCMA codes is encapsulated within a factor graph matrix denoted as $\mathbf{F} = (\mathbf{f_1}, ..., \mathbf{f_J})$, where $\mathbf{f_j} = \text{diag}\left(\mathbf{V}_j \mathbf{V}_j^T\right)$. In this graphical model, layer node j maintains connectivity with resource node k (indicating active transmission by user j over resource element kth) if and only if $\mathbf{F}_{kj} = 1$. Figure 5.18 exemplifies such a factor graph for matrix $\mathbf{F}$, featuring six symbol nodes alongside four resource nodes; and the corresponding matrix $\mathbf{F}$ is given below:

$$\mathbf{F}_{4\times 6} = \begin{bmatrix} 1\,1\,1\,0\,0\,0 \\ 1\,0\,0\,1\,1\,0 \\ 0\,1\,0\,1\,0\,1 \\ 0\,0\,1\,0\,1\,1 \end{bmatrix}. \tag{5.1}$$

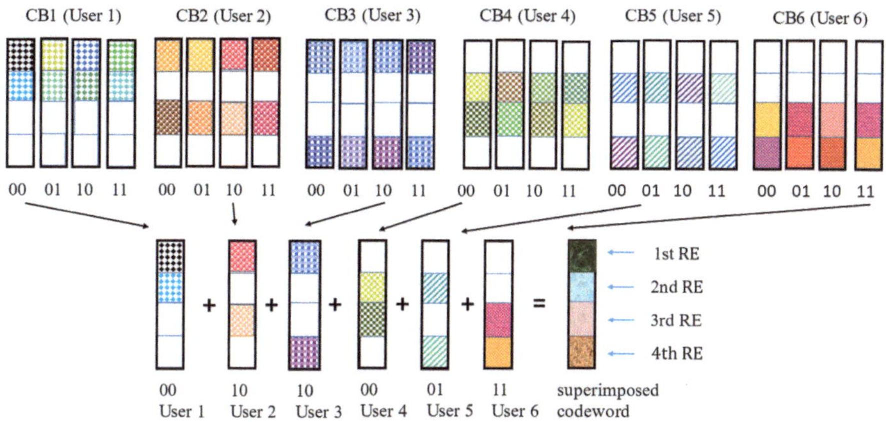

Fig. 5.15 An illustration of 4 × 6 SCMA Encoding. *Source* [49]

(a) Uplink signal model:

For an uplink SCMA system like in Fig. 5.16, let $\mathbf{h}_j = \left[h_{1j}, \ldots, h_{Kj}\right]^T$ be the effective channel fading vector for the jth user, where h_{kj} denotes the channel fading coefficient at the kth RE for the jth user. Then, the received signal at the BTS after layer multiplexing can be written as

$$\mathbf{y} = \sum_{j=1}^{J} \sqrt{P_j} \text{diag}(\mathbf{h}_j)\mathbf{x}_j + \mathbf{n}, \tag{5.2}$$

where $\mathbf{x}_j = \mathbf{V}_j \mu_j(\mathbf{b}_j)$ represents the SCMA codeword for layer j, $\mathbf{h}_j$ denotes the channel vector for layer j, P_j is the transmit power of user j, and $\mathbf{n} \in \mathbb{C}^{K\times 1}$ is the AWGN noise vector, which is Gaussian distributed, i.e. $\mathbf{n} \sim \mathcal{CN}(\mathbf{0}, N_0\mathbf{I}_K)$. If all layers are transmitted from the same geographical point, the channel vectors are identical across layers, i.e., $\mathbf{h}_j = \mathbf{h}, \forall j$. Here, we define the overloading factor of the code as $\lambda \triangleq J/K$.

Let's consider the received signal as the kth RE, which can be denoted by y_k. Due to the sparsity of $\mathbf{x}_j$, non-zero values from d_f out of J number of users overlap over the RE k. Consequently, each user's data spans only $d_v < K$ REs. Let ξ_k be the set of users transmitting data over the kth RE and ζ_j be the set of REs on which the jth user has active transmission, respectively. Then, the received signal at the kth RE can be determined as

$$\mathbf{y}_k = \sum_{j\in\xi_k} \sqrt{P_j} h_{kj} C_{k,j}(\mathbf{b}_j)\mathbf{x}_j + n_k, \forall k = 1, 2, \ldots, K. \tag{5.3}$$

(b) Downlink signal model:

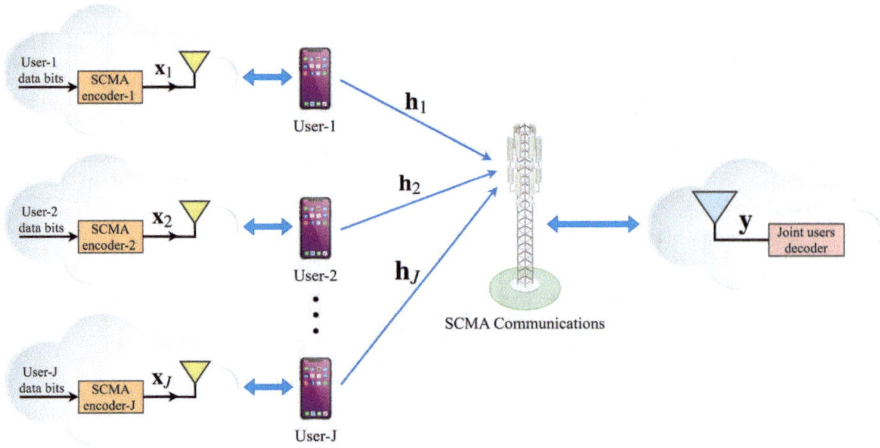

Fig. 5.16 Uplink SCMA system model

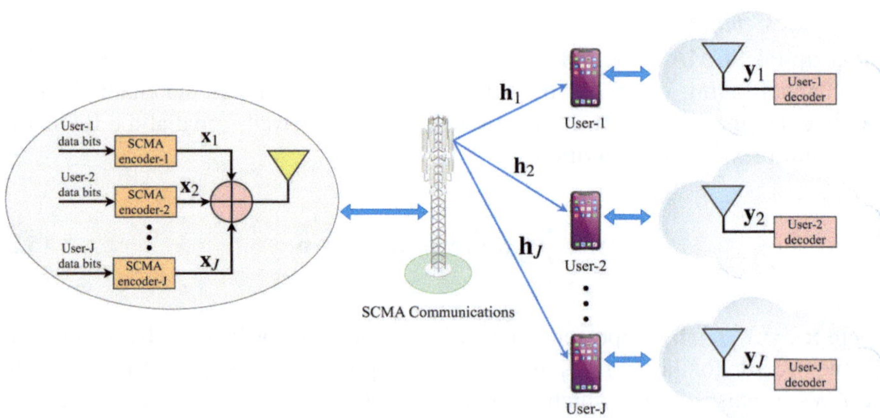

Fig. 5.17 Downlink SCMA system model

In downlink SCMA shown in Fig. 5.17, all users' information is broadcast from the BTS and each user node receives the sum of all users' codewords. The received signal $\mathbf{y}_i = [y_{i1}, \ldots, y_{iK}]$ at the ith user can be expressed as

$$\mathbf{y}_j = \text{diag}(\mathbf{h}_j) \sum_{j=1}^{J} \sqrt{P_j} \mathbf{x}_j + \mathbf{n}_j, \tag{5.4}$$

where $\mathbf{n}_j$ is the AWGN noise at the jth user, which satisfies $\mathbf{n}_j \sim \mathcal{CN}(\mathbf{0}, N_0 \mathbf{I}_K)$.

5.2.3 SCMA Codebook Design

The performance of an SCMA scheme is significantly influenced by the codebooks and their allocation to users. Currently, there are no established method in the literature for designing optimal codebooks with guaranteed results. Typically, suboptimal approaches are employed for this purpose. An SCMA scheme is generally characterized by parameters such as J, K, $\mathbf{F}$, and $\mathcal{T}j\ (Mj, \mu j, Nj, Cj)\,, \forall j$. Suppose that we focus on a specific performance indicator $\mathfrak{R}$, such as maximizing the sum rate of all users, while adhering to certain constraints. In this context, the problem of designing codebooks can be formulated accordingly as

$$\begin{aligned} &\underset{J,K,F,\mathcal{T}_j(M_j,\mu_j,N_j,C_j),\forall j}{\text{maximize}} \quad \mathfrak{R}(J, K, F, \mathcal{T}_j(M_j, \mu_j, N_j, C_j)) \\ &\text{subject to} \quad \mathfrak{C} \end{aligned} \tag{5.5}$$

where $\mathfrak{C}$ represents a set of constraints, such as $J > K, N_j \leq K\forall j, P \leq 1$, etc.

In fact, identifying a practical convex function as a performance metric in (5.5) poses a significant challenge. While metrics such as bit or packet error rate are typically considered, their analytical forms for SCMA codebooks remain elusive. Consequently, Problem (5.5) is not readily approachable without introducing certain assumptions. These may include defining arbitrary parameters, ensuring regular structural patterns among users, and decoupling dependencies across design stages, etc. To streamline the SCMA codebook design process, we assume that matrix $\mathbf{F}$ adheres to a regular design framework where each column maintains a uniform weight $N_j = N, \forall j$, and each row exhibits a consistent collision degree $d_k = d_f, \forall k$.[2]

The following multistage method for SCMA codebook design aligns with the approach proposed by Nikopour in 2013 [51] and has been adopted in subsequent research like [47].

5.2.3.1 Mapping Matrix

The mapping matrices in the set $\mathcal{V}$ are crucial for figuring out the number of interfering layers at each RE, which decide the complexity of the receiver using MPA. Sparser codewords reduce the detection complexity. For maximum overloading, the configuration for $\mathbf{V}$ can be achieved by adding $K - N$ all-zero row vectors to $\mathbf{I}_N$. This solution offers key properties: $J = \binom{K}{N}, d_f = \binom{K-1}{N-1} = \frac{JN}{K}, \forall j$, and the overloading factor $\lambda = \frac{J}{K} = \frac{df}{N}$. It also fulfills the condition $\max(0, 2N - K) \leq l \leq N - 1$, with l being the count of overlapping elements between any two different vectors $\mathbf{f}_j$. These relationships are vital for optimizing system performance within this framework.

[2] This inherently implies that every codebook C_j encompasses N complex dimensions.

5.2.3.2 Multi-dimensional Mother Constellation and Rotated Constellation

For codebooks design, some works [47, 51, 60] have proposed designing J codebooks C_j based on a Mother Constellation (MC) C combined with rotation operators. After determining the set of mapping matrices $\mathbf{V}^{+}$, the codebook optimization problem for SCMA now can be simplified to

$$\begin{aligned} \boldsymbol{\mu}^{+} &= \arg\max_{\boldsymbol{\mu}} \mathfrak{R}(J, K, M, N, \mathbf{V}^{+}, \boldsymbol{\mu}). \\ \text{subject to} &\quad \mathfrak{C}. \end{aligned} \tag{5.6}$$

The task now involves designing J distinct N-dimensional constellations, each comprising M points. To streamline the process, we model the constellation points of various layers using a mother constellation alongside some operators that are designed specifically for each layer. Formally, this is expressed as $\mu_j \equiv (\mathbf{\Delta}_j)\mu, \forall j$, where $\mathbf{\Delta} j$ represents a constellation operator. This modeling approach transforms the code optimization problem into an equivalent mathematical framework shown below, which leverages these structured operators to achieve efficient solutions across multiple layers:

$$\mu^{+}, [\mathbf{\Delta}_j^{+}]_{j=1}^{J} = \arg \max_{\mu, [\mathbf{\Delta}_j]_{j=1}^{J}} \mathfrak{R}(J, K, M, N, \mathbf{V}^{+}, \boldsymbol{\mu} = \left[\mathbf{\Delta}_j^{+}\right]_{j=1}^{J}). \tag{5.7}$$

The optimization problem (5.6) now can be solved by splitting it to two subproblems, in which the mother constellation and the designed operators are solved separately.

(a) Design Metrics and Rotated Constellations

The performance of SCMA method, particularly with a small number of layers, relies on maintaining a substantial minimum Euclidean distance in its constellation to prevent layer collisions on any RE. However, the likelihood of collision rises for large number of layers. To recover codewords that may be damaged by colliding, we can rely on the alternate REs by introducing dependencies among non-zero elements. Additionally, uneven power distribution to different codeword dimensions can create a near-far effect among the layers that are active on the same resource, facilitating the MPA process to mitigate interference between co-paired layers.

To address these challenges, one can augment a constellation that has reasonable Euclidean distances among the points by applying a unitary rotation to the base constellation. This adjustment fine-tunes dimensional dependencies and power variations of the original constellation but keeping the good Euclidean distance characteristics. Drawing inspiration from the works of [61, 62], which provided good coding for fast-fading channel communication, optimizing this rotation seeks to maximize the minimum product distance within the constellation. Consequently, the goals of our design includes minimizing both the sum and the product of the distances between mother constellation points. The product distance emerges as the important metric for evaluation of the communication performance in high-SNR regime.

(b) Rotated lattic constellations
Generally, the base constellation can be formed by any set of points in a multi-dimensional space, with the goal to maximize the minimum Euclidean distance. For low-rate scenarios, heuristic optimization methods are often sufficient for constellation design. However, as the data rate increases, a more structured design methodology becomes necessary. A lattice constellation serves as a systematic approach to designing the base constellation. Specifically, the base constellation can be constructed using orthogonal QAMs on distinct planes of complex vectors, representing a lattice structure from Z^{2N}. One key advantage of such lattice constellations is the inherent support for Gray labeling. Furthermore, unitary rotations of QAM-based lattice constellations have been explored in [61] for dimensions ranging from 2 to 4, with the objective of maximizing the minimum product distance of the resulting lattices after rotation.

(c) Shuffling multi-dimensional constellation imaginary axes
A complex constellation designed such that its real and imaginary parts are independent can significantly reduce decoding complexity while preserving the dependency across the complex dimensions of the resulted constellation. This approach decreases the complexity order of the MPA from M^{d_f} to $M^{d_{f/2}}$, offering substantial computational savings, particularly for large constellation sizes.

This section introduces a shuffling method, illustrated in Fig. 5.18, to achieve the separation of real and imaginary parts. The method constructs an N-dimensional complex mother constellation by forming the Cartesian product of two N-dimensional real constellations, each of which is generated using the technique described in the preceding section. One of these real constellations represents the real part of the complex mother constellation points, while the other represents the imaginary part.

An example of this shuffling process is depicted in Fig. 5.19, where a 16-point SCMA mother constellation (T16QAM) suitable for codebooks with $N = 2$ is constructed. The optimal rotation angle to maximize the minimum product distance is found as $\tan^{-1}\left(\frac{1+\sqrt{5}}{2}\right)$.

(d) Rotation to minimize the number of projection points
To simplify MPA decoding, it is preferable to use mother constellations with fewer projections per RE (or complex dimension). Let m represent the number of projections per complex dimension in an M-point constellation, where $m \leq M$. It can be seen that the complexity of the associated MPA detector decreases with m by a factor of m^{d_f}. In the design phase of the mother constellation, configuring the rotation matrix is crucial for minimizing the number of projected points. However, this reduction can result in a minimum product distance of zero, which negatively impacts the high-SNR performance of the SCMA system. Thus, a trade-off arises between high-SNR performance and decoding complexity. However, this configuration may lead to a zero minimum product distance, adversely affecting the performance of SCMA

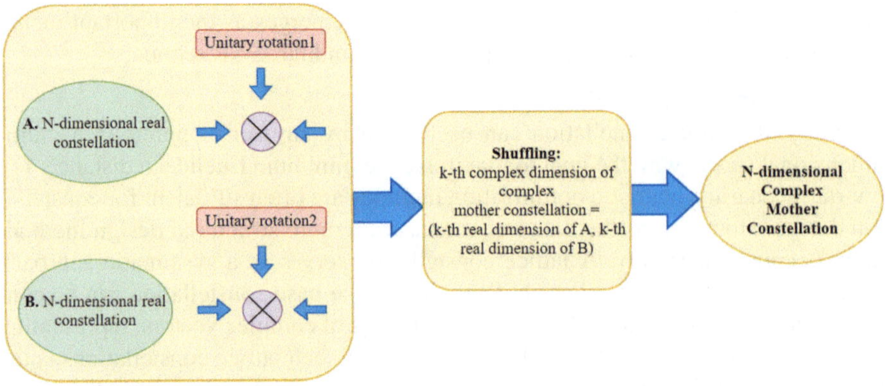

Fig. 5.18 Shuffling of real and imaginary axes to mix two orthogonal sub-constellations

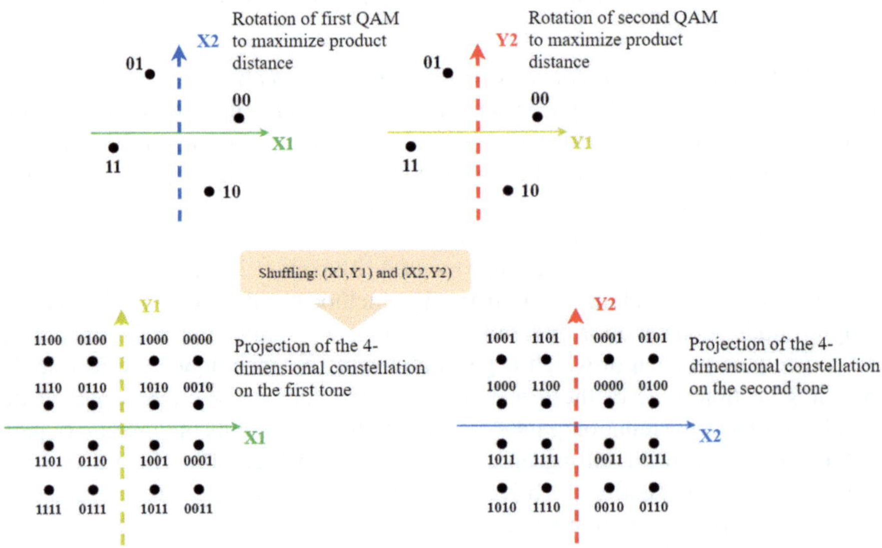

Fig. 5.19 An example of a 16-point SCMA constellation for SCMA codewords with two nonzero elements

at the high-SNR regime. This scenario presents a trade-off between the performance at high-SNR scenario and the computational workload of the system.

Figure 5.20 illustrates an example solution with $m = 9$ for 16-point constellation. This approach achieves an optimal balance by maintaining competitive performance with T16QAM in the mid-SNR regime while effectively reducing complexity from an order of 16^3 to 9^3. Such findings underscore the importance of strategic design choices in achieving lower computational complexity performance without compromising on system performance.

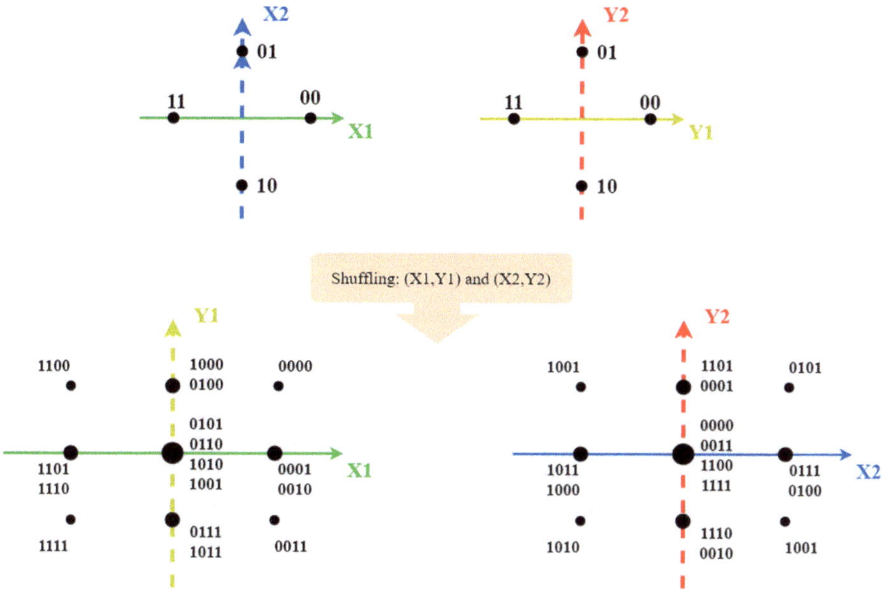

Fig. 5.20 16-point SCMA constellation with 9 projection points per complex dimension

5.2.3.3 Design of Layer-Specific Function Operators

After solving the first sub-problem for the mother constellation to obtain the solution C^+ or equivalently $\boldsymbol{\mu}^+$, the remaining task to solve the original problem (5.6) is to design an optimal operator for each layer. This problem can be formulated as

$$[\boldsymbol{\Delta}_j^+]_{j=1}^J = \arg\max_{[\boldsymbol{\Delta}_j]_{j=1}^J} \mathfrak{R}(J, K, M, \mathcal{V}^+, \boldsymbol{\mu} = [(\boldsymbol{\Delta}_j)\mu^+]_{j=1}^J)). \tag{5.8}$$

The operators in question may encompass phase rotation and power offsets that are specific to each layer. The SCMA codebooks for various layers are derived from the mother constellation μ alongside the respective operator $\boldsymbol{\Delta} j$ for the layer j. In uplink scenarios, because the layers may be transmitted on different channels, the necessity for designing the phase rotation for each layer is low. Conversely, in downlink transmission where layers may traverse on the same channel, meticulous design of phase rotation as well as power offset for each specific layer is crucial to prevent destructive codeword collisions. This strategic approach ensures that the points within the resulted super-constellation formed by combining the layers do not overlap.

There have been various SCMA codebooks design in literature. For example, in [47], the authors employed LDS-like phase signatures as referenced in [63], facilitating the construction of SCMA codebooks for multiple users from a single mother constellation without necessitating power offsets. SCMA codebooks from [47, 51] are denoted Tx-QAM, where x is the number of unique complex projections per complex-dimension of the codebook. For more information on SCMA codebooks design, the readers can refer to a survey on different codebook design methods by Vameghestahbanati et al. [64].

5.2.4 SCMA Decoding

The sparse nature of SCMA facilitates joint multiuser detection (MUD) through the application of the MPA on the factor graph, achieving performance that closely approximates that of maximum a posteriori (MAP) joint detection. The impetus for employing MPA lies in the reduced complexity order inherent in joint detection process. Specifically, the complexity order should be $O(M^J)$ when MAP detection is applied, because of brute-force search on all potential outcomes for superposed symbols at the receiver. Reversely, by integrating MPA with sparse signatures design, a reduced complexity order of $O(M^{d_f})$ can be realized for joint detection, where d_f signifies the number of branches linked to a resource node [51]. Although MPA also exhibits exponential complexity orders, it offers substantial computational cost reduction if $d_f \ll J$.

5.2.4.1 SCMA MAP Detection

Given the received signal $\mathbf{y}$ and channel information $\{\mathbf{h}_j\}_{j=1}^{J}$, the detected multiuser codeword $\hat{\mathbf{X}} = [\mathbf{x}_1, \mathbf{x}_2, \ldots, \mathbf{x}_J]$ can be written as

$$\hat{\mathbf{X}} = \underset{\mathbf{x}_j \in \mathcal{X}_j, \forall j}{\arg\max}\; p\left(\mathbf{X}|\, \mathbf{y}\right) \tag{5.9}$$

where $\mathbf{x}_j$ denotes the codeword transmitted by the jth user and $\mathcal{X}_j$ represents the jth user's codebook, $p\left(\mathbf{X}|\, \mathbf{y}\right)$ is the a-posteriori probability mass function (PMF) of the random vector $\mathbf{X}$. Then the transmitted codeword can be determined by maximizing its a-posteriori marginal PMF.

By applying Bayes' rule $p\left(\mathbf{X}|\, \mathbf{y}\right) = f\left(\mathbf{y}|\, \mathbf{X}\right) P(\mathbf{X})$ and assuming that the noise terms experienced at the K REs are i.i.d., we obtain the following equation to estimate the codeword $\mathbf{x}_j$:

$$\hat{\mathbf{x}}_j = \underset{\mathbf{x}_j \in \mathcal{X}_j}{\arg\max} \sum_{\sim \mathbf{x}_j} \left(P(\mathbf{X}) \prod_{k=1}^{K} f\left(y_k|\, \mathbf{X}\right) \right), \text{ for } j = 1, 2, \ldots, J. \tag{5.10}$$

where y_k is the received signal at the kth RE. Solving (5.10) by brute force search results in an exponential increase in computational complexity, rendering it imprac-

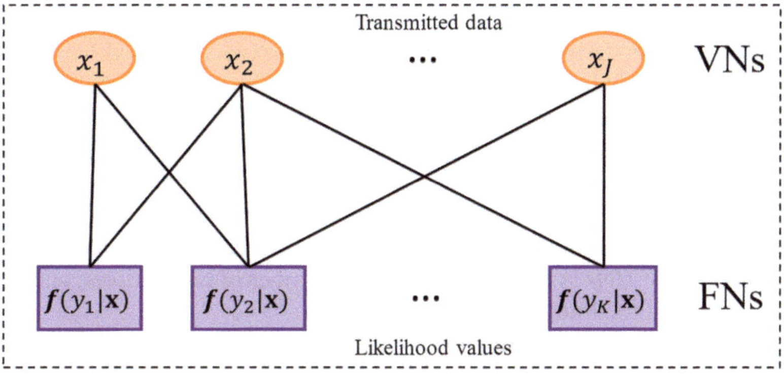

Fig. 5.21 Example factor graph for a communication channel. *Source* [49]

tical for scenarios involving a large number of users. Therefore, MPA is the popular choice for MUD in SCMA.

5.2.4.2 SCMA MAP Detection

(a) Factor graph:
To elucidate the concept of MPA, it is most effectively represented through a factor graph. Initially, we must define the factor graph: it is an undirected bipartite graph comprising two distinct sets of nodes—variable nodes (VNs) and function nodes (FNs). An edge exists between a VN and a FN if that specific variable serves as an argument for that function. Factor graphs are extensively utilized to model various communication problems due to their structural clarity and analytical utility.

Consider Fig. 5.21, which illustrates the factor graph of a communication channel. Here, $\mathbf{x} = [x_1, x_2, \ldots, x_J]$ represents the transmit vector while $\mathbf{y} = [y_1, y_2, \ldots, y_K]$ denotes the received vector. Assuming independence among noise components affecting elements of the received vector allows us to express the PDF of $\mathbf{y}$ given $\mathbf{x}$ as

$$f(\mathbf{y}|\,\mathbf{x}) = \prod_{k=1}^{K} f\,(y_k|\,\mathbf{x}). \tag{5.11}$$

It is normal that each received symbol y_k exhibits dependency solely on specific components of the vector $\mathbf{x}$. As illustrated in Fig. 5.21, when the conditional probability function $f(y_k|\,\mathbf{x})$ is contingent upon a particular element x_j, an edge will be established connecting the VN x_j to the function node representing $f(y_k|\,\mathbf{x})$. Consequently, this configuration allows us to construct a factor graph that effectively models the likelihood functions (function nodes) in relation to the transmitted data (variable nodes).

(a) SCMA detection using MPA:
The Message Passing Algorithm (MPA) serves as a pivotal method for conducting inference within graphical models by facilitating the exchange of belief messages among nodes. Especially, in SCMA, this discourse elucidates the process by which messages traverse between resource (or function) nodes (RNs) and user (or variable) nodes (UNs) within a factor graph, employing the Sum-Product Algorithm (SPA), a variant of MPA. For SCMA, each UN corresponds to an individual data layer, while each RN represents the likelihood function at a resource element (RE). Therefore, the total number of UNs is equal to J, the total number of layers/users, and the total RNs equals K, the total number of REs.

We start with the detection rule in (5.10), which can be rewritten as

$$\hat{\mathbf{x}}_j = \arg\max_{\mathbf{x}_{jm} \in \mathcal{X}_j} \sum_{\phi_j(\mathbf{x}_{jm})} . \tag{5.12}$$

where $\mathbf{x}_{jm}$ is the mth codeword of the codebook $\mathcal{X}_j$ and $\phi(\mathbf{x}_{jm})$ is the a posteriori probability that the jth user's codeword is $\mathbf{x}_{jm}$, given the received signal $\mathbf{y}$.

The detection rule outlined in equation (5.12) is executed iteratively through the application of the MPA on the factor graph depicted in Fig. 5.14. The received vector $\mathbf{y}$ acts as the input to the resource nodes. Subsequently, an iterative exchange of messages occurs between user nodes and resource nodes to decode data from various users. We define $\psi^{(l)}_{k\to j}(\mathbf{x}_{jm})$ and $\phi^{(l)}_{j\to k}(\mathbf{x}_{jm})$ as the messages transmitted by the kth resource node to the jth user node, and vice versa, during the lth iteration. These messages are length-M vectors, whose mth component is corresponding to the mth codeword of $\mathcal{X}_j$. The MPA for MUD in SCMA consists of the following steps[3]:

Step 1: Initialization: We set the initial values of the messages as $\phi^{(0)}_{j\to k}(\mathbf{x}_{jm}) = 1/M$, i.e. the messages are assumed to be equally-likely.

Step 2: Resource node update: The sum-product algorithm (SPA) starts by passing messages from leaf nodes. The received signal $\mathbf{y}$ is input to the resource nodes, which use this information to update their messages by the sum-product rule as

$$\psi^{(l)}_{k\to j}(\mathbf{x}_{jm}) = \sum_{\mathbf{c}\in C_{kj}} \frac{1}{\pi N_0} \exp\left[\frac{-1}{N_0}\left|y_k - \sqrt{P_j}h_{kj}x_{jmk} - \sum_{j'\in \mathcal{M}_k^j} \sqrt{P_{j'}}h_{ki}c_{j'k}\right|^2\right]$$
$$\times \prod_{j'\in \mathcal{M}_k^j} \phi^{(l-1)}_{j'\to k}(\mathbf{c}_{j'}) \tag{5.13}$$

where

- $\mathcal{M}_k$ is the set of the user nodes connected to the kth resource node and $\mathcal{M}_k^j \triangleq \mathcal{M}_k \backslash \{j\}$.

[3] The detailed explanation of MPA can be found in [49].

- Suppose $\mathcal{M}_k^j = \{j_1, j_2, \ldots, j_{d_f-1}\}$, then C_{kj} is the Cartesian product of the codebooks, i.e. $C_{kj} = \mathcal{X}_{j1} \times \mathcal{X}_{j2} \times \cdots \times \mathcal{X}_{j_{d_f-1}}$.
- $\mathbf{c} = \left(\mathbf{c}_{j_1}, \mathbf{c}_{j_2}, \ldots, \mathbf{c}_{j_{d_f-1}}\right)$ is a member of C_{kj} and $\mathbf{c}_{j'} = \left(c_{j'1}, c_{j'2}, \ldots, c_{j'K}\right)$.
- $\mathbf{x}_{jm} = (x_{jm1}, x_{jm2}, \ldots, x_{jmK})$.

Step 3: User node update: Next, the messages from the user nodes are updated by

$$\phi_{j\to k}^{(l)}(\mathbf{x}_{jm}) = \prod_{k'\in\mathcal{N}_j^k} \psi_{k'\to j}^{(l)}(\mathbf{x}_{jm}) \tag{5.14}$$

where $\mathcal{N}_j$ is the set of the resource nodes connected to the jth user node and $\mathcal{N}_j^k \triangleq \mathcal{N}_j \backslash \{k\}$.

Step 4: Stopping rule: When the messages converge (does not change much after a few iterations) or the maximum number of iterations κ has been reached, we come to Step 2. Otherwise, we increase l by 1 and repeat from Step 2.

Step 5: Decision: We calculate the a posterior function $\phi_j \mathbf{x}_{jm}$ as

$$\phi_j \mathbf{x}_{jm} = \prod_{k\in\mathcal{N}_j} \psi_{k\to j}^{(l)}(\mathbf{x}_{jm}). \tag{5.15}$$

Then, the trasnmitted codeword can be detected by using (5.12). In practice, to prevent the divergence due to the numerical unstability of probability-domain MPA, the messages are usually normalized after each iteration.

5.2.5 Numerical Results

The section explores the link-level performance of SCMA by simulation, highlighting its performance advantages compared to OMA methods.

Figure 5.22 illustrates the average SER performance of the SCMA system. The analysis includes a comparative evaluation using various codebooks, highlighting that the SCMA system's efficiency is intrinsically linked to each user's codebook design. Specifically, for a configuration involving nine users ($J = 9$) and six REs ($K = 6$) over an $AWGN$ channel, it is evident that the choice of codebook significantly impacts performance. Notably, employing the codebook from [48] yields an considerable gain over the others.

In Fig. 5.23, we further examine SER performance across different configurations, including 6×4, 8×4, and 9×6 SCMA systems, under Rayleigh fading condition. The analysis reveals that the 8×4 SCMA system exhibits worst performance among the three configurations due to its higher overloading factor compared to both the 6×4 and 9×6 configurations. Conversely, the superior performance of the 9×6 configuration can be attributed to reduced symbol collisions among users, thereby enhancing overall communication efficacy.

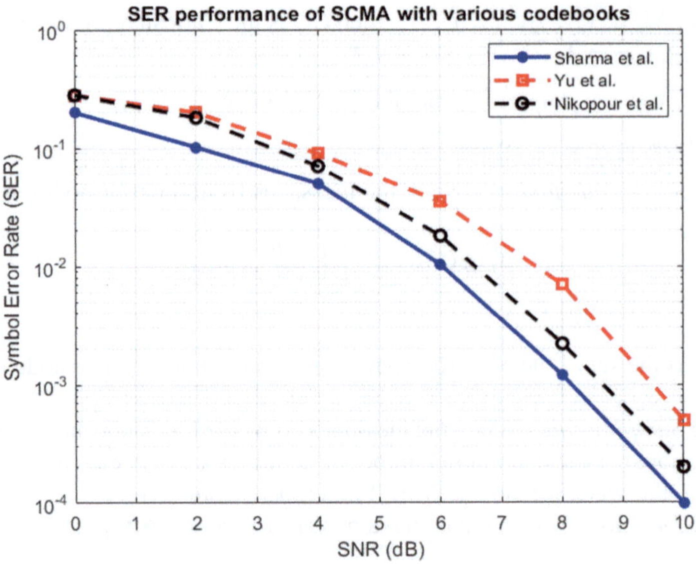

Fig. 5.22 SER performance of the SCMA system using various codebooks for $J = 9$ and $K = 6$ (overloading factor $\lambda = 150\%$) in AWGN channel

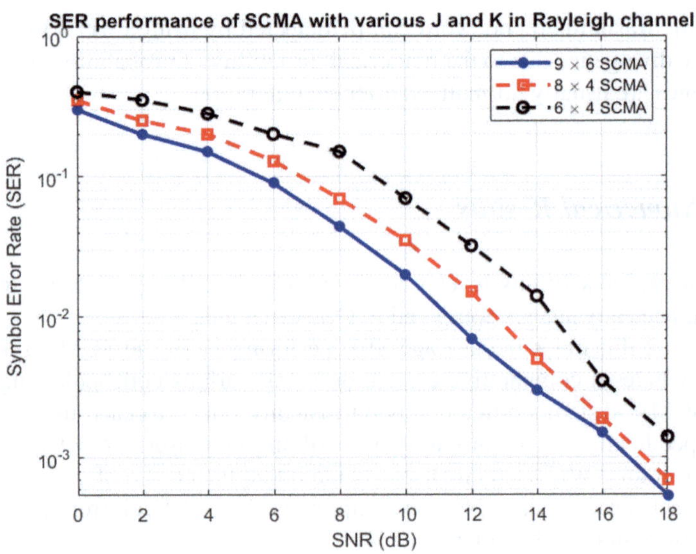

Fig. 5.23 BER performance of the SCMA system using various J and K in Rayleigh channel

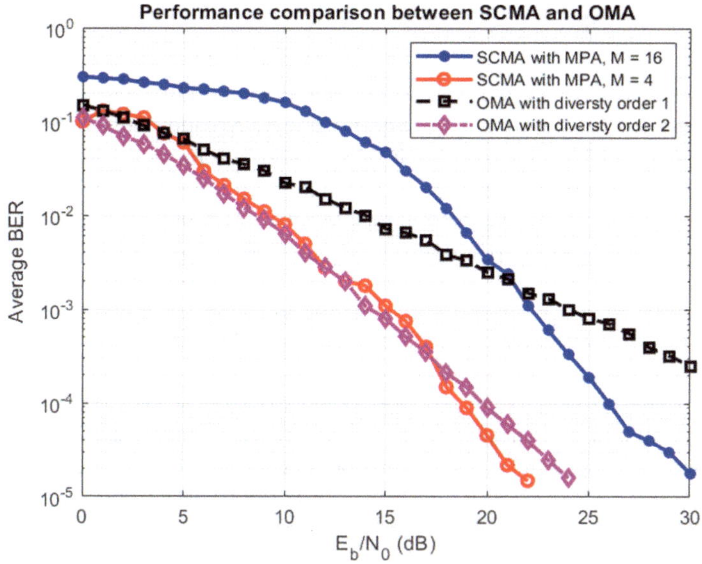

Fig. 5.24 Performance comparison of 6 × 4 SCMA in Rayleigh fading channel for M = 4, M = 16 and OMA for diversity order 1 and 2

Figure 5.24 illustrates the average BER achieved by SCMA for $M = 4$ and $M = 16$ over a Rayleigh fading channel, utilizing the SPA. In this figure, we juxtapose SCMA's BER performance against that of an OMA scheme with diversity orders 1 and 2 and without multi-user interference. Remarkably, even as SCMA transmits data from 6 users over 4 subcarriers and encounter multi-user interference, its performance remains comparable with OMA method at a diversity order of 2.

5.2.6 Conclusion

This section delivers a comprehensive tutorial on SCMA, highlighting its role as an innovative code-domain OMA scheme pivotal for massive connectivity. By employing meticulously selected sparse codebooks, SCMA achieves substantial constellation shaping gain and enhanced suppression of interference compared to earlier code-domain NOMA methods. Unlike complex MAP detector, the iterative MPA detector remarkably reduces the computational complexity. Consequently, SCMA emerges as a promising solution to enhance QoS for users with overloading factors exceeding one, while ensuring low latency and high spectral efficiency.

5.3 Conclusion

In this chapter, we provide the background and the simulation results on two typical multiple access techniques that have been considered as candidates for 5G and beyond wireless networks. The first one is NOMA, where we focus on the power-domain NOMA. Th report here is taken from [65]. The second technique is SCMA, which has been presented in [47]. We present the closed-form expressions of OP and related performance metrics and verify them by Monte Carlo simulations.

References

1. Zanella, A., Bui, N., Castellani, A., Vangelista, L., Zorzi, M.: Internet of things for smart cities. IEEE Internet Things J. **1**(1), 22–32 (2014). https://doi.org/10.1109/JIOT.2014.2306328
2. Guo, Y., Li, Y., Li, Y., Cheng, W., Zhang, H.: Swipt assisted noma for coordinated direct and relay transmission. In: 2018 IEEE/CIC International Conference on Communications in China (ICCC), pp. 111–115. IEEE (2018)
3. Li, X., Li, J., Liu, Y., Ding, Z., Nallanathan, A.: Outage performance of cooperative noma networks with hardware impairments. In: 2018 IEEE Global Communications Conference (GLOBECOM), pp. 1–6. IEEE (2018)
4. Tran, H.Q., Truong, P.Q., Phan, C.V., Vien, Q.T.: On the energy efficiency of noma for wireless backhaul in multi-tier heterogeneous cran. In: 2017 International Conference on Recent Advances in Signal Processing, Telecommunications & Computing (SigTelCom), pp. 229–234. IEEE (2017)
5. Wan, D., Wen, M., Ji, F., Yu, H., Chen, F.: Non-orthogonal multiple access for cooperative communications: challenges, opportunities, and trends. IEEE Wirel. Commun. **25**(2), 109–117 (2018)
6. Liaqat, M., Noordin, K.A., Abdul Latef, T., Dimyati, K.: Power-domain non orthogonal multiple access (pd-noma) in cooperative networks: an overview. Wireless Netw. **26**, 181–203 (2020)
7. Liu, Y., Qin, Z., Zhao, J., Wang, H., Song, L., He, A., Lao, X., Xu, B., Guo, T., Yang, J., Wang, Y., Dai, Y., Ma, Y., Zhang, X., Jiang, F., Li, A., Yao, Y.: Non-orthogonal multiple access for 5g: design and performance enhancement (2016). https://api.semanticscholar.org/CorpusID:104293341
8. Son, P.N., Duy, T.T.: Performance analysis of underlay cooperative cognitive full-duplex networks with energy-harvesting relay. Comput. Commun. **122**, 9–19 (2018)
9. Duy, T.T., Son, P.N.: Secrecy performances of multicast underlay cognitive protocols with partial relay selection and without eavesdropper's information. KSII Trans. Internet Inf. Syst. **9**(11), 4623–4643 (2015). https://doi.org/10.3837/tiis.2015.11.021
10. Huang, H., Zhu, M.: Energy efficiency maximization design for full-duplex cooperative noma systems with swipt. IEEE Access **7**, 20442–20451 (2019)
11. Al-habob, A.A., Salhab, A.M., Zummo, S.A.: A novel time-switching relaying protocol for multi-user relay networks with swipt. Arab. J. Sci. Eng. **44**, 2253–2263 (2019)
12. Duan, W., Ju, J., Hou, J., Sun, Q., Jiang, X.Q., Zhang, G.: Effective resource utilization schemes for decode-and-forward relay networks with noma. IEEE Access **7**, 51466–51474 (2019)
13. Da Costa, D.B., Yacoub, M.D.: Outage performance of two hop af relaying systems with co-channel interferers over nakagami-m fading. IEEE Commun. Lett. **15**(9), 980–982 (2011)
14. Yue, X., Liu, Y., Kang, S., Nallanathan, A., Ding, Z.: Exploiting full/half-duplex user relaying in noma systems. IEEE Trans. Commun. **66**(2), 560–575 (2017)

15. Zhou, Y., Wong, V.W., Schober, R.: Performance analysis of cooperative noma with dynamic decode-and-forward relaying. In: GLOBECOM 2017-2017 IEEE Global Communications Conference, pp. 1–6. IEEE (2017)
16. Deng, D., Yu, M., Xia, J., Na, Z., Zhao, J., Yang, Q.: Wireless powered cooperative communications with direct links over correlated channels. Phys. Commun. **28**, 147–153 (2018)
17. Vien, Q.T., Le, T.A., Barn, B., Phan, C.V.: Optimising energy efficiency of non-orthogonal multiple access for wireless backhaul in heterogeneous cloud radio access network. IET Commun. **10**(18), 2516–2524 (2016)
18. Vien, Q.T., Nguyen, H.X., Nallanathan, A.: Cooperative spectrum sensing with secondary user selection for cognitive radio networks over nakagami-m fading channels. IET Commun. **10**(1), 91–97 (2016)
19. Liaqat, M., Noordin, K.A., Latef, T.A., Dimyati, K., Ding, Z., Siddiqui, A.M., Ahmed, A., Younas, T.: Relay selection schemes for cooperative noma (c-noma) with simultaneous wireless information and power transfer (swipt). Phys. Commun. **36**, 100823 (2019)
20. Nguyen, B.C., Hoang, T.M., Tran, P.T., Nguyen, T.N.: Outage probability of noma system with wireless power transfer at source and full-duplex relay. AEU - Int. J. Electr. Commun. **116**, 152957 (2020). https://doi.org/10.1016/j.aeue.2019.152957. https://www.sciencedirect.com/science/article/pii/S1434841119314141
21. Nguyen, B.C., Hoang, T.M., Pham, X.N., Tran, P.T.: Performance analysis of energy harvesting-based full-duplex decode-and-forward vehicle-to-vehicle relay networks with nonorthogonal multiple access. Wireless Commun. Mobile Comput. **2019**(1), 6097686 (2019). https://doi.org/10.1155/2019/6097686. https://onlinelibrary.wiley.com/doi/abs/10.1155/2019/6097686
22. Liu, H., Ding, Z., Kim, K.J., Kwak, K.S., Poor, H.V.: Decode-and-forward relaying for cooperative noma systems with direct links. IEEE Trans. Wireless Commun. **17**(12), 8077–8093 (2018)
23. Duan, W., Ju, J., Sun, Q., Ji, Y., Wang, Z., Choi, J., Zhang, G.: Capacity enhanced cooperative d2d systems over rayleigh fading channels with noma (2018). arXiv:1810.06837
24. Ye, Y., Li, Y., Wang, D., Lu, G.: Power splitting protocol design for the cooperative noma with swipt. In: 2017 IEEE International Conference on Communications (ICC), pp. 1–5. IEEE (2017)
25. Pratschner, S., Tahir, B., Nissel, R., Marijanovic, L., Mussbah, M., Kirev, K., Schwarz, S., Rupp, M.: The vienna 5g link level simulator v1. 2. nt. https://www.tuwien.ac.at/wpcontent/uploads/2020/05/5gll_usermanual.pdf. Seen In **19**, 25 (2021)
26. Islam, S.R., Avazov, N., Dobre, O.A., Kwak, K.S.: Power-domain non-orthogonal multiple access (noma) in 5g systems: potentials and challenges. IEEE Commun. Surv. & Tutor. **19**(2), 721–742 (2016)
27. Ding, Z., Lei, X., Karagiannidis, G.K., Schober, R., Yuan, J., Bhargava, V.K.: A survey on non-orthogonal multiple access for 5g networks: research challenges and future trends. IEEE J. Sel. Areas Commun. **35**(10), 2181–2195 (2017)
28. Saito, Y., Benjebbour, A., Kishiyama, Y., Nakamura, T.: System-level performance evaluation of downlink non-orthogonal multiple access (noma). In: 2013 IEEE 24th Annual International Symposium on Personal, Indoor, and Mobile Radio Communications (PIMRC), pp. 611–615. IEEE (2013)
29. Liu, Y., Qin, Z., Elkashlan, M., Ding, Z., Nallanathan, A., Hanzo, L.: Nonorthogonal multiple access for 5g and beyond. Proc. IEEE **105**(12), 2347–2381 (2017)
30. Ding, Z., Yang, Z., Fan, P., Poor, H.V.: On the performance of non-orthogonal multiple access in 5g systems with randomly deployed users. IEEE Signal Process. Lett. **21**(12), 1501–1505 (2014)
31. Ding, Z., Peng, M., Poor, H.V.: Cooperative non-orthogonal multiple access in 5g systems. IEEE Commun. Lett. **19**(8), 1462–1465 (2015)
32. Aldababsa, M., Toka, M., Gökçeli, S., Kurt, G.K., Kucur, O.: A tutorial on nonorthogonal multiple access for 5g and beyond. Wirel. Commun. Mob. Comput. **2018**(1), 9713450 (2018)
33. Ding, Z., Liu, Y., Choi, J., Sun, Q., Elkashlan, M., Chih-Lin, I., Poor, H.V.: Application of non-orthogonal multiple access in lte and 5g networks. IEEE Commun. Mag. **55**(2), 185–191 (2017)

34. Sendonaris, A., Erkip, E., Aazhang, B.: Increasing uplink capacity via user cooperation diversity. In: Proceedings. 1998 IEEE International Symposium on Information Theory (Cat. No. 98CH36252), p. 156. IEEE (1998)
35. Laneman, J.N., Tse, D.N., Wornell, G.W.: Cooperative diversity in wireless networks: Efficient protocols and outage behavior. IEEE Trans. Inf. Theory **50**(12), 3062–3080 (2004)
36. Liang, X., Wu, Y., Ng, D.W.K., Zuo, Y., Jin, S., Zhu, H.: Outage performance for cooperative noma transmission with an af relay. IEEE Commun. Lett. **21**(11), 2428–2431 (2017)
37. Xiao, Y., Hao, L., Ma, Z., Ding, Z., Zhang, Z., Fan, P.: Forwarding strategy selection in dual-hop noma relaying systems. IEEE Commun. Lett. **22**(8), 1644–1647 (2018)
38. Ashraf, M., Shahid, A., Jang, J.W., Lee, K.G.: Energy harvesting non-orthogonal multiple access system with multi-antenna relay and base station. IEEE Access **5**, 17660–17670 (2017)
39. Wang, Z., Yue, X., Peng, Z.: Full-duplex user relaying for noma system with self-energy recycling. IEEE Access **6**, 67057–67069 (2018)
40. Kalokidou, V., Johnson, O., Piechocki, R.: Interference management in heterogeneous networks with blind transmitters (2016). arXiv:1601.08132
41. Yuan, Y., Xu, Y., Yang, Z., Xu, P., Ding, Z.: Energy efficiency optimization in full-duplex user-aided cooperative swipt noma systems. IEEE Trans. Commun. **67**(8), 5753–5767 (2019)
42. Choi, J.I., Jain, M., Srinivasan, K., Levis, P., Katti, S.: Achieving single channel, full duplex wireless communication. In: Proceedings of the Sixteenth Annual International Conference on Mobile Computing and Networking, pp. 1–12 (2010)
43. Zhang, Z., Long, K., Vasilakos, A.V., Hanzo, L.: Full-duplex wireless communications: challenges, solutions, and future research directions. Proc. IEEE **104**(7), 1369–1409 (2016)
44. Zhang, D., Liu, Y., Ding, Z., Zhou, Z., Nallanathan, A., Sato, T.: Performance analysis of non-regenerative massive-mimo-noma relay systems for 5g. IEEE Trans. Commun. **65**(11), 4777–4790 (2017)
45. Ding, Z., Dai, H., Poor, H.V.: Relay selection for cooperative noma. IEEE Wireless Commun. Lett. **5**(4), 416–419 (2016)
46. Huawei Technologies Co., Ltd.: 5g: a technology vision. Technical report, Huawei Technologies Co., Ltd., Shenzhen, China (2013). https://www.huawei.com/ilink/en/download/HW_314849
47. Taherzadeh, M., Nikopour, H., Bayesteh, A., Baligh, H.: Scma codebook design. In: 2014 IEEE 80th Vehicular Technology Conference (VTC2014-Fall), pp. 1–5 (2014). https://doi.org/10.1109/VTCFall.2014.6966170
48. Sharma, S., Deka, K.: Sparse code multiple access (scma) technique (2023). https://arxiv.org/abs/2309.09127
49. Chaturvedi, S., Liu, Z., Bohara, V.A., Srivastava, A., Xiao, P.: A tutorial on decoding techniques of sparse code multiple access. IEEE Access **10**, 58503–58524 (2022). https://doi.org/10.1109/ACCESS.2022.3178127
50. Hoshyar, R., Razavi, R., Al-Imari, M.: Lds-ofdm an efficient multiple access technique. In: 2010 IEEE 71st Vehicular Technology Conference, pp. 1–5. IEEE (2010)
51. Nikopour, H., Baligh, H.: Sparse code multiple access. In: 2013 IEEE 24th Annual International Symposium on Personal, Indoor, and Mobile Radio Communications (PIMRC), pp. 332–336. IEEE (2013)
52. Forney, G.D., Wei, L.F.: Multidimensional constellations. i. introduction, figures of merit, and generalized cross constellations. IEEE J. Select. Areas Commun. **7**(6), 877–892 (1989)
53. Yu, L., Fan, P., Cai, D., Ma, Z.: Design and analysis of scma codebook based on star-qam signaling constellations. IEEE Trans. Veh. Technol. **67**(11), 10543–10553 (2018). https://doi.org/10.1109/TVT.2018.2865920
54. Klimentyev, V.P., Sergienko, A.B.: Scma codebooks optimization based on genetic algorithm. In: European Wireless 2017; 23th European Wireless Conference, pp. 1–6 (2017)
55. Silva, B.F.d., Le Ruyet, D., Uch a-Filho, B.F.: Threshold-based edge selection mpa for scma. IEEE Trans. Veh. Technol. **69**(3), 2957–2966 (2020). https://doi.org/10.1109/TVT.2020.2966333
56. Yuan, W., Wu, N., Guo, Q., Li, Y., Xing, C., Kuang, J.: Iterative receivers for downlink mimo-scma: message passing and distributed cooperative detection. IEEE Trans. Wireless Commun. **17**(5), 3444–3458 (2018). https://doi.org/10.1109/TWC.2018.2813378

57. Peng, J., Chen, W., Bai, B., Guo, X., Sun, C.: Joint optimization of constellation with mapping matrix for scma codebook design. IEEE Signal Process. Lett. **24**(3), 264–268 (2017). https://doi.org/10.1109/LSP.2017.2653845
58. Xiong, G., Sun, J.: An optimal resource allocation algorithm based on sum rate maximization for uplink scma system. In: 2018 IEEE 18th International Conference on Communication Technology (ICCT), pp. 805–810 (2018). https://doi.org/10.1109/ICCT.2018.8599973
59. Salehi, J.A.: Code division multiple-access techniques in optical fiber networks. i. fundamental principles. IEEE Trans. Commun. **37**(8), 824–833 (1989)
60. Cai, D., Fan, P., Lei, X., Liu, Y., Chen, D.: Multi-dimensional scma codebook design based on constellation rotation and interleaving. In: 2016 IEEE 83rd Vehicular Technology Conference (VTC Spring), pp. 1–5 (2016). https://doi.org/10.1109/VTCSpring.2016.7504356
61. Boutros, J., Viterbo, E.: Signal space diversity: a power-and bandwidth-efficient diversity technique for the rayleigh fading channel. IEEE Trans. Inf. Theory **44**(4), 1453–1467 (1998)
62. Oggier, F., Viterbo, E., et al.: Algebraic number theory and code design for rayleigh fading channels. Found. Trends® Commun. Inf. Theory **1**(3), 333–415 (2004)
63. van de Beek, J., Popovic, B.M.: Multiple access with low-density signatures. In: GLOBECOM 2009 - 2009 IEEE Global Telecommunications Conference, pp. 1–6 (2009). https://doi.org/10.1109/GLOCOM.2009.5425243
64. Vameghestahbanati, M., Marsland, I.D., Gohary, R.H., Yanikomeroglu, H.: Multidimensional constellations for uplink scma systems–a comparative study. IEEE Commun. Surv. ys & Tutor. **21**(3), 2169–2194 (2019). https://doi.org/10.1109/COMST.2019.2910569
65. Huu, T.Q.: Non-orthogonal multiple access (noma) for energy harvesting based wireless cooperative relay networks. Ph.D. dissertation, Ho Chi Minh City University of Technology and Education, Thu Duc City, Ho Chi Minh City, Vietnam (2023)

Chapter 6
Massive MIMO

6.1 Optimal Beamforming Design for Massive MIMO

Massive MIMO (multiple-input multiple-output), a pivotal technology in advancing spectral efficiency (SE) for next-generation cellular networks, leverages sophisticated beamforming and spatial multiplexing techniques [1]. By utilizing a base transceiver station (BTS) equipped with an extensive array of steerable antennas—often numbering in the hundreds—Massive MIMO facilitates the concurrent servicing of multiple user equipments (UEs) within identical time-frequency resources. Typically operating in time-division duplex (TDD) mode, this system acquires CSI through uplink pilot signaling and capitalizes on channel reciprocity [2]. The acquired CSI is indispensable for coherent uplink receiver combining and downlink transmitter precoding.

The optimization of achievable SE in Massive MIMO systems under conditions of imperfect CSI has been rigorously explored across various fading environments, including those modeled by spatially uncorrelated [3] or correlated Rayleigh fading [2, 4]. Additionally, scenarios devoid of fading, such as line-of-sight propagation, have been thoroughly analyzed [2, 5]. However, practical wireless channels often embody both deterministic line-of-sight components and small-scale multipath-induced fading characteristics effectively captured by the Rician fading model [6].

Despite its practical significance, the performance analysis of Massive MIMO under Rician fading channels has not been as extensively investigated compared to Rayleigh models. In single-cell contexts with spatially uncorrelated Rician channels employing zero forcing (ZF) processing, studies have derived approximate SE expressions for both uplink [7] and downlink scenarios [8, 9]. In multi-cell environments where each cell assumes spatially uncorrelated Rician fading while inter-cell interference follows a Rayleigh model, SE approximations were developed using ZF combining for uplinks [10] and either ZF [11] or regularized ZF precoding for downlinks [12].

P. T. Tran et al., *Developing MATLAB Comprehensive Simulation for 5G/6G Wireless Communications*, Signals and Communication Technology,
https://doi.org/10.1007/978-981-95-1117-4_6

In this chapter, we introduce the results presented in [13], which was the first work that overcame the above shortcomings, and illustrate how these results are verified by our simulation module of Massive MIMO. In [13], the authors examine a multi-cell scenario with spatially correlated Rician fading channels between BTSs and UEs, while other pairs experience Rayleigh fading. This channel model has previously been applied only to single-cell scenarios with perfect CSI [14, 15]. They derive the minimum mean-squared error (MMSE), element-wise MMSE (EW-MMSE), and least-square (LS) channel estimators, characterizing their statistics. Using these estimates for MR combining and precoding, the authors compute closed-form UL and DL SEs and discuss their structure. They analyze the asymptotic behavior of UL and DL SEs under spatially correlated Rician fading using these estimators. Lastly, they numerically compare the UL and DL SEs across MMSE, EW-MMSE, and LS estimation for both correlated and uncorrelated Rician and Rayleigh fading.

It is crucial to acknowledge that these investigations predominantly address scenarios characterized by imperfect CSI—a condition reflecting real-world applicability—while earlier research assuming perfect CSI can be referenced from prior works cited within these studies [7, 12].

6.1.1 System Model

In this analysis, we delve into the Massive MIMO system as introduced by [13], which is structured with L cells, each housing a BTS equipped with M_j antennas. These stations serve K single-antenna user equipments (UEs) as depicted in Fig. 6.1. The system operates under time-division duplexing (TDD), ensuring channel responses remain stable over a coherence block of size τ_c. Notably, these channel realizations are assumed to be independent across different coherence blocks.

The coherence block size, denoted by τ_c is influenced by factors such as carrier frequency and external conditions like propagation environments and UE mobility [3]. Within each block, samples are allocated for three core functions: UL pilot signaling (τ_p), UL data transmission (τ_u), and DL data transmission (τ_d), satisfying the equation $\tau_c = \tau_p + \tau_u + \tau_d$. Channel estimates for both uplink and downlink are derived using uplink pilot signals due to the channel reciprocity feature inherent in TDD systems, facilitated by fully synchronized hardware.

The channel response between UE k in cell l and the BTS in cell j is represented mathematically as $\mathbf{h}_{lk}^j \in \mathbb{C}^{M_j}$, where each element of $\mathbf{h}_{lk}^j$ describes the propagation path from the UE to an individual antenna at the BTS. Herein, superscripts of $\mathbf{h}_{lk}^j$ denote BTS indices while its subscripts indicate specific cell and UE indices. Although UL and DL channels share identical responses within a single coherence block for simplicity's sake, we use notation where UL channels are represented as $\mathbf{h}_{lk}^j$ and DL channels are denoted by its Hermitian transpose $(\mathbf{h}_{lk}^j)^H$.

Our focus shifts toward spatially correlated Rician fading channels where every channel vector follows a circularly symmetric complex Gaussian distribution:

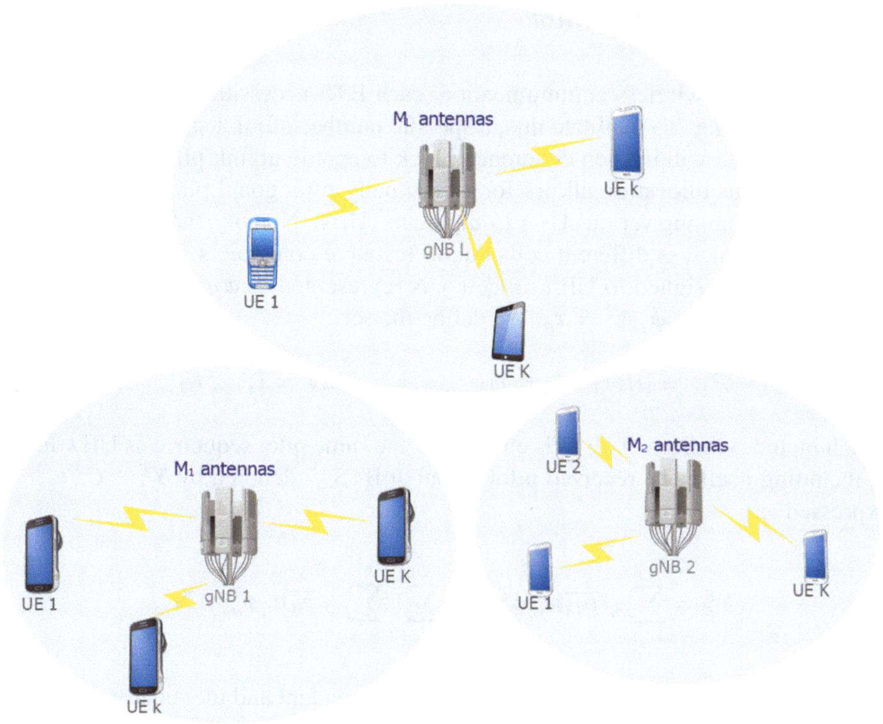

Fig. 6.1 System module of the considered beamforming design for Massive MIMO systems

$$\mathbf{h}_{lk}^{j} \sim \mathcal{N}_C\left(\mathbf{h}_{lk}^{-j}, \mathbf{R}_{lk}^{j}\right), \tag{6.1}$$

as per [13]. Herein lies an intricate balance between LoS components captured through mean vectors $\mathbf{h}_{lk}^{-j} \in C^{M_j}$ alongside non-line-of-sight (NLoS) elements characterized via positive semi-definite covariance matrices $\mathbf{R}_{lk}^{j} \in C^{M_j \times M_j}$. Small-scale fading adheres closely to Gaussian distributions, while macroscopic effects—pathlosses, shadow-fading, antenna gains, and radiation patterns—are encapsulated within these covariance structures at both transmitter/receiver ends ($\mathbf{R}_{lk}^{j}$ and $\mathbf{h}_{lk}^{-j}$).

To quantify average gains from one of the antennas at BTS j toward specific UE k across cell l effectively requires normalized traces—a mathematical operation providing insights into signal strength variations amidst complex wireless landscapes:

$$\beta_{lk}^{j} = \frac{1}{M_j}\mathrm{tr}(\mathbf{R}_{lk}^{j}). \tag{6.2}$$

where β_{lk}^{j} is called the large-scale fading coefficient.

6.1.2 Channel Estimation

In the domain of wireless communication, each BTS necessitates CSI for effective receive processing. To facilitate this, a specific number of pilot samples, denoted as τ_p, are allocated within each coherence block to enable uplink pilot-based channel estimation. This allocation allows for τ_p mutually orthogonal pilot sequences that are distributed among various User Equipments (UEs). Notably, these sequences are reused by UEs across different cells due to resource constraints. The deterministic pilot sequence assigned to UE k in cell j is represented as $\phi_{jk} \in \mathbb{C}^{\tau_p}$, adhering to the norm condition $\|\boldsymbol{\phi}_{jk}\|^2 = \tau_p$. We define the set

$$\mathcal{P}_{jk} = \{(l, i) : \phi_{\mathbf{li}} = \phi_{jk}, l = 1, ..., L, i = 1, ..., K_l\}, \tag{6.3}$$

which includes indices of all UEs employing the same pilot sequence as UE k in cell j, including itself. The received pilot signal at BTS j, denoted by $\mathbf{Y}_j^p \in C^{M_j \times \tau_p}$, is expressed as

$$\mathbf{Y}_j^p = \sum_{k=1}^{K_j} \sqrt{p_{jk}}\mathbf{h}_{jk}^j \phi_{jk}^T + \sum_{l=1, l \neq j}^{L} \sum_{i=1}^{K_l} \sqrt{p_{li}}\mathbf{h}_{li}^j \phi_{li}^T + \mathbf{N}_j^p, \tag{6.4}$$

where noise matrix $\mathbf{N}_j^p \in C^{M_j \times \tau_p}$ comprises independent and identically distributed elements following a complex Gaussian distribution $\mathcal{N}_C(0, \sigma_{ul}^2)$ with variance σ_{ul}^2. To estimate channel coefficients $\mathbf{h}_{li}^j$, BTS j multiplies its received signal $\mathbf{Y}_j^p$ with the corresponding UE's pilot sequence $\phi_{\mathbf{li}}^*$, yielding:

$$\mathbf{y}_{jli}^p = \mathbf{Y}_j^p \phi_{li}^* = \sqrt{p_{li}}\tau_p \mathbf{h}_{li}^j + \sum_{(l',i') \in \mathcal{P}_{li} \setminus (l,i)} \sqrt{p_{l'i'}}\tau_p \mathbf{h}_{l'i'}^j + \mathbf{N}_j^p \phi_{li}^*. \tag{6.5}$$

The processed received pilot signal vector $\mathbf{y}_{jli}^p \in C^{M_j}$ serves as a sufficient statistic for estimating h_{lij} [2]. In this analysis, the authors of [13] explore three distinct channel estimation methodologies that leverage varying degrees of statistical channel knowledge. These statistical attributes—mean vectors and covariance matrices—can be effectively approximated through sample mean and covariance matrices [2, 16, 17].

6.1.2.1 MMSE Channel Estimator

By employing the processed received pilot signal as delineated in equation (6.5), the BTS can utilize the MMSE estimation technique to derive an estimate of $\mathbf{h}_{li}^j$, as articulated in the subsequent lemma [13]. It is crucial to highlight that the Bayesian MMSE estimator presupposes comprehensive knowledge of statistical distributions.

Lemma 6.1 *The MMSE estimate of channel from BS j to UE i in cell l is*

$$\hat{\mathbf{h}}_{li}^{j} = \bar{\mathbf{h}}_{li}^{-j} + \sqrt{p_{li}}\mathbf{R}_{li}^{j}\Psi_{li}^{j}(\mathbf{y}_{jli}^{p} - \mathbf{y}_{jli}^{-p}), \tag{6.6}$$

where $\mathbf{y}_{jli}^{-p} = \sum_{(l',i')\in\mathcal{P}_{li}} \sqrt{p_{l'i'}}\tau_p\bar{\mathbf{h}}_{l'i'}^{-j}$ *and*

$$\Psi_{li}^{j} = \tau_p Cov\{\mathbf{y}_{jli}^{p}\}^{-1} = \left(\sum_{(l',i')\in\mathcal{P}_{li}} p_{l'i'}\tau_p\mathbf{R}_{l'i'}^{j} + \sigma^2\mathbf{I}_{M_j}\right)^{-1}. \tag{6.7}$$

The estimation error $\tilde{\mathbf{h}}_{li}^{j} = \mathbf{h}_{li}^{j} - \hat{\mathbf{h}}_{li}^{j}$ *has the covariance matrix*

$$\mathbf{C}_{li}^{j} = \mathbf{R}_{li}^{j} - p_{li}\tau_p\mathbf{R}_{li}^{j}\Psi_{li}^{j}\mathbf{R}_{li}^{j} \tag{6.8}$$

and the mean-squared error is $MSE = E\{\|\mathbf{h}_{li}^{j} - \hat{\mathbf{h}}_{li}^{j}\|^2\} = tr(\mathbf{C}_{li}^{j})$.
The MMSE estimate $\hat{\mathbf{h}}_{li}^{j}$ *and the estimation error* $\tilde{\mathbf{h}}_{li}^{j}$ *are independent random variables and distributed as*

$$\hat{\mathbf{h}}_{li}^{j} \sim \mathcal{N}_C(\bar{\mathbf{h}}_{li}^{j}, \mathbf{R}_{li}^{j} - \mathbf{C}_{li}^{j}), \tag{6.9}$$

$$\tilde{\mathbf{h}}_{li}^{j} \sim \mathcal{N}_C(\mathbf{0}_M, \mathbf{C}_{li}^{j}). \tag{6.10}$$

Proof The proof follows from standard MMSE estimation techniques for Gaussian random variables observed under Gaussian noise conditions [2, 18]. □

Importantly, it should be noted that the covariance matrix of estimation errors, $\mathbf{C}_{li}^{j}$, remains unaffected by mean values. In essence, this implies that LoS components do not influence these errors since they are known quantities and subject to be subtracted from received signals. Moreover, the channel estimates for UEs within set $\mathcal{P}_{li}$ are not independent, although the channels are assumed independent, due to shared pilot sequences among UEs, resulting in what's termed "pilot contamination."

6.1.2.2 Element-Wise MMSE Channel Estimator

In scenarios where the BTS does not have full access to covariance matrices, the EW-MMSE estimator emerges as a robust alternative [2, 19]. This method uniquely utilizes only the diagonal elements of these matrices, effectively ignoring inter-element correlations. Consequently, it circumvents matrix inversions, thereby significantly reducing computational complexity compared to the MMSE estimator outlined in Lemma 6.1.

Lemma 6.2 *The EW-MMSE estimate of channel from BS j to UE i in cell l is*

$$\hat{\mathbf{h}}_{li}^{j} = \bar{\mathbf{h}}_{li}^{j} + \sqrt{p_{li}}\mathbf{D}_{li}^{j}\Lambda_{li}^{j}(\mathbf{y}_{jli}^{p} - \bar{\mathbf{y}}_{jli}^{p}), \tag{6.11}$$

where $\mathbf{D}_{li}^{j} \in C^{M_j \times M_j}$ *and* $\Lambda_{li}^{j} \in C^{M_j \times M_j}$ *are diagonal matrices with*

$$\mathbf{D}_{li}^{j} = diag([\mathbf{R}_{li}^{j}]_{mm} : m = 1, ..., M_j)$$

and

$$\Lambda_{li}^{j} = diag\left(\left[\sum_{(l',i')\in\mathcal{P}_{li}} p_{l'i'}\tau_p \mathbf{R}_{l'i'}^{j} + \sigma^2 \mathbf{I}_{M_j}\right]_{mm} : m = 1, ..., M_j\right)^{-1}$$

The distributions of EW-MMSE estimate $\hat{\mathbf{h}}_{li}^{j}$ *and the estimation error* $\tilde{\mathbf{h}}_{li}^{j}$ *are correlated and distributed as*

$$\hat{\mathbf{h}}_{li}^{j} \sim \mathcal{N}_C(\bar{\mathbf{h}}_{li}^{j}, \sum_{li}^{j}), \tag{6.12}$$

$$\tilde{\mathbf{h}}_{li}^{j} \sim \mathcal{N}_C(\mathbf{0}_M, \tilde{\sum}_{li}^{j}), \tag{6.13}$$

where

$$\sum_{li}^{j} = p_{li}\tau_p \mathbf{D}_{li}^{j}\Lambda_{li}^{j}(\Psi_{li}^{j})^{-1}\Lambda_{li}^{j}\mathbf{D}_{li}^{j} \tag{6.14}$$

and

$$\tilde{\sum}_{li}^{j}) = \mathbf{R}_{li}^{j} - p_{li}\tau_p \mathbf{R}_{li}^{j}\Lambda_{li}^{j}\mathbf{D}_{li}^{j} - p_{li}\tau_p \mathbf{D}_{li}^{j}\Lambda_{li}^{j}\mathbf{R}_{li}^{j} + \sum_{li}^{j} \tag{6.15}$$

Proof Proof follows by applying similar techniques used for deriving the MMSE estimator, but focuses on estimating each element independently using only the signal received at the considered antenna while computing resulting statistics. Unlike traditional MMSE estimation, vectors $\hat{\mathbf{h}}_{li}^{j}$ and $\tilde{\mathbf{h}}_{li}^{j}$ are correlated with $E\{\hat{\mathbf{h}}_{li}^{j}(\tilde{\mathbf{h}}_{li}^{j})^H\} = p_{li}\tau_p \mathbf{D}_{li}^{j}\Lambda_{li}^{j}\mathbf{R}_{li}^{j} - \sum_{li}^{j}$

We can easily get the desired result using the same approach as was employed to derive the MMSE estimator, but by estimating each element separately using only the signal obtained at the antenna and then computing the resulting statistics. In contrast to MMSE estimation, $\hat{\mathbf{h}}_{li}^{j}$ and $\tilde{\mathbf{h}}_{li}^{j}$ are correlated with $E\{\hat{\mathbf{h}}_{li}^{j}(\tilde{\mathbf{h}}_{li}^{j})^H\} = p_{li}\tau_p \mathbf{D}_{li}^{j}\Lambda_{li}^{j}\mathbf{R}_{li}^{j} - \sum_{li}^{j}$, except when all covariance matrices are strictly diagonal. □

6.1.2.3 LS Channel Estimator

In scenarios where the BTS lacks prior knowledge of $\mathbf{R}_{li}^{j}$ and $\bar{\mathbf{h}}_{li}^{j}$, the application of a non-Bayesian LS estimator becomes pertinent for estimating the propagation channel $\mathbf{h}_{li}^{j}$. The LS estimate is mathematically characterized as the value of $\hat{\mathbf{h}}_{li}^{j}$ that minimizes the Euclidean norm $\|\mathbf{y}_{jli}^{p} - \sqrt{p_{li}\tau_p}\hat{\mathbf{h}}_{li}^{j}\|^2$. This minimization leads to:

$$\hat{\mathbf{h}}_{li}^{j} = \frac{1}{\sqrt{p_{li}\tau_p}}\mathbf{y}_{jli}^{p}. \tag{6.16}$$

Lemma 6.3 *The LS estimator and estimation error are correlated random variables and distributed as*

$$\hat{\mathbf{h}}_{li}^{j} \sim \mathcal{N}_C\left(\frac{1}{\sqrt{p_{li}\tau_p}}\bar{\mathbf{y}}_{jli}^{p}, \frac{1}{p_{li}\tau_p}(\Psi_{li}^{j})^{-1}\right), \tag{6.17}$$

$$\tilde{\mathbf{h}}_{li}^{j} \sim \mathcal{N}_C\left(\bar{\mathbf{h}}_{li}^{j} - \frac{1}{\sqrt{p_{li}\tau_p}}\bar{\mathbf{y}}_{jli}^{p}, \frac{1}{p_{li}\tau_p}(\Psi_{li}^{j})^{-1} - \mathbf{R}_{li}^{j}\right). \tag{6.18}$$

Proof The proof is given in Appendix C of [13]. □

It is crucial to note that these statistical properties are more intricate compared to those derived from MMSE and EW-MMSE estimators. In particular, the mean of estimation error is not zero, which must be considered when evaluating the performance metrics.

6.1.3 *Uplink Spectral Efficiency with MR Combining*

During data transmission, the signal received at BS j, which is denoted as $\mathbf{y}_j \in C^{M_j}$, is expressed by

$$\mathbf{y}_j = \sum_{k=1}^{K_j}\mathbf{h}_{jk}^{j}s_{jk} + \sum_{l=1,l\neq j}^{L}\sum_{i=1}^{K_l}\mathbf{h}_{li}^{j}s_{li} + \mathbf{n}_j, \tag{6.19}$$

where $\mathbf{n}_j \sim \mathcal{N}_C(\mathbf{0}_{M_j}, \sigma_{ul}^2\mathbf{I}_{M_j})$ is the AWGN. The UL signal from UE k in cell l, represented by $s_{lk} \in C$ with power $p_{lk} = E\{|s_{lk}|^2\}$, contributes to this expression. The first term (6.19) represents the desired signal while subsequent terms account for interference and noise.

BTS j employs the receive combining vector $\mathbf{v}_{jk} \in \mathbb{C}^{M_j}$ based on its CSI, which it uses to process $\mathbf{y}_j$ and isolate the desired signal from UE k from interference. According to [2, Theorem 4.4], the ergodic UL capacity for UE k in cell j has a lower bound given by

$$SE_{jk}^{ul} = \frac{\tau_u}{\tau_c} \log_2\left(1 + \gamma_{jk}^{ul}\right) \qquad [bit/s/Hz], \tag{6.20}$$

where

$$\gamma_{jk}^{ul} = \frac{p_{jk}|E\{\mathbf{v}_{jk}^H \mathbf{h}_{jk}^j\}|^2}{\sum_{l=1}^{L}\sum_{i=1}^{K_l} p_{li} E\{|\mathbf{v}_{jk}^H \mathbf{h}_{li}^j|^2\} - p_{jk}|E\{\mathbf{v}_{jk}^H \mathbf{h}_{jk}^j\}|^2 + \sigma_{ul}^2 E\{\|\mathbf{v}_{jk}\|^2\}}, \tag{6.21}$$

denotes the effective SINR. Herein, expectations are taken over all random variables involved. Since SE_{jk}^{ul} falls below capacity limits, it represents an achievable ergodic SE. The effective SINR γ_{jk}^{ul} can be determined numerically across various combining schemes and channel estimators. Notably, closed-form computation is feasible when employing MR, utilizing any of the three channel estimators discussed in Sect. 6.1.2.

6.1.3.1 Uplink Spectral Efficiency with MMSE estimator

The utilization of the MMSE estimator, as outlined in equation (6.6), facilitates the derivation of a closed-form expression for the SE in (6.20), as demonstrated in Theorem 6.1 [13].

Theorem 6.1 *If MR combining with* $\mathbf{v}_{jk} = \hat{\mathbf{h}}_{jk}^j$ *is used based on the MMSE estimator, then*

$$E\{\mathbf{v}_{jk}^H \mathbf{h}_{jk}^j\} = p_{jk}\tau_p tr(\mathbf{R}_{jk}^j \Psi_{jk}^j \mathbf{R}_{jk}^j) + \|\bar{\mathbf{h}}_{jk}^j\|^2, \tag{6.22}$$

$$E\{\|\mathbf{v}_{jk}\|^2\} = p_{jk}\tau_p tr(\mathbf{R}_{jk}^j \Psi_{jk}^j \mathbf{R}_{jk}^j) + \|\bar{\mathbf{h}}_{jk}^j\|^2, \tag{6.23}$$

and $E\{|\mathbf{v}_{jk}^H \mathbf{h}_{li}^j|^2\}$ *is given in* (6.24).

$$\begin{aligned} E\{|\mathbf{v}_{jk}^H \mathbf{h}_{li}^j|^2\} &= p_{jk}\tau_p tr(\mathbf{R}_{li}^j \mathbf{R}_{jk}^j \Psi_{jk}^j \mathbf{R}_{jk}^j) + p_{jk}\tau_p (\bar{\mathbf{h}}_{li}^j)^H \mathbf{R}_{jk}^j \Psi_{jk}^j \mathbf{R}_{jk}^j \bar{\mathbf{h}}_{li}^j \\ &+ (\bar{\mathbf{h}}_{jk}^j)^H \mathbf{R}_{li}^j \bar{\mathbf{h}}_{jk}^j + |(\bar{\mathbf{h}}_{jk}^j)^H \bar{\mathbf{h}}_{li}^j|^2 \\ &+ \begin{cases} p_{jk} p_{li} \tau_p^2 |tr(\mathbf{R}_{li}^j \Psi_{jk}^j \mathbf{R}_{jk}^j)|^2 + 2\sqrt{p_{jk} p_{li}}\tau_p \Re\{tr(\mathbf{R}_{li}^j \Psi_{jk}^j \mathbf{R}_{jk}^j)(\bar{\mathbf{h}}_{li}^j)^H \bar{\mathbf{h}}_{jk}^j\}, & if\,(l,i) \in \mathcal{P}_{jk}, \\ 0, & if\,(l,i) \notin \mathcal{P}_{jk}. \end{cases} \end{aligned} \tag{6.24}$$

Plugging these expression into the SINR in (6.21) *yields*

$$\gamma_{jk}^{ul,mmse} = \frac{p_{jk}^2 \tau_p tr\left(\mathbf{R}_{jk}^j \Psi_{jk}^j \mathbf{R}_{jk}^j\right) + p_{jk}\|\bar{\mathbf{h}}_{jk}^j\|^2}{\sum\limits_{l=1}^{L}\sum\limits_{i=1}^{K_l} p_{li}\xi_{li}^{ul} + \sum\limits_{(l,i)\in\mathcal{P}_{jk}\backslash(j,k)} p_{li}\Gamma_{li}^{ul} - p_{jk}\upsilon_{jk}^{ul} + \sigma_{ul}^2}, \tag{6.25}$$

where $v_{jk}^{ul} = \frac{\|\bar{\mathbf{h}}_{jk}^{j}\|^4}{p_{jk}\tau_p tr\left(\mathbf{R}_{jk}^{j}\Psi_{jk}^{j}\mathbf{R}_{jk}^{j}\right)+\|\bar{\mathbf{h}}_{jk}^{j}\|^2}$, ξ_{li}^{ul} *and* Γ_{li}^{ul} *correspond to LoS-related interference, non-coherent interference, respectively.*

$$\xi_{li}^{ul} = \frac{p_{jk}\tau_p tr\left(\mathbf{R}_{li}^{j}\mathbf{R}_{jk}^{j}\Psi_{jk}^{j}\mathbf{R}_{jk}^{j}\right) + p_{jk}\tau_p\left(\bar{\mathbf{h}}_{li}^{j}\right)^H \mathbf{R}_{jk}^{j}\Psi_{jk}^{j}\mathbf{R}_{jk}^{j}\bar{\mathbf{h}}_{li}^{j} + \left(\bar{\mathbf{h}}_{jk}^{j}\right)^H \mathbf{R}_{li}^{j}\bar{\mathbf{h}}_{jk}^{j} + \left|\left(\bar{\mathbf{h}}_{jk}^{j}\right)^H \bar{\mathbf{h}}_{li}^{j}\right|^2}{p_{jk}\tau_p tr\left(\mathbf{R}_{jk}^{j}\Psi_{jk}^{j}\mathbf{R}_{jk}^{j}\right) + \left\|\bar{\mathbf{h}}_{jk}^{j}\right\|^2} \tag{6.26}$$

$$\Gamma_{li}^{ul} = \frac{p_{jk}p_{li}\tau_p^2\left|tr\left(\mathbf{R}_{li}^{j}\Psi_{jk}^{j}\mathbf{R}_{jk}^{j}\right)^2\right| + 2\sqrt{p_{jk}p_{li}}\tau_p\Re\left\{tr\left(\mathbf{R}_{li}^{j}\Psi_{jk}^{j}\mathbf{R}_{jk}^{j}\right)\left(\bar{\mathbf{h}}_{li}^{j}\right)^H\bar{\mathbf{h}}_{jk}^{j}\right\}}{p_{jk}\tau_p tr\left(\mathbf{R}_{jk}^{j}\Psi_{jk}^{j}\mathbf{R}_{jk}^{j}\right) + \left\|\bar{\mathbf{h}}_{jk}^{j}\right\|^2} \tag{6.27}$$

Proof The proof is given in Appendix D of [13]. □

The rigorous closed-form SINR expression provided in (6.25) offers critical insights into how Massive MIMO systems behave under Rician fading conditions. The signal terms within the numerator are influenced by both estimation quality and the LoS component. Pilot contamination diminishes estimation quality since $p_{jk}^2\tau_p \mathrm{tr}\left(\mathbf{R}_{jk}^{j}\Psi_{jk}^{j}\mathbf{R}_{jk}^{j}\right) = p_{jk}\mathrm{tr}\left(\mathbf{R}_{jk}^{j} - \mathbf{C}_{jk}^{j}\right)$, which represents the multiplication of transmit power and the trace of the channel estimate covariance matrix from (6.9).

In contrast, denominator interference magnitude is dictated by relations between covariance matrix $\mathbf{R}_{jk}^{j}$ and $\mathbf{R}_{li}^{j}$, alongside inner product of LoS components $\bar{\mathbf{h}}_{jk}^{j}$ and $\bar{\mathbf{h}}_{li}^{j}$. If covariance matrices occupy distinct subspaces or exhibit minimal eigenvalues due to weak large-scale fading effects, interference stemming from LoS propagation remains negligible when $\bar{\mathbf{h}}_{jk}^{j}$ and $\bar{\mathbf{h}}_{li}^{j}$ are nearly orthogonal.

Non-coherent interference term ξ_{li}^{ul} is not an increasing function with respect to M, unless $\bar{\mathbf{h}}_{jk}^{j}$ and $\bar{\mathbf{h}}_{li}^{j}$ are nearly parallel, while coherent interference term Γ_{li}^{ul} involves pilot-contaminating UEs $(l, i) \in \mathcal{P}_{jk}\backslash(j, k)$. It grows linearly alongside M. The term v_{jk}^{ul}, dependent upon the desired UE's norm component, increases with M_j.

6.1.3.2 Uplink Spectral Efficiency with EW-MMSE Estimator

In the context of employing the EW-MMSE estimator as outlined in Equation (6.11), the authors in [13] derived a closed-form expression for the SE presented in (6.20), as articulated in Theorem 6.2.

Theorem 6.2 *If MR combining with* $v_{jk} = \hat{\mathbf{h}}_{jk}^{j}$ *is used based on the EW-MMSE estimator, then]*

$$E\left\{\mathbf{v}_{jk}^{H}\mathbf{h}_{jk}^{j}\right\} = p_{jk}\tau_p tr\left(\mathbf{D}_{jk}^{j}\Lambda_{jk}^{j}\mathbf{D}_{jk}^{j}\right) + \left\|\bar{\mathbf{h}}_{jk}^{j}\right\|^2, \tag{6.28}$$

$$E\left\{\|\mathbf{v}_{jk}\|^2\right\} = tr\left(\sum_{jk}^{j}\right) + \|\bar{\mathbf{h}}_{jk}^{j}\|^2, \tag{6.29}$$

and $E\{|\mathbf{v}_{jk}^H \mathbf{h}_{li}^j|^2\}$ *is given in* (6.30).

$$\begin{aligned} E\{|\mathbf{v}_{jk}^H \mathbf{h}_{li}^j|^2\} = \chi_{li}^{ul} &= tr\left(\mathbf{R}_{li}^j \sum_{jk}^{j}\right) + (\bar{\mathbf{h}}_{jk}^j)^H \mathbf{R}_{li}^j \bar{\mathbf{h}}_{jk}^j + (\bar{\mathbf{h}}_{li}^j)^H \sum_{jk}^{j} \bar{\mathbf{h}}_{li}^j + |(\bar{\mathbf{h}}_{jk}^j)^H \bar{\mathbf{h}}_{li}^j|^2 \\ &+ \begin{cases} p_{jk} p_{li} \tau_p^2 \left(tr\left(\mathbf{D}_{li}^j \Lambda_{jk}^j \mathbf{D}_{jk}^j\right)\right)^2 + 2\sqrt{p_{jk} p_{li}}\tau_p tr\left(\mathbf{D}_{li}^j \Lambda_{jk}^j \mathbf{D}_{jk}^j\right) \Re\left\{\left(\bar{\mathbf{h}}_{jk}^j\right)^H \bar{\mathbf{h}}_{li}^j\right\}, & \text{if } (l,i) \in \mathcal{P}_{jk}, \\ 0, & \text{if } (l,i) \notin \mathcal{P}_{jk}. \end{cases} \end{aligned} \tag{6.30}$$

Plugging these expressions into the SINR in (6.21) *gives* $\gamma_{jk}^{ul,ew}$ *in* (6.31).

$$\gamma_{jk}^{ul,ew} = \frac{p_{jk}\left(p_{jk}\tau_p tr\left(\mathbf{D}_{jk}^j \Lambda_{jk}^j \mathbf{D}_{jk}^j\right) + \|\bar{\mathbf{h}}_{jk}^j\|^2\right)^2}{\sum_{l=1}^{L}\sum_{i=1}^{K_l} p_{li}\chi_{li}^{ul} - p_{jk}\left(p_{jk}\tau_p tr\left(\mathbf{D}_{jk}^j \Lambda_{jk}^j \mathbf{D}_{jk}^j\right) + \|\bar{\mathbf{h}}_{jk}^j\|^2\right)^2 + \sigma_{ul}^2\left(tr\left(\Sigma_{jk}^j\right) + \|\bar{\mathbf{h}}_{jk}^j\|^2\right)}. \tag{6.31}$$

Proof The proof is given in Appendix E of [13]. □

The SINR delineated in Theorem 6.2, while more intricate than that derived using the MMSE estimator, retains an analogous interpretative framework.

6.1.3.3 Uplink Spectral Efficiency with LS Estimator

The application of the LS estimator, as delineated in equation (6.16), facilitates the derivation of a closed-form expression for the SE in (6.20), as stated in Theorem 6.3.

Theorem 6.3 *If MR combining with* $\mathbf{v}_{jk} = \frac{1}{\sqrt{p_{jk}\tau_p}}\mathbf{y}_{jjk}^p$ *is used based on the LS estimator, then*

$$E\left\{\mathbf{v}_{jk}^H \mathbf{h}_{jk}^j\right\} = tr\left(\mathbf{R}_{jk}^j\right) + \sum_{(l,i)\in\mathcal{P}_{jk}} \frac{\sqrt{p_{li}}}{\sqrt{p_{jk}}}\left(\bar{\mathbf{h}}_{li}^j\right)^H \bar{\mathbf{h}}_{jk}^j, \tag{6.32}$$

$$E\left\{\|\mathbf{v}_{jk}^H\|^2\right\} = \frac{1}{p_{jk}\tau_p} tr\left((\Psi_{jk}^j)^{-1}\right) + \frac{1}{p_{jk}\tau_p^2}\|\bar{\mathbf{y}}_{jjk}^p\|^2, \tag{6.33}$$

and $E\{|\mathbf{v}_{jk}^H \mathbf{h}_{li}^j|^2\}$ *is given in* (6.34), *where* $\bar{\mathbf{x}}_{jk} = \bar{\mathbf{y}}_{jjk}^p - \sqrt{p_{li}}\tau_p \bar{\mathbf{h}}_{li}^j$ *and* $(\Omega_{jk}^j)^{-1} = (\Psi_{jk}^j)^{-1} - p_{li}\tau_p \mathbf{R}_{li}^j$.

$$
p_{jk}\tau_p^2 E\left\{|\mathbf{v}_{jk}^H \mathbf{h}_{li}^j|^2\right\} = p_{jk}\tau_p^2 \chi_{li}^{ul,ls} = \tau_p tr\left(\mathbf{R}_{li}^j (\Psi_{jk}^j)^{-1}\right) + \left|(\bar{\mathbf{y}}_{jjk}^p)^H \bar{\mathbf{h}}_{li}^j\right|^2
$$
$$
+ \begin{cases} (\bar{\mathbf{y}}_{jjk}^p)^H \mathbf{R}_{li}^j \bar{\mathbf{y}}_{jjk}^p + \tau_p (\bar{\mathbf{h}}_{li}^j)^H (\Psi_{jk}^j)^{-1} \bar{\mathbf{h}}_{li}^j, & if\,(l,i) \notin \mathcal{P}_{jk}, \\ p_{li}\tau_p^2 |tr(\mathbf{R}_{li}^j)|^2 + \bar{\mathbf{x}}_{jk}^H \mathbf{R}_{li}^j \bar{\mathbf{x}}_{jk} + \tau_p (\bar{\mathbf{h}}_{li}^j)^H (\Omega_{jk}^j)^{-1} \bar{\mathbf{h}}_{li}^j & \\ +2\sqrt{p_{li}}\tau_p \Re\left\{(\bar{\mathbf{y}}_{jjk}^p)^H \bar{\mathbf{h}}_{li}^j tr\left(\mathbf{R}_{li}^j\right) + (\bar{\mathbf{y}}_{jjk}^p)^H \mathbf{R}_{li}^j \bar{\mathbf{h}}_{li}^j + \bar{\mathbf{x}}_{jk}^H \bar{\mathbf{h}}_{li}^j \|\bar{\mathbf{h}}_{li}^j\|^2\right\}, & if\,(l,i) \in \mathcal{P}_{jk}. \end{cases} \tag{6.34}
$$

By substituting these into (6.21), *we get* $\gamma_{jk}^{ul,ls}$ *in* (6.35), *where* $\chi_{li}^{ul,ls}$ *is defined in* (6.34).

$$
\gamma_{jk}^{ul,ls} = \frac{p_{jk}\left|tr(\mathbf{R}_{jk}^j) + \sum_{(l,i)\in\mathcal{P}_{jk}} \frac{\sqrt{p_{li}}}{\sqrt{p_{jk}}} (\bar{\mathbf{h}}_{li}^j)^H \bar{\mathbf{h}}_{jk}^j\right|^2}{\sum\limits_{l=1}^{L}\sum\limits_{i=1}^{K_l} p_{li}\chi_{li}^{ul,ls} - p_{jk}\left|tr(\mathbf{R}_{jk}^j) + \sum\limits_{(l,i)\in\mathcal{P}_{jk}} \frac{\sqrt{p_{li}}}{\sqrt{p_{jk}}} (\bar{\mathbf{h}}_{li}^j)^H \bar{\mathbf{h}}_{jk}^j\right|^2 + \sigma_{ul}^2\left(\frac{tr\left((\Psi_{jk}^j)^{-1}\right)}{p_{jk}\tau_p} + \frac{\|\bar{\mathbf{y}}_{jjk}^p\|^2}{p_{jk}\tau_p^2}\right)}, \tag{6.35}
$$

Proof The proof is given in Appendix F of [13]. □

It is noteworthy that under Rayleigh fading conditions—achieved by setting all mean vectors to zero—the difference between SE using MMSE and LS/EW-MMSE estimators becomes negligible [2]. However, under Rician fading conditions where LoS paths are dominant, employing an LS estimator can result in substantial SE loss due to larger interference terms compared to the MMSE estimator [13].

6.1.3.4 Uplink Spectral Efficiency with Mean Only (MO) Estimator

In scenarios where the BTS is limited to knowledge of only the mean values of UEs' channels, these mean values can serve as effective channel estimates without necessitating pilot transmissions. This approach leverages the fact that mean values exhibit slower variations compared to small-scale fading coefficients. Consequently, employing a Mean Only (MO) estimator significantly reduces computational complexity by eliminating the need for channel estimation in each coherence block. However, this method proves beneficial primarily when UEs possess dominant LoS (LoS) paths since it disregards NLoS components.

If the BS only knows the mean values of the UEs' channels but not the covariance matrices, we can use this information as a channel estimate, without the need for sending pilots. The mean values vary much slower compared to the small-scale fading coefficients. Thus, using the MO estimator may greatly reduce the computational complexity since the channels are not estimated for each coherence block. However, it is only meaningful when the UEs have dominant LoS paths since it ignores the NLoS paths. If MR combining is applied using $\mathbf{v}_{jk} = \bar{\mathbf{h}}_{jk}^j$ based on such an MO estimator, we can derive:

$$
E\left\{\mathbf{v}_{jk}^H \mathbf{h}_{jk}^j\right\} = E\left\{\|\mathbf{v}_{jk}^H\|^2\right\} = \|\bar{\mathbf{h}}_{jk}^j\|^2, \tag{6.36}
$$

$$E\left\{|\mathbf{v}_{jk}^{H}\mathbf{h}_{li}^{j}|^2\right\} = (\bar{\mathbf{h}}_{jk}^{j})^H \mathbf{R}_{li}^{j} \bar{\mathbf{h}}_{jk}^{j} + |(\bar{\mathbf{h}}_{jk}^{j})^H \bar{\mathbf{h}}_{li}^{j}|^2. \tag{6.37}$$

By substituting these expressions into the SINR in (6.21), $\gamma_{jk}^{ul,mo}$ is obtained as

$$\frac{p_{jk}\|\bar{\mathbf{h}}_{jk}^{j}\|^2}{\sum_{l=1}^{L}\sum_{i=1}^{K_l} \frac{p_{li}}{\|\bar{\mathbf{h}}_{jk}^{j}\|^2}\left((\bar{\mathbf{h}}_{jk}^{j})^H \mathbf{R}_{li}^{j} \bar{\mathbf{h}}_{jk}^{j} + |(\bar{\mathbf{h}}_{jk}^{j})^H \bar{\mathbf{h}}_{li}^{j}|^2\right) - p_{jk}\|\bar{\mathbf{h}}_{jk}^{j}\|^2 + \sigma_{ul}^2}. \tag{6.38}$$

It is crucial to note that if UE (j, k) lacks an LoS component, its channel information remains unknown and consequently renders the SINR in (6.38) null. Therefore, the MO estimator's utility is confined to UEs with an identifiable LoS path.

6.1.4 Downlink Spectral Efficiency with MR Precoding

In each coherence block, there are τd downlink (DL) data transmissions. The signal transmitted from BTS l is represented as

$$\mathbf{x}_l = \sum_{k=1}^{K_l} \mathbf{w}_{lk} \varsigma_{lk}, \tag{6.39}$$

where $\varsigma_{lk} \sim \mathcal{N}_C(0, p_{lk})$ represents the DL data signal intended for UE k, and p_{lk} denotes the signal power. The transmit precoding vector $\mathbf{w}_{lk}$ is crucial in determining the spatial directivity of transmission and satisfies $E\{\|\mathbf{w}_{lk}\|^2\} = 1$. Consequently, the expected transmit power allocated to this UE is given by $E\{\|\mathbf{w}_{lk}\varsigma_{lk}\|^2\} = p_{lk}$.

The received signal at UE k in cell j, denoted as $y_{jk} \in C$, can be expressed as

$$\begin{aligned} y_{jk} &= (\mathbf{h}_{jk}^{l})^H \mathbf{w}_{jk}\varsigma_{jk} + \sum_{i=1, i\neq k}^{K_j} (\mathbf{h}_{jk}^{l})^H \mathbf{w}_{ji}\varsigma_{ji} \\ &+ \sum_{l=1, l\neq j}^{L} \sum_{i=1}^{K_l} (\mathbf{h}_{jk}^{l})^H \mathbf{w}_{li}\varsigma_{li} + n_{jk}, \end{aligned} \tag{6.40}$$

where $n_{jk} \sim \mathcal{N}_C(0, \sigma_{dl}^2)$ represents i.i.d. additive receiver noise with variance σ_{dl}^2. The first term in this equation corresponds to the desired signal; the second term accounts for intra-cell interference, while inter-cell interference is captured by the third term.

The ergodic DL capacity of UE k within cell j has a lower bound [2, Theorem 4–6]:

$$SE_{jk}^{dl} = \frac{\tau_d}{\tau_c} \log_2\left(1 + \gamma_{jk}^{dl}\right) \qquad [bit/s/Hz] \tag{6.41}$$

with

$$\gamma_{jk}^{dl} = \frac{p_{jk} \left| E\{\mathbf{w}_{jk}^H \mathbf{h}_{jk}^j\} \right|^2}{\sum_{l=1}^{L} \sum_{i=1}^{K_l} p_{li} E\left\{ \left| \mathbf{w}_{li}^H \mathbf{h}_{jk}^l \right|^2 \right\} - p_{jk} \left| E\left\{ \mathbf{w}_{jk}^H \mathbf{h}_{jk}^j \right\} \right|^2 + \sigma_{dl}^2}, \tag{6.42}$$

where expectations are taken over all random variables. Subsequent subsections will compute the effective SINR γ_{jk}^{dl} for MR precoding by using different channel estimators.

6.1.4.1 Downlink Spectral Efficiency with MMSE Estimator

The application of the MMSE estimator, as delineated in equation (6.6), facilitates the derivation of a closed-form expression for the DL SE, as articulated in Theorem 6.4.

Theorem 6.4 *If MR precoding with* $\mathbf{w}_{jk} = \frac{\hat{\mathbf{h}}_{jk}^j}{\sqrt{E\{\|\hat{\mathbf{h}}_{jk}^j\|^2\}}}$ *is used based on the MMSE estimator, then*

$$E\{\mathbf{w}_{jk}^H \mathbf{h}_{jk}^j\} = \sqrt{p_{jk}\tau_p tr\left(\mathbf{R}_{jk}^j \Psi_{jk}^j \mathbf{R}_{jk}^j\right) + \|\bar{\mathbf{h}}_{jk}^j\|^2}, \tag{6.43}$$

and $E\left\{ \left| \mathbf{w}_{li}^H \mathbf{h}_{jk}^l \right|^2 \right\}$ *is given in* (6.44), *where* $E\{\|\hat{\mathbf{h}}_{li}^l\|^2\} = p_{li}\tau_p tr\left(\mathbf{R}_{li}^l \Psi_{li}^l \mathbf{R}_{li}^l\right) + \|\bar{\mathbf{h}}_{li}^l\|^2$.

$$\begin{aligned} E\{\|\hat{\mathbf{h}}_{li}^l\|^2\} E\left\{ \left| \mathbf{w}_{li}^H \mathbf{h}_{jk}^l \right|^2 \right\} &= p_{li}\tau_p tr\left(\mathbf{R}_{jk}^l \mathbf{R}_{li}^l \Psi_{li}^l \mathbf{R}_{li}^j\right) + p_{li}\tau_p (\bar{\mathbf{h}}_{jk}^l)^H \mathbf{R}_{li}^l \Psi_{li}^l \mathbf{R}_{li}^l \bar{\mathbf{h}}_{jk}^l \\ &+ \left| (\bar{\mathbf{h}}_{jk}^l)^H \bar{\mathbf{h}}_{li}^l \right|^2 + \left(\bar{\mathbf{h}}_{li}^l\right)^H \mathbf{R}_{jk}^l \bar{\mathbf{h}}_{li}^l \\ &+ \begin{cases} p_{jk} p_{li} \tau_p^2 \left| tr\left(\mathbf{R}_{jk}^l \Psi_{li}^l \mathbf{R}_{li}^l\right) \right|^2 + 2\sqrt{p_{jk} p_{li}}\tau_p \Re\left\{ tr\left(\mathbf{R}_{jk}^l \Psi_{li}^l \mathbf{R}_{li}^l\right) \left(\bar{\mathbf{h}}_{li}^l\right)^H \bar{\mathbf{h}}_{jk}^l \right\}, & \text{if } (l,i) \in \mathcal{P}_{jk}, \\ 0, & \text{if } (l,i) \notin \mathcal{P}_{jk}. \end{cases} \end{aligned} \tag{6.44}$$

By incorporating these into the DL SINR in (6.42), *we obtain*

$$\gamma_{jk}^{dl,mmse} = \frac{\rho_{jk} p_{jk} \tau_p tr\left(\mathbf{R}_{jk}^j \Psi_{jk}^j \mathbf{R}_{jk}^j\right) + \rho_{jk} \|\bar{\mathbf{h}}_{jk}^j\|^2}{\sum_{l=1}^{L} \sum_{i=1}^{K_l} \rho_{li} \xi_{li}^{dl} + \sum_{(l,i) \in \mathcal{P}_{jk} \setminus (j,k)} \rho_{li} \Gamma_{li}^{dl} - \rho_{jk} \upsilon_{jk}^{d} l + \sigma_{dl}^2}, \tag{6.45}$$

where $v_{jk}^{dl} = \|\bar{\mathbf{h}}_{jk}^{l}\|^4 / \left(p_{jk}\tau_p tr\left(\mathbf{R}_{jk}^{l}\Psi_{jk}^{l}\mathbf{R}_{jk}^{l}\right) + \|\bar{\mathbf{h}}_{jk}^{l}\|^2\right)$, ξ_{li}^{dl} *and* Γ_{li}^{dl} *correspond to the non-coherent interference in* (6.46) *and the coherent interference in* (6.47), *respectively.*

$$\xi_{li}^{dl} = \frac{p_{li}\tau_p tr(\mathbf{R}_{jk}^{l}\mathbf{R}_{li}^{l}\Psi_{li}^{l}\mathbf{R}_{li}^{l}) + p_{li}\tau_p \left(\bar{\mathbf{h}}_{jk}^{l}\right)^H \mathbf{R}_{li}^{l}\Psi_{li}^{l}\mathbf{R}_{li}^{l}\bar{\mathbf{h}}_{jk}^{l} + \left(\bar{\mathbf{h}}_{li}^{l}\right)^H \mathbf{R}_{jk}^{l}\bar{\mathbf{h}}_{li}^{l} + \left|(\bar{\mathbf{h}}_{jk}^{l})^H\bar{\mathbf{h}}_{li}^{l}\right|^2}{p_{li}\tau_p tr\left(\mathbf{R}_{li}^{l}\Psi_{li}^{l}\mathbf{R}_{li}^{l}\right) + \|\bar{\mathbf{h}}_{li}^{l}\|^2}, \tag{6.46}$$

$$\Gamma_{li}^{dl} = \frac{p_{jk}p_{li}\tau_p^2\left|tr\left(\mathbf{R}_{jk}^{l}\Psi_{li}^{l}\mathbf{R}_{li}^{l}\right)\right|^2 + 2\sqrt{p_{jk}p_{li}}\tau_p\Re\left\{tr\left(\mathbf{R}_{jk}^{l}\Psi_{li}^{l}\mathbf{R}_{li}^{l}\right)\left(\bar{\mathbf{h}}_{li}^{l}\right)^H\bar{\mathbf{h}}_{jk}^{l}\right\}}{p_{li}\tau_p tr\left(\mathbf{R}_{li}^{l}\Psi_{li}^{l}\mathbf{R}_{li}^{l}\right) + \|\bar{\mathbf{h}}_{li}^{l}\|^2}. \tag{6.47}$$

Proof The proof is similar to the uplink case in Appendix D of [13]. □

Notably, this SINR expression mirrors its UL counterpart from Theorem 6.1, albeit with distinct notations for transmit power and noise variance alongside index permutations within interference terms—a reflection of UL-DL duality principles [2, Sect. 4.3]. Consequently, its qualitative interpretation remains consistent with that observed within uplink contexts.

6.1.4.2 Downlink Spectral Efficiency with EW-MMSE Estimator

In the context of employing the EW-MMSE estimator as delineated in (6.11), the authors in [13] derive a closed-form expression for the DL SE as presented in (6.41), which is stated in Theorem 6.5.

Theorem 6.5 *If MR precoding with* $\mathbf{w}_{jk} = \frac{\hat{\mathbf{h}}_{jk}^{j}}{\sqrt{E\{\|\hat{\mathbf{h}}_{jk}^{j}\|^2\}}}$ *is used based on the EW-MMSE estimator, then*

$$E\{\mathbf{w}_{jk}^H\mathbf{h}_{jk}^{j}\} = \frac{p_{jk}\tau_p tr\left(\mathbf{D}_{jk}^{j}\Lambda_{jk}^{j}\mathbf{D}_{jk}^{j}\right) + \|\bar{\mathbf{h}}_{jk}^{j}\|^2}{\sqrt{tr\left(\sum_{jk}^{j}\right) + \|\bar{\mathbf{h}}_{jk}^{j}\|^2}}, \tag{6.48}$$

and $E\left\{\left|\mathbf{w}_{li}^H\mathbf{h}_{jk}^{l}\right|^2\right\}$ *is given in* (6.49), *where* $E\{\|\hat{\mathbf{h}}_{li}^{l}\|^2\} = tr\left(\sum_{li}^{l}\right) + \|\bar{\mathbf{h}}_{li}^{l}\|^2$. *By substituting these expressions into the DL SINR in* (6.42), *we achieve* $\gamma_{jk}^{dl,ew}$ *in* (6.50).

$$E\{\|\hat{\mathbf{h}}_{li}^{l}\|^2\}E\left\{\left|\mathbf{w}_{li}^{H}\mathbf{h}_{jk}^{l}\right|^2\right\} = \chi_{li}^{dl} = tr\left(\mathbf{R}_{jk}^{l}\sum_{li}^{l}\right) + (\bar{\mathbf{h}}_{li}^{l})^H\mathbf{R}_{jk}^{l}\bar{\mathbf{h}}_{li}^{l} + (\bar{\mathbf{h}}_{jk}^{l})^H\sum_{li}^{l}\bar{\mathbf{h}}_{jk}^{l} + |(\bar{\mathbf{h}}_{li}^{l})^H\bar{\mathbf{h}}_{jk}^{l}|^2$$
$$+\begin{cases} p_{jk}p_{li}\tau_p^2\left(tr\left(\mathbf{D}_{jk}^{l}\Lambda_{li}^{l}\mathbf{D}_{li}^{l}\right)\right)^2 + 2\sqrt{p_{jk}p_{li}}\tau_p tr\left(\mathbf{D}_{jk}^{l}\Lambda_{li}^{l}\mathbf{D}_{li}^{l}\right)\Re\left\{(\bar{\mathbf{h}}_{li}^{l})^H\bar{\mathbf{h}}_{jk}^{l}\right\}, & if\ (l,i)\in\mathcal{P}_{jk}, \\ 0, & if\ (l,i)\notin\mathcal{P}_{jk}. \end{cases} \tag{6.49}$$

$$\gamma_{jk}^{dl,ew} = \frac{\rho_{jk}\left(p_{jk}\tau_p tr\left(\mathbf{D}_{jk}^{j}\Lambda_{jk}^{j}\mathbf{D}_{jk}^{j}\right) + \|\bar{\mathbf{h}}_{jk}^{j}\|^2\right)/\left(tr\left(\sum_{jk}^{j}\right) + \|\bar{\mathbf{h}}_{jk}^{j}\|^2\right)}{\sum_{l=1}^{L}\sum_{i=1}^{K_l}\rho_{li}\frac{\chi_{li}^{dl}}{tr\left(\sum_{li}^{l}\right) + \|\bar{\mathbf{h}}_{li}^{l}\|^2} - \rho_{jk}\frac{\left(p_{jk}\tau_p tr\left(\mathbf{D}_{jk}^{j}\Lambda_{jk}^{j}\mathbf{D}_{jk}^{j}\right) + \|\bar{\mathbf{h}}_{jk}^{j}\|^2\right)^2}{tr\left(\sum_{jk}^{j}\right) + \|\bar{\mathbf{h}}_{jk}^{j}\|^2} + \sigma_{dl}^2}. \tag{6.50}$$

Proof The proof parallels those used for the UL scenario. □

Theorem 6.5 provides an analytical framework for understanding how MR precoding can be effectively utilized within this estimation paradigm to optimize DL performance metrics such as SE and SINR.

6.1.4.3 Downlink Spectral Efficiency with LS Estimator

Utilizing the LS estimator as defined in (6.16), we can derive a closed-form formula for the DL SE presented in (6.41), which is encapsulated by Theorem 6.6.

Theorem 6.6 *If MR precoding with* $\mathbf{w}_{jk} = \frac{\hat{\mathbf{h}}_{jk}^{j}}{\sqrt{E\{\|\hat{\mathbf{h}}_{jk}^{j}\|^2\}}}$ *is used based on the LS estimator, then*

$$E\left\{\mathbf{w}_{jk}^{H}\mathbf{h}_{jk}^{j}\right\} = \frac{E\left\{\mathbf{v}_{jk}^{H}\mathbf{h}_{jk}^{j}\right\}}{\sqrt{E\left\{\|\mathbf{v}_{jk}\|^2\right\}}} = \frac{tr(\mathbf{R}_{jk}^{j}) + \sum_{(l,i)\in\mathcal{P}_{jk}}\frac{\sqrt{p_{li}}}{\sqrt{p_{jk}}}(\bar{\mathbf{h}}_{li}^{j})^H\bar{\mathbf{h}}_{jk}^{j}}{\sqrt{\frac{1}{p_{jk}\tau_p^2}\left(\tau_p tr\left((\Psi_{jk}^{j})^{-1}\right) + \|\bar{\mathbf{y}}_{jjk}^{p}\|^2\right)}} \tag{6.51}$$

and $E\left\{\left|\mathbf{w}_{li}^{H}\mathbf{h}_{jk}^{l}\right|^2\right\}$ *is given in* (6.52), *where* $E\{\|\hat{\mathbf{h}}_{li}^{l}\|^2\} = \frac{1}{p_{jk}\tau_p^2}\left(\tau_p tr\left((\Psi_{jk}^{j})^{-1}\right) + \|\bar{\mathbf{y}}_{jjk}^{p}\|^2\right)$. *By substituting these expressions into the DL SINR in* (6.42), *we obtain* $\gamma_{jk}^{dl,ls}$ *in* (6.53), *where the interference term* $\chi_{li}^{dl,ls}$ *is defined in* (6.52).

$$E\left\{\|\hat{\mathbf{h}}_{li}^{l}\|^2\right\}E\left\{|\mathbf{w}_{li}^{H}\mathbf{h}_{jk}^{l}|^2\right\} = p_{li}\tau_p^2\chi_{li}^{dl,ls} = \tau_p tr\left(\mathbf{R}_{jk}^{l}(\Psi_{li}^{l})^{-1}\right) + |(\bar{\mathbf{y}}_{jli}^{p})^H\bar{\mathbf{h}}_{jk}^{l}|^2$$
$$+\begin{cases} (\bar{\mathbf{y}}_{jli}^{p})^H\mathbf{R}_{jk}^{l}\bar{\mathbf{y}}_{jli}^{p} + \tau_p(\bar{\mathbf{h}}_{jk}^{l})^H(\Psi_{li}^{l})^{-1}\bar{\mathbf{h}}_{jk}^{l}, & if\ (l,i)\notin\mathcal{P}_{jk}, \\ p_{jk}\tau_p^2\left|tr(\mathbf{R}_{jk}^{l})\right|^2 + \bar{\mathbf{x}}_{li}^{H}\mathbf{R}_{jk}^{l}\bar{\mathbf{x}}_{li} + \tau_p(\bar{\mathbf{h}}_{jk}^{l})^H(\Omega_{li}^{l})^{-1}\bar{\mathbf{h}}_{jk}^{l} & \\ +2\sqrt{p_{jk}}\tau_p\Re\left\{(\bar{\mathbf{y}}_{jli}^{p})^H\bar{\mathbf{h}}_{jk}^{l}tr\left(\mathbf{R}_{jk}^{l}\right) + (\bar{\mathbf{y}}_{jli}^{p})^H\mathbf{R}_{jk}^{l}\bar{\mathbf{h}}_{jk}^{l} + \bar{\mathbf{x}}_{li}^{H}\bar{\mathbf{h}}_{jk}^{l}\|\bar{\mathbf{h}}_{jk}^{l}\|^2\right\}, & if\ (l,i)\in\mathcal{P}_{jk}. \end{cases} \tag{6.52}$$

$$\gamma_{jk}^{dl,ls} = \frac{\rho_{jk} p_{jk} \tau_p^2 \left| tr(\mathbf{R}_{jk}^j) + \sum_{(l,i)\in\mathcal{P}_{jk}} \frac{\sqrt{p_{li}}}{p_{jk}} (\bar{\mathbf{h}}_{li}^j)^H \bar{\mathbf{h}}_{jk}^j \right|^2 / \left(\tau_p tr\left((\Psi_{jk}^j)^{-1} \right) + \|\bar{\mathbf{y}}_{jjk}^p\|^2 \right)}{\sum_{l=1}^{L} \sum_{i=1}^{K_l} \rho_{li} \frac{p_{li} \tau_p^2 \chi_{li}^{dl,ls}}{\tau_p tr\left((\Psi_{li}^l)^{-1} \right) + \|\bar{\mathbf{y}}_{lli}^p\|^2} - \rho_{jk} \frac{p_{jk} \tau_p^2 \left| tr(\mathbf{R}_{jk}^j) + \sum_{(l,i)\in\mathcal{P}_{jk}} \frac{\sqrt{p_{li}}}{\sqrt{p_{jk}}} (\bar{\mathbf{h}}_{li}^j)^H \bar{\mathbf{h}}_{jk}^j \right|^2}{\tau_p tr\left((\Psi_{jk}^j)^{-1} \right) + \|\bar{\mathbf{y}}_{jjk}^p\|^2} + \sigma_{dl}^2}, \tag{6.53}$$

Proof The proof is similar to the uplink case. □

6.1.4.4 Downlink Spectral Efficiency with Mean Only Estimator

When MR precoding employs $\mathbf{w}_{jk} = \frac{\bar{\mathbf{h}}_{jk}^j}{\|\bar{\mathbf{h}}_{jk}^j\|}$ via MO estimator, we have:

$$E\left\{ \mathbf{w}_{jk}^H \mathbf{h}_{jk}^j \right\} = \|\bar{\mathbf{h}}_{jk}^j\|, \tag{6.54}$$

$$E\left\{ |\mathbf{w}_{li}^H \mathbf{h}_{jk}^l|^2 \right\} = \left((\bar{\mathbf{h}}_{li}^l)^H \mathbf{R}_{jk}^l \bar{\mathbf{h}}_{li}^l + \left| (\bar{\mathbf{h}}_{li}^l)^H \bar{\mathbf{h}}_{jk}^l \right|^2 \right) / \|\bar{\mathbf{h}}_{li}^l\|^2. \tag{6.55}$$

By substituting these equations into the DL SINR in (6.42), $\gamma_{jk}^{dl,mo}$ can be computed as

$$\frac{p_{jk} \|\bar{\mathbf{h}}_{jk}^j\|^2}{\sum_{l=1}^{L} \sum_{i=1}^{K_l} \frac{\rho_{li}}{\|\bar{\mathbf{h}}_{li}^l\|^2} \left((\mathbf{h}_{li}^l)^H \mathbf{R}_{jk}^l \bar{\mathbf{h}}_{li}^l + |(\bar{\mathbf{h}}_{li}^l)^H \mathbf{h}_{jk}^l|^2 \right) - p_{jk} \|\mathbf{h}_{jk}^j\|^2 + \sigma_{dl}^2}. \tag{6.56}$$

6.1.5 Asymptotic Analysis

In this section, we delve into the asymptotic behavior of Rician fading channels when employing the MR combining technique, utilizing various channel estimators. The foundational work by [13] is built upon several technical assumptions:

- **Assumption 1:** For $l, j = 1, ..., L$ and $i = 1, ..., K_l$, the spatial covariance matrix $\mathbf{R}_{li}^j$ satisfies $\lim\sup_{M_j} \|\mathbf{R}_{li}^j\|^2 < \infty$ and $\lim\inf_{M_j} \frac{1}{M_j} \mathrm{tr}\left(\mathbf{R}_{li}^j \right) > 0$.
- **Assumption 2:** For $l, j = 1, ..., L$ and $i = 1, ..., K_l$, the LoS component $\bar{\mathbf{h}}_{li}^j$ satisfies $\lim\sup_{M_j} \frac{1}{M_j} \|\bar{\mathbf{h}}_{li}^j\|^2 < \infty$.
- **Assumption 3:** For $l, j = 1, ..., L$, $i = 1, ..., K_l$ and $k = 1, ..., K_j$, the LoS component $\bar{\mathbf{h}}_{jk}^j$ and $\bar{\mathbf{h}}_{li}^j$ satisfy $\lim_{M_j} \frac{1}{M_j} \left| (\bar{\mathbf{h}}_{jk}^j)^H \bar{\mathbf{h}}_{li}^j \right| \to 0$ if $(j, k) \neq (l, i)$.

The first assumption aligns with standard practices in Massive MIMO asymptotic analysis [4], indicating that the gathered signal energy scales with antenna count due to diverse spatial directions contributing to it. Further discussions on these assumptions are provided in Sect. 6.1.5.5.

We also revisit spatially orthogonal matrices as defined by Bjornson et al. [2]:

Definition 6.1 For $l = 1, ..., L$ and $i = 1, ..., K_l$, two spatial covariance matrices $\mathbf{R}^j_{li}$ and $\mathbf{R}^j_{jk}$ are considered asymptotically spatially orthogonal if their cross-product's trace normalized by array size approaches zero as array size becomes infinite:

$$\frac{1}{M_j}\text{tr}\left(\mathbf{R}^j_{li}\mathbf{R}^j_{jk}\right) \to 0 \quad \text{as} \quad M_j \to \infty. \tag{6.57}$$

Definition 6.1 implies significant rank deficiency of both matrices involved. In scenarios involving spatial correlation fading—particularly when the BTS employs uniform linear arrays—the condition holds if UE channels exhibit non-overlapping multipath angular distributions [20, Theorem 1]. However, such ideal angular separations are rare in practical sub-6 GHz cellular networks.

6.1.5.1 Asymptotic Analysis of Spectral Efficiency with MMSE Estimator

Theorem 6.7 *Under Assumptions 1–3, it is deduced that the uplink SINR, denoted as* $\gamma^{ul,mmse}_{jk}$, *tends to infinity as* $M_j \to \infty$, *provided that the matrix* $\mathbf{R}^j_{jk}$ *is asymptotically spatially orthogonal to* $\mathbf{R}^j_{li}\forall(l,i) \in \mathcal{P}_{jk}\backslash(j,k)$. *Otherwise, as* $M_j \to \infty$, *we have:*

$$\gamma^{ul,mmse}_{jk} - \frac{p^2_{jk}\tau_p tr(\mathbf{R}^j_{jk}\Psi^j_{jk}\mathbf{R}^j_{jk}) + p_{jk}\|\bar{\mathbf{h}}^j_{jk}\|^2}{\sum\limits_{(l,i)\in\mathcal{P}_{jk}\backslash(j,k)} p^2_{li}\dfrac{p_{jk}\tau^2_p|tr(\mathbf{R}^j_{li}\Psi^j_{jk}\mathbf{R}^j_{jk})|^2}{p_{jk}\tau_p tr\left(\mathbf{R}^j_{jk}\Psi^j_{jk}\mathbf{R}^j_{jk}\right) + \|\bar{\mathbf{h}}^j_{jk}\|^2}} \to 0. \tag{6.58}$$

Proof The proof is given in Appendix G of [13]. □

Theorem 6.8 *Under Assumption 1–3, it follows that* $\gamma^{dl,mmse}_{jk}$ *grows without bound as* $M_1 = ... = M_L \to \infty$ *if* $\mathbf{R}^l_{jk}$ *is asymptotically spatially orthogonal to* $\mathbf{R}^l_{li}\forall(l,i) \in \mathcal{P}_{jk}\backslash(j,k)$. *Otherwise, as* $M_1 = ... = M_L \to \infty$, *it follows that*

$$\gamma^{dl,mmse}_{jk} - \frac{\rho_{jk}p_{jk}\tau_p tr\left(\mathbf{R}^j_{jk}\Psi^j_{jk}\mathbf{R}^j_{jk}\right) + \rho_{jk}\|\bar{\mathbf{h}}^j_{jk}\|^2}{\sum\limits_{(l,i)\in\mathcal{P}_{jk}\backslash(j,k)} \rho_{li}\dfrac{p_{jk}p_{li}\tau^2_p|tr(\mathbf{R}^l_{jk}\Psi^l_{li}\mathbf{R}^l_{li})|^2}{p_{li}\tau_p tr\left(\mathbf{R}^l_{li}\Psi^l_{li}\mathbf{R}^l_{li}\right) + \|\bar{\mathbf{h}}^l_{li}\|^2}} \to 0. \tag{6.59}$$

Proof The proof is given in Appendix G of [13]. □

It is shown by these theorems that in general the SINRs are upper bounded by simplified asymptotic SINR formulas, which depend on the covariance matrices and mean values. One exception is that the covariance matrices are asymptotically spatially orthogonal. In this case, the SE grows unboundedly with rate $\log_2(M)$, similar to the results for correlated Rayleigh fading in [2, 20].

6.1.5.2 Asymptotic Analysis of Spectral Efficiency with EW-MMSE Estimator

Theorem 6.9 *Under Assumptions 1–3, it can be rigorously demonstrated that* $\gamma_{jk}^{ul,ew}$ *tends toward infinity as* $M_j \to \infty$*, provided that* $\mathbf{D}_{jk}^{j}$ *is asymptotically spatially orthogonal to* $\mathbf{D}_{li}^{j} \forall (l,i) \in \mathcal{P}_{jk} \backslash (j,k)$*. Conversely, if this condition does not hold, under the same assumptions and as* $M_j \to \infty$*, we obtain:*

$$\gamma_{jk}^{ul,ew} - \frac{p_{jk}^2 \tau_p tr(\mathbf{D}_{jk}^{j} \Lambda_{jk}^{j} \mathbf{D}_{jk}^{j}) + p_{jk} \|\bar{\mathbf{h}}_{jk}^{j}\|^2}{\sum_{(l,i)\in\mathcal{P}_{jk}\backslash(j,k)} p_{li}^2 \frac{p_{jk} p_{li} \tau_p^2 \left(tr\left(\mathbf{D}_{li}^{j} \Lambda_{jk}^{j} \mathbf{D}_{jk}^{j}\right)\right)^2}{p_{jk}\tau_p tr\left(\mathbf{D}_{jk}^{j} \Lambda_{jk}^{j} \mathbf{D}_{jk}^{j}\right) + \|\bar{\mathbf{h}}_{jk}^{j}\|^2}} \to 0. \tag{6.60}$$

Proof The proof is given in Appendix H of [13]. □

Theorem 6.10 *Under Assumptions 1–3, it follows that* $\gamma_{jk}^{ul,ew}$ *grows without bound as* $M_j \to \infty$ *if* $\mathbf{D}_{li}^{l}$ *is asymptotically spatially orthogonal to* $\mathbf{D}_{jk}^{l} \forall (l,i) \in \mathcal{P}_{jk} \backslash (j,k)$*. Otherwise, with the same Assumptions 1, as* $M_j \to \infty$*, it follows that*

$$\gamma_{jk}^{dl,ew} - \frac{\frac{\rho_{jk}\left(p_{jk}\tau_p tr\left(\mathbf{D}_{jk}^{j} \Lambda_{jk}^{j} \mathbf{D}_{jk}^{j}\right) + \|\bar{\mathbf{h}}_{jk}^{j}\|^2\right)^2}{tr\left(\Sigma_{jk}^{j}\right) + \|\bar{\mathbf{h}}_{jk}^{j}\|^2}}{\sum_{(l,i)\in\mathcal{P}_{jk}\backslash(j,k)} \rho_{li} \frac{p_{jk} p_{li} \tau_p^2 \left(tr\left(\mathbf{D}_{jk}^{l} \Lambda_{li}^{l} \mathbf{D}_{li}^{l}\right)\right)^2}{tr\left(\Sigma_{li}^{l}\right) + \|\bar{\mathbf{h}}_{li}^{l}\|^2}} \to 0. \tag{6.61}$$

Proof The proof is given in Appendix H of [13]. □

The implications from these theorems are similar to the MMSE estimation case. However, it requires a more stringent condition that the diagonals of covariance matrices (rather than the matrices themselves) achieve asymptotic spatial orthogonality to ensure an unbounded SE.

6.1.5.3 Asymptotic Analysis of Spectral Efficiency with LS Estimator

Theorem 6.11 *Under Assumption 1–3, as $M_j \to \infty$, we have*

$$\gamma_{jk}^{ul,ls} - \frac{p_{jk}^2 \tau_p^2 \left(tr(\mathbf{R}_{jk}^j) + \|\bar{\mathbf{h}}_{jk}^j\|^2 \right)}{\sum_{(l,i)\in\mathcal{P}_{jk}\setminus(j,k)} \frac{P_{li}^2 \tau_p^2 \left(tr(\mathbf{R}_{li}^j) + \|\bar{\mathbf{h}}_{li}^j\|^2 \right)^2}{tr(\mathbf{R}_{jk}^j) + \sum_{(l,i)\in\mathcal{P}_{jk}} \frac{\sqrt{p_{jk}}}{\sqrt{p_{li}}} \left| (\bar{\mathbf{h}}_{li}^j)^H \bar{\mathbf{h}}_{jk}^j \right|}} \to 0. \tag{6.62}$$

Theorem 6.12 *Under Assumptions 1–3, as $M_1 = \ldots = M_L \to \infty$, it follows that*

$$\gamma_{jk}^{ul,ls} - \frac{p_{jk}^2 \left(tr(\mathbf{R}_{jk}^j) + \|\bar{\mathbf{h}}_{jk}^j\|^2 \right)}{\sum_{(l,i)\in\mathbf{P}_{jk}\setminus(j,k)} \frac{p_{li}^2 \tau_p^2 \left(tr(\mathbf{R}_{jk}^j) + \|\bar{\mathbf{h}}_{jk}^j\|^2 \right)}{tr(\mathbf{R}_{jk}^j) + \sum_{(l,i)\in\mathcal{P}_{jk}} \frac{p_{li} p_{jk} \tau_p^2 \left(tr(\mathbf{R}_{jk}^j) + \|\bar{\mathbf{h}}_{jk}^l\|^2 \right)^2}{\tau_p tr\left(\left(\Psi_{li}^l\right)^{-1} \right) + \|\bar{\mathbf{y}}_{lli}^p\|^2}}} \to 0. \tag{6.63}$$

Proof The proof is given in Appendix I of [13]. □

There is an important contrast with scenarios involving MMSE and EW-MMSE estimator. Specifically, when employing the LS estimator, SE does not exhibit unbounded growth.

6.1.5.4 Asymptotic Analysis of Spectral Efficiency with Mean Only Estimator

Under Assumptions 1–3, if UE (j, k) possesses a LoS path, the SINR $\gamma_{jk}^{ul,mo}$ tends to infinity as $M_j \to \infty$. Similarly, in the DL scenario under the same assumptions, if UE (j, k) has an LoS path, then $\gamma_{jk}^{ul,mo}$ also grows unbounded as $M_1, \ldots, M_L \to \infty$.

Proof In the UL, the LoS-related term $\frac{p_{li} \left| \left(\bar{\mathbf{h}}_{jk}^j\right)^H \bar{\mathbf{h}}_{li}^j \right|^2}{\|\bar{\mathbf{h}}_{jk}^j\|^2}$ in (6.38) approaches zero as $M_j \to \infty$ for $(l, i) \neq (j, k)$ due to Assumption 2 and 3. Noting that the $\frac{\left(\bar{\mathbf{h}}_{jk}^j\right)^H \mathbf{R}_{li}^j \bar{\mathbf{h}}_{jk}^j}{\|\bar{\mathbf{h}}_{jk}^j\|^2} \leq \|\mathbf{R}_{li}^j\|^2$ remains finite as $M_j \to \infty$ due to Assumption 1, which completes the proof. In the DL, the expressions for $\gamma_{jk}^{dl,mo}$ mirrors those of $\gamma_{jk}^{ul,mo}$, except that indices (l, i) and (j, k) are swapped in the interference terms. □

It is important to note that while mean vectors are assumed perfectly known herewith, minor phase shifts still can significantly impact the system performance—a topic that will be further explored in Sect. 6.1.2.

6.1.5.5 Discussion on the Asymptotic Behaviors of Rician Fading Channels

To elucidate Assumptions 2–3 and Definition 1's rationale within this context, a uniform linear array (ULA) equipped with omni-directional antennas is considered, in which the LoS component is written as [2, Sect. 1.3]

$$\bar{\mathbf{h}}_{li}^{j} = \sqrt{\beta_{li}^{j,LoS}} \left[1 e^{j2\pi d_H sin(\varphi_{jk}^{j})} \ldots e^{j2\pi d_H (M-1) sin(\varphi_{jk}^{j})} \right]^T, \tag{6.64}$$

where $\beta_{li}^{j,LoS}$ is the large-scale fading coefficient, $d_H \leq 0.5$ is the antenna spacing parameter (in fractions of the wavelength), and φ_{jk}^{j} is the angle of arrival (AoA) to the UE seen from the BTS. By utilizing this model, we get $\frac{1}{M_j} \|\bar{\mathbf{h}}_{li}^{j}\|^2 = \beta_{li}^{j,LoS}$, which is finite for any M_j and consistent with Assumption 2.

The magnitude of the inner product of LoS components of two different UEs can be written as

$$\begin{aligned} \left| (\bar{\mathbf{h}}_{jk}^{j})^H \bar{\mathbf{h}}_{li}^{j} \right| &= \sqrt{\beta_{jk}^{j} \beta_{li}^{j}} \left| \sum_{m=0}^{M_j - 1} \left(e^{j2\pi d_H (\sin(\varphi_{jk}^{j}) - \sin(\varphi_{li}^{j}))} \right)^m \right| \\ &= \sqrt{\beta_{jk}^{j} \beta_{li}^{j}} g_1(\varphi_{jk}^{j}, \varphi_{li}^{j}), \end{aligned} \tag{6.65}$$

where

$$g_1(\varphi_{jk}^{j}, \varphi_{li}^{j}) = \begin{cases} \dfrac{\sin\left(\pi d_H M_j (\sin(\varphi_{jk}^{j}) - \sin(\varphi_{li}^{j}))\right)}{\sin\left(\pi d_H (\sin(\varphi_{jk}^{j}) - \sin(\varphi_{li}^{j}))\right)}, & \text{if } \sin(\varphi_{jk}^{j}) \notin \sin(\varphi_{li}^{j}), \\ M_j, & \text{if } \sin(\varphi_{jk}^{j}) \in \sin(\varphi_{li}^{j}). \end{cases} \tag{6.66}$$

Hence, if the LoS components $\bar{\mathbf{h}}_{jk}^{j}$ and $\bar{\mathbf{h}}_{li}^{j}$ do not have the same AoA (more precisely, if $\sin(\varphi_{jk}^{j}) \notin \sin(\varphi_{li}^{j})$) then, $\lim_{M_j} \frac{1}{M_j} \left| (\bar{\mathbf{h}}_{jk}^{j})^H \bar{\mathbf{h}}_{li}^{j} \right| \to 0$ holds [2, Sect. 1.3]. This implies that Assumption 3 holds, except in the extreme case when AoAs are exactly the same.

6.1.6 *Numerical Results*

In this section, we rigorously validate and evaluate the closed-form SE formulas derived earlier by simulating a Massive MIMO cellular network. The simulation involves a 16-cell configuration, each cell corresponds to a square area of 250 m ×

250 m. This network employs a wrap-around topology to ensure uniform interference reception from all directions across BTSs. Each cell contains $K = 10$ UEs, which are uniformly and independently distributed at distances exceeding 35 m from the BTS. Each UE is assigned to the BTS that offers the highest channel gain among all possible UE-BS combinations within the system. The spatial positioning of each UE is crucial for calculating large-scale fading and determining nominal angles between UEs and BTSs.

Each BTS uses a Uniform Linear Array (ULA) featuring half-wavelength antenna spacing, as detailed in [2, Sect. 1]. Consequently, the LoS component from UE i in cell l to BS j can be represented by (6.64). Regarding covariance matrices, there are $N = 6$ scattering clusters, and the covariance matrix of each cluster follows an approximate Gaussian local scattering model [2, Sect. 2.6] as

$$\left[\mathbf{R}_{li}^{j}\right]_{s,m} = \frac{\beta_{li}^{j,NLoS}}{N} \sum_{n=1}^{N} e^{j\pi(s-m)\sin(\varphi_{li,n}^{j})} e^{\frac{\sigma_{\varphi}^{2}}{2}\left(\pi(s-m)\cos(\varphi_{li,n}^{j})\right)^{2}}, \tag{6.67}$$

where $\beta_{li}^{j,NLoS}$ is the large-scale fading coefficient and $\varphi_{li,n}^{j} \sim \mathcal{U}\left[\varphi_{li,n}^{j} - 40°, \varphi_{li}^{j} + 40°\right]$ is the nominal AoA for the n cluster. The multipath components of a cluster have Gaussian distributed AoAs, distributed around the nominal AoA with the angular standard deviation (ASD) $\sigma_{\varphi} = 5°$. Communication occurs over a bandwidth of 20 MHz with total receiver noise power set at −94 dBm. Each coherence block comprises $\tau_c = 200$ samples while maintaining identical pilot allocation across cells with $\tau_p = 10$ pilots per cell. These pilots are assigned randomly such that any kth UE in two cells sharing a pilot group utilizes identical pilots.

The probability of LoS between UE (l, i) and the BTS j is governed by distance as per the established in the 3GPP model in [21]:

$$\Pr(LoS) = \begin{cases} \frac{300-d_{li}^{j}}{300}, & 0 < d_{li}^{j} < 300\,\text{m}, \\ 0, & d_{li}^{j} > 300\,\text{m}. \end{cases} \tag{6.68}$$

For channels where an LoS path exists, the large-scale fading coefficient (in dB) can be expressed as

$$\beta_{li}^{j} = -30.18 - 26\log_{10}\left(d_{li}^{j}\right) + F_{li}^{j}, \tag{6.69}$$

where $F_{li}^{j} \sim \mathcal{N}(0, \sigma_{sf}^{2})$ represents shadow fading with a standard deviation $\sigma_{sf} = 10$. In this scenario, we set $\beta_{li}^{j,LoS} = 0$ and maintain $\beta_{li}^{j,NLoS} = \beta_{li}^{j}$ (in linear scale).

The heuristic UL power control policy proposed in [2, Sect. 7.3], in which the transmit power for UE k in cell j can be found as

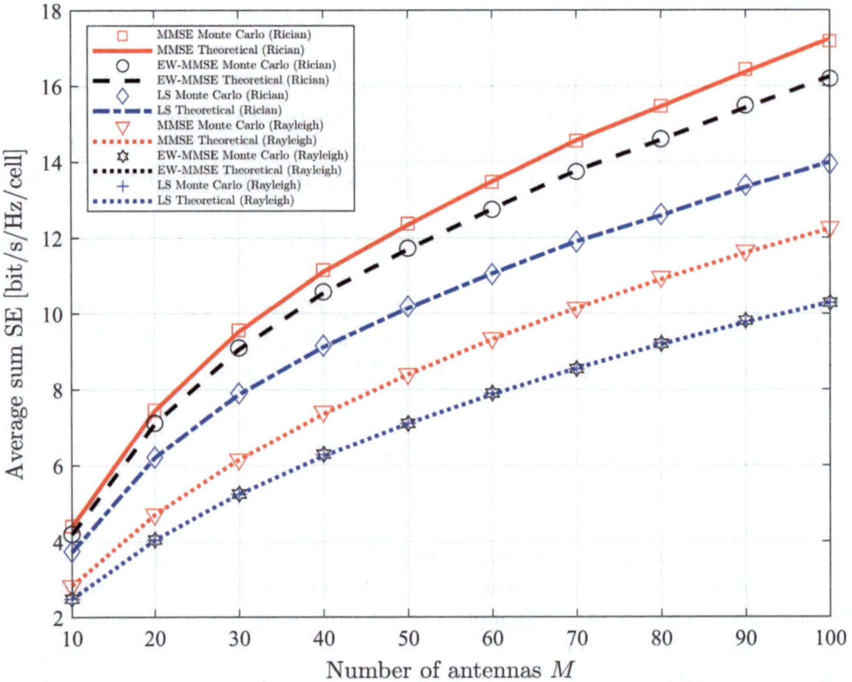

Fig. 6.2 Average UL sum SE for $K = 10$ as a function of the number of BS antennas for different channel estimators

$$p_{jk} = \begin{cases} p_{\max}^{ul}, & \Delta > \frac{\beta_{jk}^{j}}{\beta_{j,\min}^{j}}, \\ p_{\max}^{ul}\Delta\frac{\beta_{j,\min}^{j}}{\beta_{jk}^{j}}, & \Delta \leq \frac{\beta_{jk}^{j}}{\beta_{j,\min}^{j}}, \end{cases} \tag{6.70}$$

where the maximum power is $p_{\max}^{ul} = 10\,\text{dBm}$ and $\beta_{j,\min}^{j} = \min(\beta_{j1}^{j}, ..., \beta_{jk}^{j}, ..., \beta_{jK}^{j})$. This strategy ensures optimal resource allocation by allowing UEs with weaker channels to transmit at full capacity while others adjust their output to maintain the difference in UL SNRs in a threshold of $\Delta = 10$. For simplicity's sake, we set DL transmit power equal to the UL one, i.e., $\rho_{jk} = p_{jk}$ for each UE.

Figure 6.2 presents the sum UL SE averaged across various UE locations and shadow-fading realizations, by utilizing Maximum Ratio (MR) combining with either MMSE, LS, or EW-MMSE estimators. For comparative purposes, the authors in [13] also include curves for Rayleigh fading using identical covariance matrices, representing scenarios where all LoS components are obstructed but small-scale fading persists. This results in a reduced average channel gain $E\{\|\mathbf{h}_{li}^{j}\|^2\}$, as would typically occur in practical environments. The curves are derived from closed-form expressions detailed in Sect. 6.1.3, while the "□" markers stem from Monte Carlo simulations.

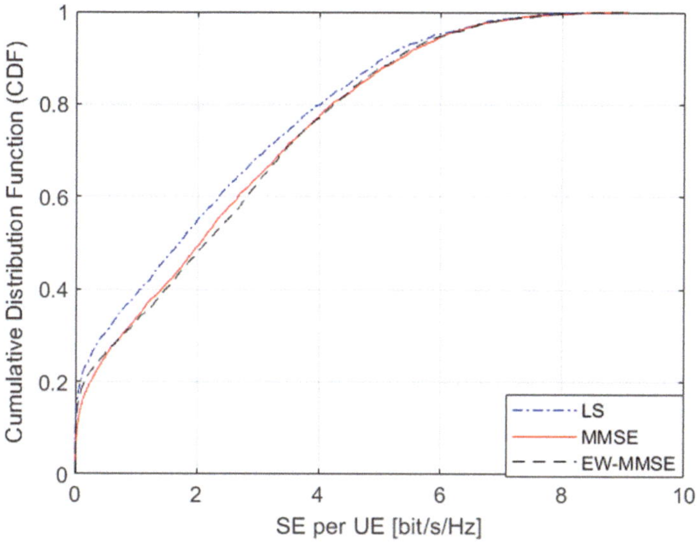

Fig. 6.3 CDF of the UL SE per UE with M = 100 for different channel

The congruence of these markers with the analytical curves substantiates our theoretical findings. Notably, MMSE estimation yields the highest uplink SE due to its utilization of known LoS components and spatial correlation data. In Rician fading conditions, MMSE and EW-MMSE perform comparably well; however, EW-MMSE surpasses LS by leveraging knowledge of channel mean values. Under Rayleigh fading conditions, MMSE remains superior while LS and EW-MMSE yield identical sum SEs due to equivalent estimates up to a scaling factor.

Figure 6.3 illustrates the CDFs of SE per UE, which are influenced by random UE placements and shadow-fading variations. For UEs experiencing favorable channels, both MMSE and EW-MMSE estimators achieve similar SEs due to minimal estimation errors. Conversely, significant disparities between MMSE and EW-MMSE arise for UEs under the worst channel conditions.

In Fig. 6.4, we observe average DL sum SE across different UE locations and shadow-fading scenarios, using MR precoding based on various estimators, namely, MSE, LS, or EW-MMSE. The results are similar to the uplink case. Figure 6.5 examines average UL SE amidst uncorrelated Rician versus Rayleigh fading with respect to the number of antennas. Here, both EW-MMSS & MMSE overlap with each other due to their diagonal spatial covariance matrices, but both of them outperform LS when dealing with Rician fading because of the incorporation of mean vectors in the estimates. Reversely, all vectors exhibit uniform performance under Rayleigh fading, because the estimates are only different by the same constant scaling factor, which is canceled out in SINR.

Figure 6.6 explores spatially correlated Rician fading channels between BTSs and UEs, a scenario unlikely to be feasible practically, yet demonstrates the increase in

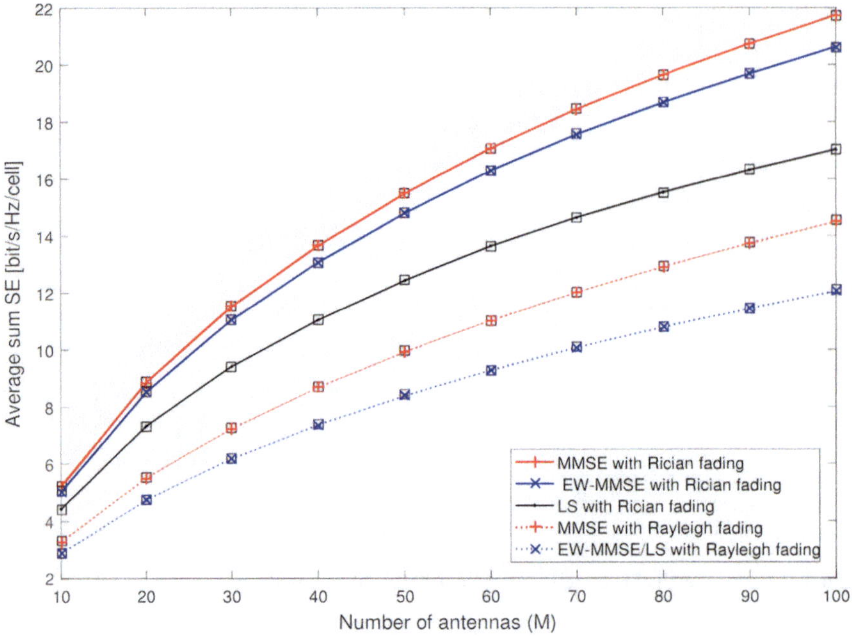

Fig. 6.4 Average DL sum SE for K = 10 as a function of the number of BS antennas for different channel estimators

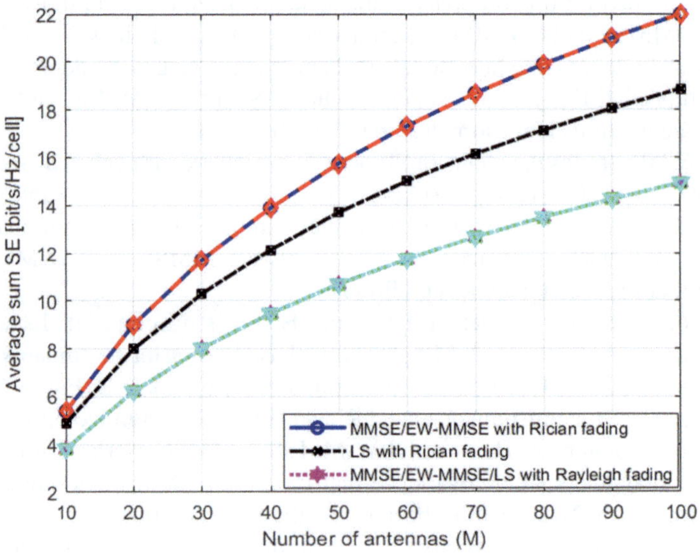

Fig. 6.5 Average UL SE with spatially uncorrelated Rician and Rayleigh fading with $\mathbf{R}_{li}^{j} = \beta_{li}^{j,NLoS}\mathbf{I}_M$

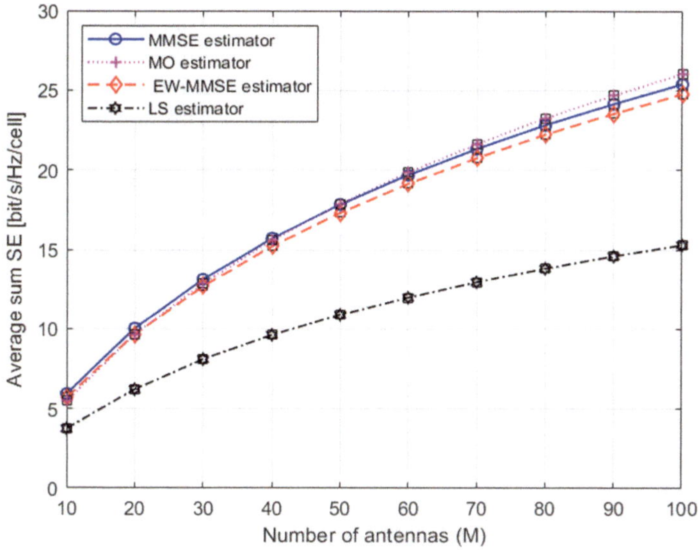

Fig. 6.6 Average UL sum SE for $K = 10$ as a function of BS antennas for different channel estimators where all UE-BS pairs have an LoS path

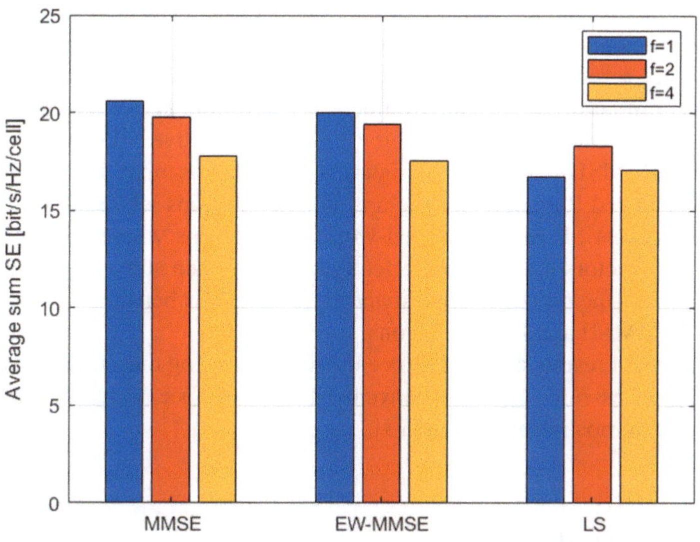

Fig. 6.7 Average UL sum SE for different pilot reuse factors and different estimators

average UL sum SEs compared with the case that only a portion of LoS components are available (e.g., Fig. 6.2). Note that in the case that some channels are Rayleigh fading and the MO estimator is employed, the corresponding SEs would become zero.

Figure 6.7 illustrates the impact of three distinct methods for reusing pilot sequences across cells on the average sum UL SEs for various estimators. The integer f represents the pilot reuse factor, defined as $\tau p = f\mathbf{K}$, indicating that each subset of pilots is being reused in a fraction $1/f$ of the cells. This configuration leads to a trade-off. Specifically, increasing of the reuse factor can help reduce the pilot contamination (by increasing the instantaneous SINR in (6.21)), but also cause a larger penalty in the pre-log factor in (6.20). The LS estimator is particularly susceptible to pilot contamination due to its inability to mitigate interference through spatial processing. In contrast, MMSE estimators can effectively suppress such interference even with tighter reuse factors. Thus, optimal reuse factors vary among different estimators.

6.1.7 Conclusion

This section rigorously analyzed both UL and DL SEs within multi-cell Massive MIMO systems characterized by spatially correlated Rician fading channels. We introduce the closed-form SE formulas derived by [13] when employing MMSE, EW-MMSE, or LS estimation techniques. These formulations offer exact insights into operational dynamics and interference impact under Rician fading condition and are instrumental for optimizing user scheduling, pilot allocation, power control, and other resource management tasks in practical scenarios. It is highlighted that spatial correlation significantly influences the system performance; notably, the LoS components can enhance the achievable SE in Massive MIMO configurations. Furthermore, MMSE estimation consistently outperforms other methods under both spatially correlated Rayleigh and Rician fading conditions while the LS estimation yields the weakest SE results. In real-world applications where covariance matrices and mean vectors may not be perfectly known - often subject to random phase shifts or errors—the practical performance typically falls between that predicted by MMSE/EW-MMSE and LS estimations.

The analytical results of [13] has been verified by using our proposed simulation framework successfully. With the convenient user interface, we can cover all of the simulation scenarios introduced in [13].

6.2 Spatial Modulation

In this section, a simulation framework for a typical modulation technique in MIMO systems, i.e., spatial modulation (SM), is introduced. For SM, only a single antenna is activated during MIMO transmission. The antenna index is selected based on the transmitted message symbol. One of the major problems that makes SM difficult to implement in practice is that channel state information (CSI) is required at the receiver end. To overcome this difficulty, a differential SM scheme (DSM) has been proposed to completely bypass any CSI estimation at the transmitter or receiver. This scheme can be applied to any constant energy constellation, such as phase-shift keying (PSK), and to systems with arbitrary numbers of transmit and receive antennas.

6.2.1 Introduction

While MIMO systems enable high data rate transmission due to their large number of antennas, this advantage comes at the cost of high energy consumption and the complexity of multi-stream decoding at the receiver. To address some of these challenges, a single-RF chain transmission scheme called Spatial Modulation (SM) was introduced. In SM, information bits are divided into two groups: one modulates the transmitted symbol, and the other determines which transmit antenna is activated. This design eliminates inter-channel interference and avoids synchronization issues among transmit antennas. The SM scheme relies on pilot signals to satisfy the requirement for CSI. Although SM is more robust than earlier approaches like V-BLAST, CSI estimation becomes increasingly complex and costly in highly mobile environments where mobility rates approach the data transmission rates.

A common approach to overcome these challenges is differential modulation. Numerous techniques have been proposed, including differential detection for the Alamouti scheme, which avoids the need for CSI. Building on the principles of SM, a novel technique called Space-Time Shift Keying (STSK) emerged, demonstrating the feasibility of a differential version. This gave rise to Differential Space-Time Shift Keying (DSTSK). However, existing documentation on DSTSK highlights two key limitations and lacks clear design details.

This section introduces a differential scheme tailored specifically for SM. This Differential Spatial Modulation (DSM) scheme eliminates the need for CSI at both the transmitter and receiver. Moreover, DSM exploits both spatial and temporal dimensions to facilitate differential modulation and demodulation. Unlike DSTSK, DSM allows for the systematic design of transmission matrices based on the order of activated transmit antennas, enabling multiple symbols to be transmitted per block. This overcomes the constraints of DSTSK and significantly enhances spectral efficiency (Fig. 6.8).

6.2.2 Spectral Efficiency

The set G contains $N_t!2^{N_t b}$ blocks, where N_t represents the number of antennas and b signifies 2^b-PSK constellation. Based on Stirling's formula, the theoretical spectral efficiency of DSM can be given as

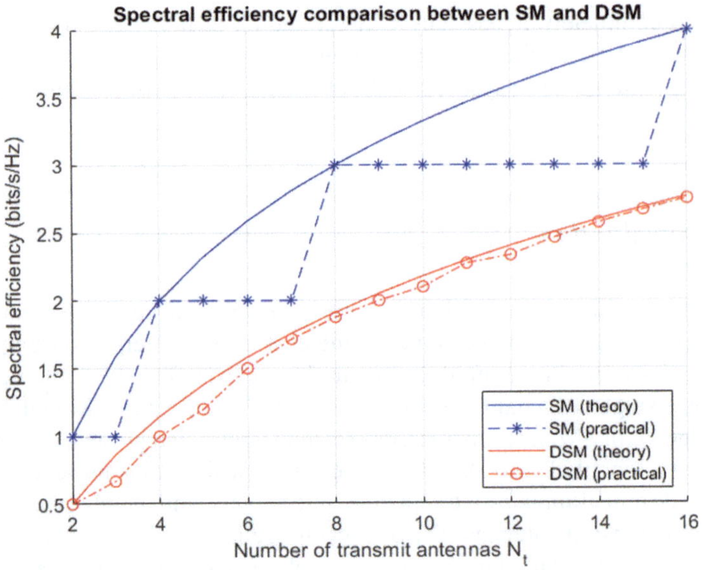

Fig. 6.8 Spectral comparison of SM and DSM

$$R_{DSM,theory} = \frac{1}{N_t}\log_2(N_t!) + b, \tag{6.71}$$

$$R_{DSM,theiry} \approx \log_2(N_t) - \left(\log_2(e) - \frac{1}{N_t}\log_2(\sqrt{2\pi N_t})\right) + b, \tag{6.72}$$

where the middle term indicates the loss of spectral efficiency, which is upper bounded by $\log_2(e)$. While the spectral efficiency of a conventional SM is given as

$$R_{SM,theiry} = \log_2(N_t) + b. \tag{6.73}$$

Usually, $N_t!$ is not an integral power of 2, which is usually tackled by fractional bit encoding (FBE). However, FBE leads to longer decoding delay and worse bit error rate (BER) performance.

So, in practice, we assume that N_t is an integer power of 2. Hence, the spectral efficiency of SM and DSM is given respectively as

$$\overline{R}_{SM} = \lfloor \log_2(N_t) \rfloor + b, \tag{6.74}$$

$$\overline{R}_{DSM} = \frac{1}{N_t}\lfloor \log_2(N_t!) \rfloor + b. \tag{6.75}$$

6.2.3 System Model

The system under consideration comprises N_t transmitting antennas and N_r receiving antennas (see Fig. 6.9). In a standard SM system, the constellation vector $\mathbf{s}_t$, of dimension N_t, contains only a single non-zero symbol since only one antenna is active at any given time. Consequently, the wireless link between transmitter and receiver exhibits time-hopping behavior, complicating differential operations in SM.

To address this, the system design utilizes an $N_t * N_t$ matrix constructed by stacking constellation vectors transmitted over the time interval $[k, k + N_t - 1]$ where k is an integer. This matrix, referred to as the *space-time block* and denoted by S, is structured such that each antenna is activated exactly once per block. This design enables differential operations as long as the wireless channel remains stable across two consecutive blocks.

Let us define a set G containing all possible combinations of matrices S for a system with N_t transmitting antennas and 2^b-PSK constellation. The total number of such matrices is $N_t! * 2^{N_t b}$. However, for $N_t > 3$, $N_t!$ is not an integer power of 2. Therefore, practical implementations of the DSM system employ a subset of G, denoted as G_M, comprising $2^{\lfloor \log_2(N_t!) \rfloor} * 2^{N_t b}$ transmitting blocks. This allows each transmitting block to be encoded with $\lfloor \log_2 (N_t!) \rfloor + N_t b$ bits.

6.2.3.1 Algorithmic Details

The transmission process begins with input data bits and an initial transmission block, S_0, which is defined as the identity matrix. This is a natural choice since the current system employs a $2^b - PSK$ constellation for symbol encoding, where the symbol "1" is always included in the set, regardless of the value of b.

As outlined earlier, the DSM scheme encodes the bits into two parts. The first $\lfloor \log_2 (N_t!) \rfloor$ are encoded spatially using specific mapping algorithms (refer to Sect. 6.1.2), while the remaining $N_t * b$ bits are mapped onto N_t constellation symbols in the signal domain. Specifically, the first set of bits determines the activation order of the transmit antennas within a space-time block through an index-mapping procedure. The second set of bits specifies which symbols each transmitting antenna should send. By combining these two mappings, a unique transmission block, X_t, is generated. Multiplying X_t with S_{t-1} yields the transmission block S_t, which is then transmitted. This process continues iteratively until all input data bits are transmitted. For the final transmission, if the remaining data bits are insufficient to complete the encoding, zero-padding is applied to the message data.

At the receiver, the transmitted block is received after being affected by channel fading and noise. The receiver initializes with Y_0 as the identity matrix. For the purposes of this discussion, we assume $N_r = N_t$. The differential detection can be explained as follows:

$$Y_{t-1} = H_{t-1} S_{t-1} + N_{t-1}, \tag{6.76}$$

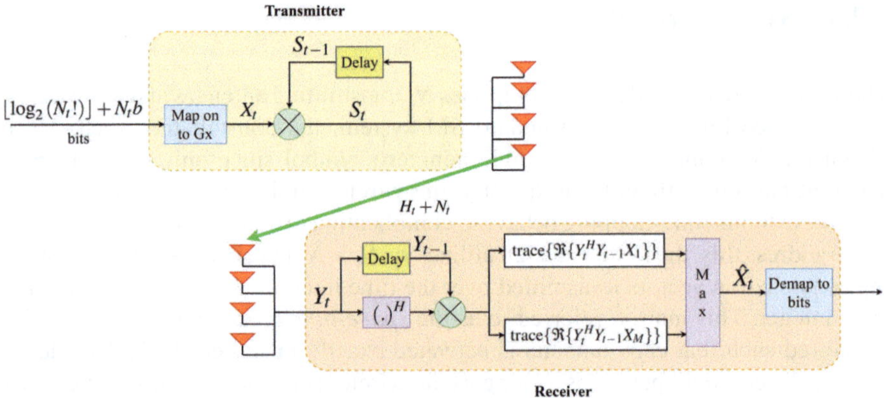

Fig. 6.9 System overview

$$Y_t = H_t S_t + N_t. \tag{6.77}$$

For DSM to work as equivalently good as the theoretical spectral efficiency methods, the channel is assumed to be quasi-static. The fading coefficients remain constant over two adjacent DSM transmission blocks. With the given assumption, we could solve (6.76) and (6.77) to get

$$Y_t = Y_{t-1} X_t - N_{t-1} X_t + N_t. \tag{6.78}$$

The optimal maximum likelihood detector can be derived as

$$X_{predicted} = \underset{X \in G_M}{\arg\max}\, trace\left[\mathrm{Re}\left\{Y_t^H Y_{t-1} X\right\}\right]. \tag{6.79}$$

This explains the iterative procedure used by the receiver in Sect. 6.2.3.2.

6.2.3.2 Algorithm

The differential process starts with sending $S_0 = I_{N_t}$ as the first transmission block sent as a default. The algorithms mentioned below assume the process during time t, when the block S_{t-1} is transmitted.

Algorithm 1: Transmitter algorithm
Initialize N_t, b for 2^b-PSK constellation.
Initialize the transmission with $S_0 = I_{N_t}$.
Initialize sequence of bits as inputs.
for *time* $\leftarrow$ 0 ***to*** *till transmission is complete* **do**
  S_{t-1} = The previous transmitted block;
  Bits = Accumulate $\lfloor \log_2 (N_t!) \rfloor + N_t b$ bits from the incoming sequence;
  $X_t = mappingAlgo(Bits, N_t, b)$;
  $S_t = S_{t-1} X_t$;
end

Algorithm 2: Receiver algorithm
Assuming the algorithm has saved Y_{t-1}.
for *time* $\leftarrow$ 0 ***to*** *till receiving is complete* **do**
  $X_{predicted} = \arg\max_{X \in G_M} trace\left[\mathrm{Re}\left\{Y_t^H Y_{t-1} X\right\}\right]$.
end

Algorithm 3: Mapping algorithm used in transmission
Given a number m = first $\lfloor \log_2 (N_t!) \rfloor$ bits in decimal.
$b^{(m)} = LehnerCode(m, N_t)$;
Define $\theta = (1, 2, 3, \ldots, N_t)$ and container $a^{(m)}$.
for $l \leftarrow 1$ ***to*** N_t **do**
  $a_l^{(m)} = \theta_{b_l^{(m)}}$; $b_l^{(m)}$ is removed from θ;
end

6.2.3.3 Mapping Algorithm

There are two index-based mapping methods that could be used for relating the first $\lfloor \log_2 (N_t!) \rfloor$ bits to the transmission block set G_M.

- Table Lookup method
- Permutation method

Table 6.1 Lookup table

Bits	Permutation of indices	Blocks
00	(1 2 3)	$\begin{bmatrix} s_{11} & 0 & 0 \\ 0 & s_{22} & 0 \\ 0 & 0 & s_{33} \end{bmatrix}$
01	(1 3 2)	$\begin{bmatrix} s_{11} & 0 & 0 \\ 0 & 0 & s_{23} \\ 0 & s_{32} & 0 \end{bmatrix}$
10	(2 1 3)	$\begin{bmatrix} 0 & s_{12} & 0 \\ s_{21} & s_{22} & 0 \\ 0 & 0 & s_{33} \end{bmatrix}$
11	(2 3 1)	$\begin{bmatrix} 0 & 0 & s_{13} \\ s_{21} & 0 & 0 \\ 0 & s_{32} & 0 \end{bmatrix}$

In the first method, a lookup table is created to provide the corresponding permutations for the incoming bits (Table 6.1). The selection of permutations may affect the overall performance. For ease of implementation of permutation index mapping, the algorithm discards the lexicographically larger permutations in this development. The main problem is that with increasing transmitting antennas, the table size increases exponentially.

The second method is based on a one-to-one mapping between integers and a permutation of N_t elements in lexicographic order, known as the Lehner Code. The detailed algorithm is mentioned in Algorithm 3. The algorithm first converts a number m to its factorial representation $b^{(m)}$ and then maps the factorial representation to a permutation $a^{(m)}$, which is an indexed representation of a matrix with one-hot encoding. For example, if the matrix we seek is $I_{N_t*N_t}$, then $a^{(m)} = [1; 2; 3]$.

While the second method is better than the first one, assuming the reduced time and space complexity, all the elements of the set G_M are required on the receiver side. Thus, the advantage of the Lehner Code approach lies only on the transmission side and not the receiving end.

6.2.4 Simulation Results

In our experiments, the mapping algorithm encodes bits in $N_t * N_t$ square matrices, where N_t denotes the number of transmitting antennas, and each value of the matrix is determined by the encoding constellation given by 2^b. The dimensions of the quasi-static Rayleigh flat fading channel used in the experiment are $N_r * N_t$, where N_r denotes the number of receiving antennas.

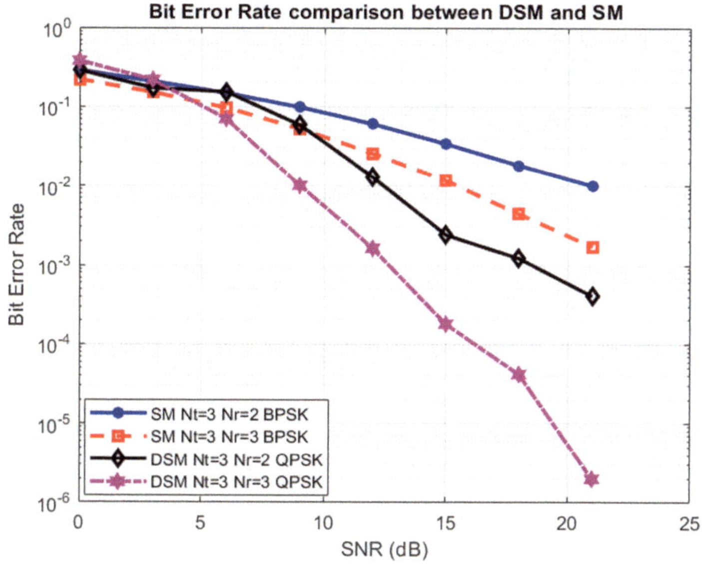

Fig. 6.10 BER performance of differential detection DSM versus coherent detection SM

We have conducted various Bit Error Rate(BER) simulations under various system configurations to examine DSM performance.

1. **DSM Versus SM:**
   In Fig. 6.10, we have compared the BER performance between DSM and SM using $N_t = 3$ and different values of N_r ($N_r = 2; 3$), transmit antennas. In our experiments, we have used N_t greater than N_r. We have used quaternary PSK (QPSK) for DSM and BPSK for SM.
2. **DSM With Multiple Transmit Antennas:**
   In Fig. 6.11, we have compared the DSM BER performance with multiple transmit antennas. For a distinct number of transmit antennas, we have experimented with different number of receiver antennas. For all the experiments, the DSM always uses QPSK, and the SM always uses BPSK.

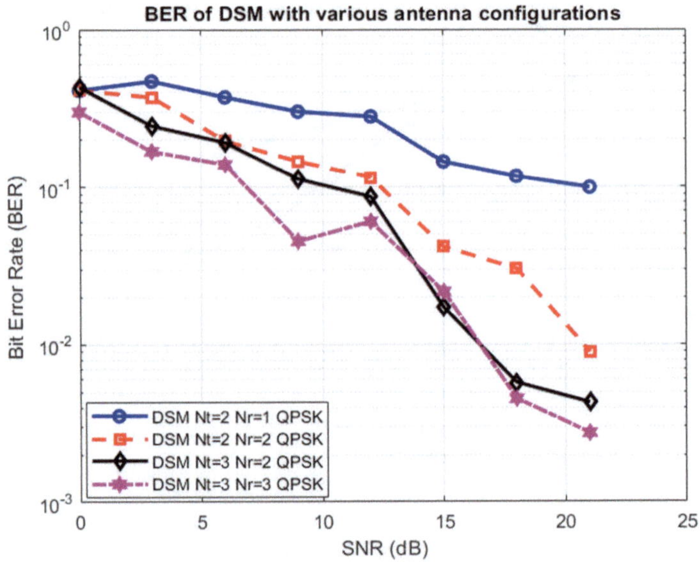

Fig. 6.11 BER performance of differential detection DSM

6.2.5 *Conclusion and Future Work*

In this section, we introduce the concept of DSM [22] and its clear advantages with respect to spectral efficiency. Using both time and space dimensions without the need for CSI at both the transmitter and the receiver makes the method highly beneficial. In comparison with coherent SM [23], DSM exhibits a spectral efficiency loss that is upper bounded by $\log_2(e)$ and a performance loss of no more than 3 dB.

The major problems that the method faces is the exponential order complexity on the receiver side. The receiver side algorithm iterates on every possible transmission block to search for the block with the maximum trace value. The number of elements in such a set can be calculated as $2^{\log_2(N_t!)+N_t*b}$, which increases exponentially with the increase in N_t and b. Future research must focus upon a reduced-complexity detection process at the receiver side.

References

1. Larsson, E.G., Edfors, O., Tufvesson, F., Marzetta, T.L.: Massive mimo for next generation wireless systems. IEEE Commun. Mag. **52**(2), 186–195 (2014)
2. Björnson, E.: Hoydis j sanguinetti l. Massive MIMO networks: spectral, energy, and hardware efficiency. Found. Trends® Signal Proc. **11**, 3–4 (2017)
3. Marzetta, T.L., Larsson, E.G., Yang, H., Ngo, H.Q.: Fundamentals of Massive MIMO. Cambridge University Press (2016)

4. Hoydis, J., Ten Brink, S., Debbah, M.: Massive mimo in the ul/dl of cellular networks: How many antennas do we need?" vol. 31, no. 2 (2013)
5. Yang, H., Marzetta, T.L.: Massive mimo with max-min power control in line-of-sight propagation environment. IEEE Trans. Commun. **65**(11), 4685–4693 (2017)
6. David Tse, P.V.: Fundamentals of Wireless Communication. Cambridge University Press (2005). http://gen.lib.rus.ec/book/index.php?md5=7010d0a3a40f9335b63735114f2ed017
7. Zhang, Q., Jin, S., Wong, K.K., Zhu, H., Matthaiou, M.: Power scaling of uplink massive mimo systems with arbitrary-rank channel means. IEEE J. Select. Top. Signal Proc. **8**(5), 966–981 (2014)
8. Kong, C., Zhong, C., Matthaiou, M., Zhang, Z.: Performance of downlink massive mimo in ricean fading channels with zf precoder. In: 2015 IEEE International Conference on Communications (ICC), pp. 1776–1782. IEEE (2015)
9. Hu, Y., Hong, Y., Evans, J.: Angle-of-arrival-dependent interference modeling in rician massive mimo. IEEE Trans. Veh. Technol. **66**(7), 6171–6183 (2016)
10. Wu, L., Zhang, Z., Dang, J., Wang, J., Liu, H., Wu, Y.: Channel estimation for multicell multiuser massive mimo uplink over rician fading channels. IEEE Trans. Veh. Technol. **66**(10), 8872–8882 (2017)
11. Zhao, L., Yang, T., Geraci, G., Yuan, J.: Downlink multiuser massive mimo in rician channels under pilot contamination. In: 2016 IEEE International Conference on Communications (ICC), pp. 1–6. IEEE (2016)
12. Sanguinetti, L., Kammoun, A., Debbah, M.: Asymptotic analysis of multicell massive mimo over rician fading channels. In: 2017 IEEE International Conference on Acoustics, Speech and Signal Processing (ICASSP), pp. 3539–3543. IEEE (2017)
13. Özdogan, O., Bj rnson, E., Larsson, E.G.: Massive mimo with spatially correlated rician fading channels. IEEE Trans. Commun. **67**(5), 3234–3250 (2019). https://doi.org/10.1109/TCOMM.2019.2893221
14. Zhang, J., Wen, C.K., Jin, S., Gao, X., Wong, K.K.: On capacity of large-scale mimo multiple access channels with distributed sets of correlated antennas. IEEE J. Sel. Areas Commun. **31**(2), 133–148 (2013). https://doi.org/10.1109/JSAC.2013.130203
15. Tataria, H., Smith, P.J., Greenstein, L.J., Dmochowski, P.A., Matthaiou, M.: Impact of line-of-sight and unequal spatial correlation on uplink mu-mimo systems. IEEE Wireless Commun. Lett. **6**(5), 634–637 (2017). https://doi.org/10.1109/LWC.2017.2726062
16. Björnson, E., Sanguinetti, L., Debbah, M.: Massive mimo with imperfect channel covariance information. In: 2016 50th Asilomar Conference on Signals, Systems and Computers, pp. 974–978. IEEE (2016)
17. Haghighatshoar, S., Caire, G.: Massive mimo pilot decontamination and channel interpolation via wideband sparse channel estimation. IEEE Trans. Wireless Commun. **16**(12), 8316–8332 (2017)
18. Kay, S.M.: Statistical Signal Processing: Estimation Theory, vol. 1. Prentice Hall, Chapter–3 (1993)
19. Shariati, N., Björnson, E., Bengtsson, M., Debbah, M.: Low-complexity polynomial channel estimation in large-scale mimo with arbitrary statistics. IEEE J. Sel. Top. Signal Proc. **8**(5), 815–830 (2014)
20. Yin, H., Gesbert, D., Filippou, M., Liu, Y.: A coordinated approach to channel estimation in large-scale multiple-antenna systems. IEEE J. Sel. Areas Commun. **31**(2), 264–273 (2013)
21. 3GPP: 3rd generation partnership project; technical specification group radio access network. study on lte time division duplex (tdd)-frequency division duplex (fdd) joint operation including carrier aggregation (ca)(release 12) (2016)
22. Bian, Y., Cheng, X., Wen, M., Yang, L., Poor, H.V., Jiao, B.: Differential spatial modulation. IEEE Trans. Veh. Technol. **64**(7), 3262–3268 (2014)
23. Mesleh, R.Y., Haas, H., Sinanovic, S., Ahn, C.W., Yun, S.: Spatial modulation. IEEE Trans. Veh. Technol. **57**(4), 2228–2241 (2008)

Chapter 7
Energy-Saving Solutions in 5G/6G Networks

7.1 Wireless Energy Harvesting (EH)

Recently, the use of RF signals for both energy transmission and information signal transmission has been widely accepted by the scientific community. Simultaneous wireless information and power transfer (SWIPT) [1] is a protocol proposed to transmit RF energy and information signals at the same time through the wireless medium. There are two conventional wireless energy harvesting (EH) protocols that have been proposed, namely, time-switching relaying (TSR) and power-splitting relaying (PSR) [2]. Additionally, the research group of the author [3] has also proposed a new EH protocol, known as the hybrid protocol between time-switching and power-splitting.

For the energy saving solutions simulation module, in this book, a simulation tool is constructed for the aforementioned wireless EH protocols, thereby integrating with other modules to perform simulations for various scenarios.

In [4], the OP and network throughput of the dual-mode decode-and-forward relay network, where the relay node is equipped with EH capability based on time-switching, are rigorously established. The analytical results provide theoretical insights into various system parameters such as time-switching coefficient, source data rate, and power ratio of the transmit power to the system noise power. The analysis in the paper is validated by Monte Carlo simulations and can serve as guidelines for designing practical EH relay systems. This section is written based on the result of [4, 5] to simulate the solution of EH via RF waves.

Moreover, in practice, the transceiver hardware at the nodes is imperfect due to phase noise, I/Q imbalance, and nonlinearity of the amplifier. Both aforementioned papers investigate the impact of hardware impairments at both source and relay nodes in analyzing the OP and throughput of the wireless information. The effects of other parameters such as power supply and time-splitting coefficient on system performance considering hardware errors are also studied. Furthermore, the optimal time-switching coefficient to maximize system throughput is found using numerical methods [4].

P. T. Tran et al., *Developing MATLAB Comprehensive Simulation for 5G/6G Wireless Communications*, Signals and Communication Technology,
https://doi.org/10.1007/978-981-95-1117-4_7

7.1.1 System Model

Consider a dual-mode relay system as illustrated in Fig. 7.1, where the source S sends information to the destination D with the assistance of the relay node R. Assume that the direct connection between the source and the destination is weak. Therefore, a relay R is deployed to enhance this connection without needing any separate data to transmit. Relay R does not have any power supply other than the energy harvested from the source.

To perform EH at the relay node, there are three protocols that can be chosen: Time Switching Relay (TSR), Power Splitting Relay (PSR), or a combination of both. These protocols are described in Fig. 7.2, where T represents the total time to transmit one data symbol.

7.1.2 Performance Analysis

Assuming h and g correspondingly represent the channel gains from the source to the relay (R) and from the relay (R) to the destination (D), and both follow Rayleigh

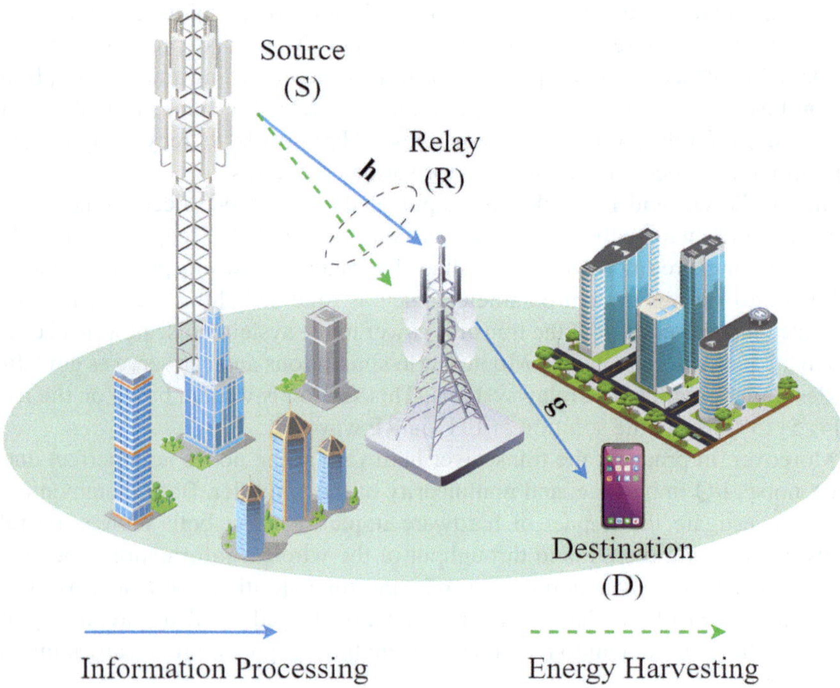

Fig. 7.1 System model of an energy harvesting-based relay network

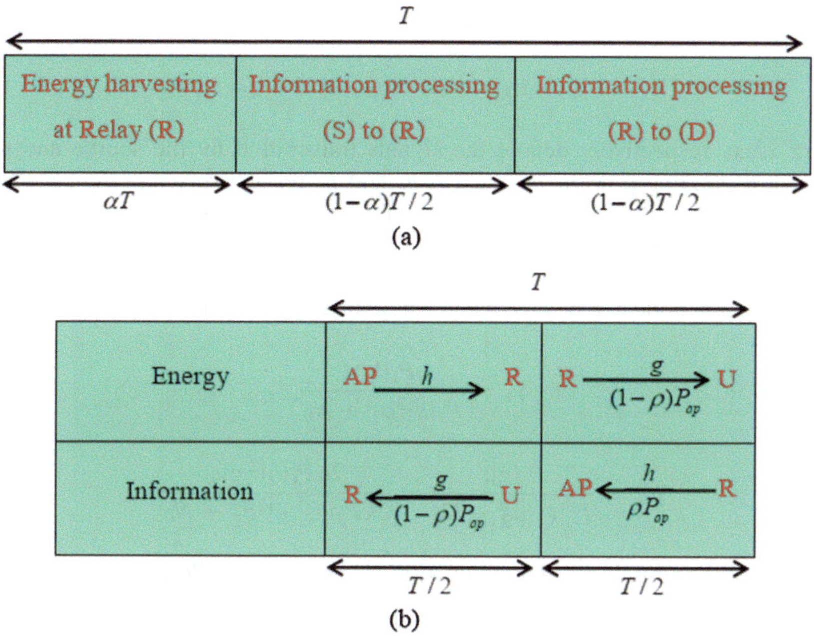

Fig. 7.2 EH Protocols: **a** Time switching relaying (TSR); **b** power-splitting relaying (PSR)

distributions. Consequently, $|h|^2$ is a random exponential variable with a mean value λ_h, and $|g|^2$ is distributed according to the exponential function with a mean value λ_g.

The following section illustrates the analysis process for the case of time-switching EH protocol applied to the decoding and forwarding network. Other protocols are analyzed similarly.

If the linear EH model is applied, the energy harvested from the source is calculated as

$$\lambda_h = \mu P_s |h|^2 \alpha T, \tag{7.1}$$

where μ represents the energy conversion efficiency, and P_s is the power of the signal transmitted from the source.

Consequently, the transmit power of the relay can be calculated as [4]

$$P_r = \frac{E_h}{(1-\alpha)T/2} = \frac{2\mu P_s |h|^2 \alpha}{1-\alpha} = 2k P_s |h|^2, \tag{7.2}$$

where $k = \dfrac{\mu\alpha}{1-\alpha}$.

Due to hardware impairments, the received signal at the relay (R) and at the destination node can be respectively represented by

$$y_r = h(x_s + \eta_s) + n_r, \tag{7.3}$$

$$y_D = h(x_r + \eta_r) + n_d. \tag{7.4}$$

where x_s, x_r respectively denote the signals transmitted by the source and relay with corresponding powers P_s and P_r; η_s and η_r respectively represent hardware impairments at the source and relay; n_r and n_d respectively represent AWGN noise at the relay node and at the destination point, assumed to have the same power N_0.

From there, the signal-to-interference-plus-noise ratio (SINR) at the relay node and destination node can be calculated using the following formulas:

$$\gamma_1 = \frac{P_s|h|^2}{P_s|h|^2\sigma_1^2 + N_0}, \tag{7.5}$$

$$\gamma 2 = \frac{P_r|g|^2}{P_r|g|^2\sigma_2^2 + N_0} = \frac{2kP_s|h|^2 g|^2}{2kP_s|h|^2 g|^2\sigma_2^2 + N_0}, \tag{7.6}$$

where σ_1 and σ_2 represent the attenuation at the source node and the relay node, respectively, i.e., $P_s\sigma_1^2$ is the variance of η_s, and $P_r\sigma_2^2$ is the variance of η_r.

Assuming that the source transmits at a constant rate R, thus $z = 2^R - 1$ represents the minimum threshold for SINR required for successful reception at either the relay node or the destination node. Suppose that all transmitters have the same hardware impairment structure, so $\sigma_1 = \sigma_2 = \sigma$. The system OP can be expressed as

$$P_{out} = P_r(min(\gamma_1, \gamma_2) < z) = 1 - P_r\{\gamma_1 \geq z, \gamma_2 \geq z\} \tag{7.7}$$

By probability theory, we obtain Lemma 1 [4].

Lemma 7.1 *For the dual-function relay system DF with EH based on time-switching, satisfying the condition $z * \sigma^2 < 1$, the OP of the system can be expressed as*

$$P_{out} = 1 - e^{-\frac{x_0}{\lambda_h} - \frac{1}{2k\lambda_g}} - \frac{1}{\lambda_g}\int_0^{\frac{1}{2k}} e^{-\frac{y}{\lambda_g} - \frac{x_0}{2\lambda_h ky}} dy \tag{7.8}$$

where $x_0 = \dfrac{N_0 z}{P_s(1 - \sigma^2 z)}$.

Proof Refer to [4]. □

Next, the average throughput can be calculated based on the OP:

$$R_{DL}(\alpha) = (1 - P_{out})R\frac{(1-\alpha)}{2} = \frac{R(1-\alpha)}{2}.(e^{-\frac{x_0}{\lambda_h} - \frac{1}{2k\lambda_g}} + \frac{1}{\lambda_g}\int_0^{\frac{1}{2k}} e^{-\frac{y}{\lambda_g} - \frac{x_0}{2\lambda_h ky}} dy) \tag{7.9}$$

Moreover, the optimal time-switching conversion coefficient $\alpha*$ can be attained by solving the following optimization problem:

$$\alpha^* = \arg\max_{\alpha} \{R_{DL}(\alpha)\} \tag{7.10}$$

This optimization problem does not have an explicit solution. Instead, it can be solved numerically using the Golden search algorithm.

7.1.3 MATLAB Coding

The following is a sample script in MATLAB to implement the simulation of RF EH to verify the above analysis

```
x = eval(app.DistancematrixEditField.Value);
PL = str2double(app.PathlossexponentEditField.Value);
lambda1 = x.^PL;
lambda_h = lambda1(1); % channel power gain of source-relay path
lambda_g = lambda1(2); % channel power gain of relay-destination path
t = app.NoofsamplesEditField.Value; % number of channel samples
eta = app.EnergyconversionefficiencyKnob.Value/100; % energy conversion efficiency
PsN0dB = app.TransmitpowerEditField.Value - app.NoisevarianceEditField.Value; % Transmit power
R = app.RequireddatarateEditField.Value; % Required data rate

switch app.SweepparameterDropDown.Value
    case 'Transmit SNR (dB)'
        PsN0dB = eval(app.SweepvaluesEditField.Value);
        if strcmp(app.EHProtocolDropDown.Value, 'Power-splitting')
            beta1 = app.PowersplittingfactorKnob.Value; % power-splitting factor
            [P_out_ana, P_out_sim, throughput_ana, throughput_sim] =
```

```
            EH_AF_PSR(lambda_h,lambda_g,eta,beta1,PsN0dB,R,t);
        a1 = app.UIAxes;
        axes(a1);
        if strcmp(app.SelectresulttoplotDropDown.Value,'Outage probability')
            h1 = semilogy(PsN0dB,P_out_ana,'r--','LineWidth',2); hold on;
            h2 = semilogy(PsN0dB,P_out_sim,'r*','LineWidth',2); hold on;
        else
            h3 = semilogy(PsN0dB,throughput_ana,'b-','LineWidth',2); hold on;
            h4 = semilogy(PsN0dB,throughput_sim,'bd','LineWidth',2); hold on;
        end
        grid on;
        title(['OP versus transmit SNR for \beta = ',num2str(beta1),', R = ',num2str(R),', PSR model']);
        xlabel('Transmit SNR (dB)');
        if strcmp(app.SelectresulttoplotDropDown.Value,'Outage probability')
            ylabel('Outage probability');
            legend([h1,h2],'OP (Ana)','OP (Sim)');
        else
            ylabel('Throughput');
            legend([h3,h4],'Throughput (Ana)','Throughput (Sim)');
        end
        app.UITable.ColumnName = {app.SweepparameterDropDown.Value; 'Outage Probability';'Throughput'};
        app.UITable.Data = table(PsN0dB',P_out_ana',throughput_ana');
```

```
        elseif strcmp(app.EHProtocolDropDown.Value,'Time-switching')
            alpha1 = app.TimeswitchingfactorKnob.Value;
            if strcmp(app.RelayingstrategyDropDown.Value,'DF')
                sigma1 = 0.1; % hardware impairment level --> add to GUI later
                [P_out_ana, P_out_sim, throughput_ana, throughput_sim] = EH_DF_TSR(lambda_h,lambda_g,eta,alpha1,PsN0dB,R,t,sigma1);
            else
                if strcmp(app.RelayingstrategyDropDown.Value,'AF')
                    [P_out_ana, P_out_sim, throughput_ana, throughput_sim] = EH_AF_TSR(lambda_h,lambda_g,eta,alpha1,PsN0dB,R,t);
                end
            end
            a1 = app.UIAxes;
            axes(a1);
            if strcmp(app.SelectresulttoplotDropDown.Value,'Outage probability')
                h1 = semilogy(PsN0dB,P_out_ana,'r--','LineWidth',2); hold on;
                h2 = semilogy(PsN0dB,P_out_sim,'r*','LineWidth',2); hold on;
            else
                h3 = semilogy(PsN0dB,throughput_ana,'b-','LineWidth',2); hold on;
```

```
                    h4 = semilogy(PsN0dB, throughput_sim,'bd','LineWidth',2); hold on;
                end
                grid on;
                title(['OP versus transmit SNR for \alpha = ',num2str(alpha1),', R = ',num2str(R),', TSR model']);
                xlabel('Transmit SNR (dB)');
                if strcmp(app.SelectresulttoplotDropDown.Value,'Outage probability')
                    ylabel('Outage probability');
                    legend([h1,h2],'OP (Ana)','OP (Sim)');
                else
                    ylabel('Outage probability');
                    legend([h3,h4],'Throughput (Ana)','Throughput (Sim)');
                end
                app.UITable.ColumnName = {app.SweepparameterDropDown.Value; 'Outage Probability';'Throughput'};
                app.UITable.Data = table(PsN0dB',P_out_ana',throughput_ana');
            end
        end
        % Other cases should follow after this ...
end
```

```
function [P_out_ana, P_out_sim,throughput_ana, throughput_sim] = EH_AF_PSR(lambda_h, lambda_g,eta,beta1,PsN0dB,R,t)

gamma = 2.^(2*R)-1;
PsN0 = 10.^(PsN0dB/10);
% Channel generation
```

```
h = sqrt(1/2/lambda_h)*(randn(1,t) + 1i*randn(1,t)); %Rayleigh fading for channel from transmitter 1 to relay
g = sqrt(1/2/lambda_g)*(randn(1,t) + 1i*randn(1,t)); %Rayleigh fading for channel from transmitter 2 to relay
X = abs(h).^2;   %Fading power of h1
Y = abs(g).^2 ;  %Fading power of g2


if (length(beta1) > 1) || ((length(PsN0) == 1) && (length(beta1) == 1) && (length(R) == 1))
    P_out_ana = zeros(1,length(beta1));
    P_out_sim = zeros(1,length(beta1));
    throughput_sim = zeros(1,length(beta1));
    throughput_ana = zeros(1,length(beta1));

    for p = 1: length(beta1)
        a = eta.^3.*(beta1(p).^2).*(1-beta1(p)).*PsN0.^2;
        b = eta.*beta1(p).*PsN0;
        c = eta.^2.*beta1(p).*(1-beta1(p)).*PsN0;
        num = a.*X.^3.*Y.^2;
        den = b.*X.^2+c.*X.*Y.^2+1;
        SNR_AF = num ./ den;
        P_out_sim(p) = sum(SNR_AF < gamma) / t;
        throughput_sim(p) = (1 - P_out_sim(p)) * (R./2);

        k = sqrt(gamma./(eta.*beta1(p).*PsN0));
        f = @(x)(exp(-(x.*lambda_h + (sqrt((gamma.*b.*x.^2 +gamma)./(a.*x.^3-c.*gamma.*x))).*lambda_g))).*lambda_h;
        l = integral(f, k,inf);

        P_out_ana(p) = 1 - l;
        throughput_ana(p) = (1 - P_out_ana(p)).* (R./2);
    end
elseif length(PsN0) > 1
    P_out_ana = zeros(1,length(PsN0));
    P_out_sim = zeros(1,length(PsN0));
```

```
    throughput_sim = zeros(1,length(PsN0));
    throughput_ana = zeros(1,length(PsN0));

    for p = 1: length(PsN0)
        a = eta.^3.*(beta1.^2).*(1-beta1).*PsN0(p).^2;
        b = eta.*beta1.*PsN0(p);
        c = eta.^2.*beta1.*(1-beta1).*PsN0(p);
        num = a.*X.^3.*Y.^2;
        den = b.*X.^2+c.*X.*Y.^2+1;
        SNR_AF = num ./ den;
        P_out_sim(p) = sum(SNR_AF < gamma) / t;
        throughput_sim(p) = (1 - P_out_sim(p)) * (R./2);

        k = sqrt(gamma./(eta.*beta1.*PsN0(p)));
        f = @(x)(exp(-(x.*lambda_h + (sqrt((gamma.*b.*x.^2 +gamma)./(a.*x.^3-c.*gamma.*x))).*lambda_g))).*lambda_h;
        l = integral(f, k,inf);

        P_out_ana(p) = 1 - l;
        throughput_ana(p) = (1 - P_out_ana(p)) .* (R./2);
     end
elseif length(R) > 1
    P_out_ana = zeros(1,length(R));
    P_out_sim = zeros(1,length(R));
    throughput_sim = zeros(1,length(R));
    throughput_ana = zeros(1,length(R));

    for p = 1: length(R)
        a = eta.^3.*(beta1.^2).*(1-beta1).*PsN0.^2;
        b = eta.*beta1.*PsN0;
        c = eta.^2.*beta1.*(1-beta1).*PsN0;
        num = a.*X.^3.*Y.^2;
        den = b.*X.^2+c.*X.*Y.^2+1;
        SNR_AF = num ./ den;
        P_out_sim(p) = sum(SNR_AF < gamma(p)) / t;
        throughput_sim(p) = (1 - P_out_sim(p)) * (R./2);
```

Table 7.1 Parameters for EH simulation

Symbol	Parameter name	Value
R	Transmission data rate	3 bps/ Hz
Z	SINR threshold	7
μ	Energy conversion efficiency	0.8
λ_h	Average of $\|h\|^2$	0.5
λ_g	Average of $\|g\|^2$	0.5
P_s/N_0	Signal-to-noise ratio at the transmitter	0–30 dB

```

            k = sqrt(gamma(p)./(eta.*beta1.*delta))
                ;
            f = @(x)(exp(-(x.*lambda_h + (sqrt((
                gamma(p).*b.*x.^2 +gamma(p))./(a.*x
                .^3-c.*gamma(p).*x))).*lambda_g))).*
                lambda_h;
            l = integral(f, k,inf);

            P_out_ana(p) = 1 - l;
            throughput_ana(p) = (1 - P_out_ana(p))
                .* (R./2);
        end
else
    msgbox('Error: more than one vector input!'
        );
end
```

7.1.4 Simulation Results

This section illustrates simulation results using the Monte Carlo method to validate the analyses above. For simplicity, the distances from the source to the relay and from the relay to the destination are both normalized by the source-destination distance (meaning the maximum distance is equal to 1). Other simulation parameters are listed in Table 7.1.

The results demonstrate that the analytical solution aligns perfectly with the simulation results. The OP and throughput graphs as a function of P_s/N_0 and the time-switching coefficient are presented in Figs. 7.3, 7.4 and 7.5, respectively.

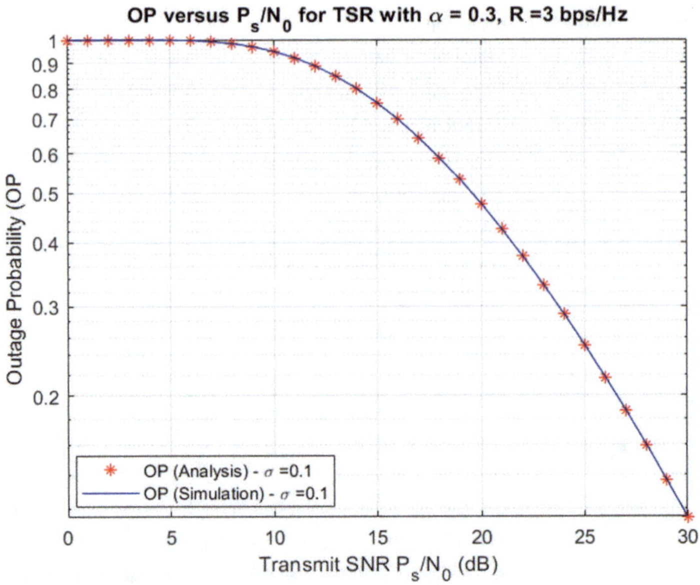

Fig. 7.3 OP plotted against the transmit power of the source

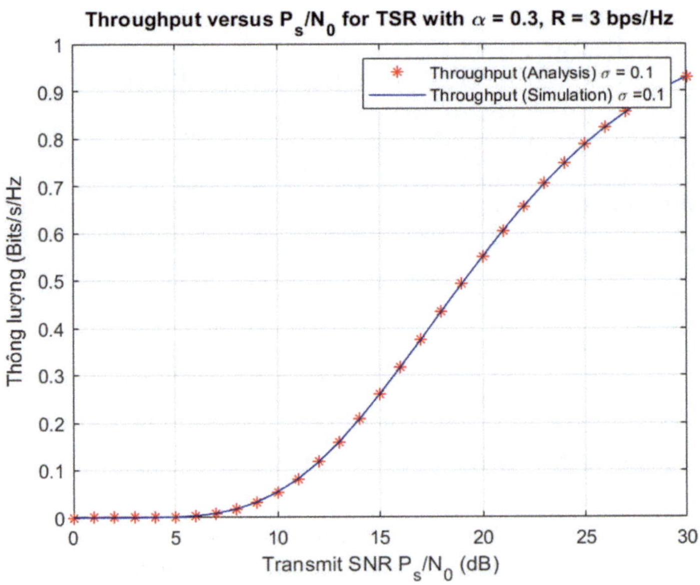

Fig. 7.4 Throughput plotted against the transmit power of the source

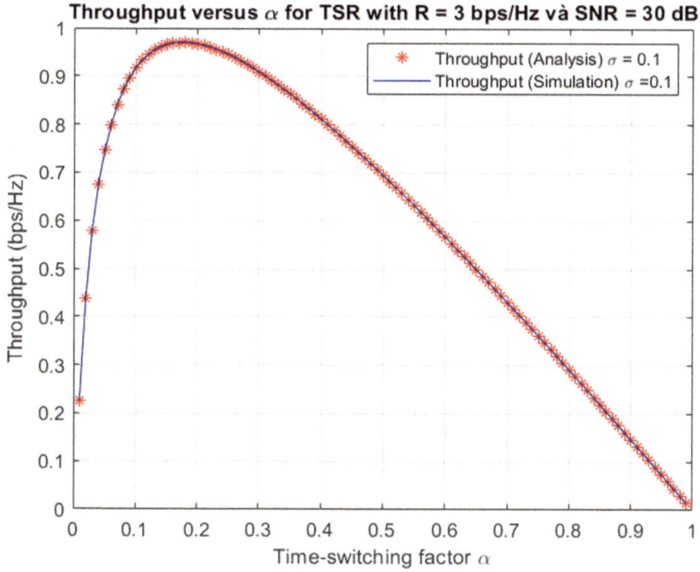

Fig. 7.5 Throughput plotted against the time-switching coefficient

7.2 Self-energy Recycling

Another proposed solution for energy saving in wireless relay networks is energy reuse through self-interference cancellation at relays, along with the application of relay selection algorithms to enhance network performance [6].

7.2.1 Introduction

A relay with full-duplex (FD) capability can transmit and receive simultaneously on the same frequency band, providing twice the spectral efficiency. Therefore, FD communication is expected to become one of the key technologies for 5G and beyond wireless networks. However, this increase in throughput is limited by the presence of unavoidable interfering signals because the strong transmitted signal travels back to the receiver in the same device. Through effective self-interference signal suppression (SIC) strategies, FD has developed rapidly over the past decade, especially in the sub-6 GHz band. Specifically, SIC can be suppressed by several methods in different domains, such as the antenna domain (polarization or isolation), analog domain (using low noise amplification), and digital domain (interfering with the baseband signal after the ADC converter). However, in reality, the aforementioned methods still cannot completely eliminate self-interference signals.

Although self-interference is harmful to signal decoding at the relay, it can be energetically beneficial. Based on radio frequency (RF) EH technology, the concept of energy reuse (S-ER) in FD networks has been proposed, where energy from self-interfering signals can be harvested, accumulated, and reused. Consequently, it has been applied to wireless FD networks to enhance energy performance (EE), especially for FD networks with support for radio EH. Researchers have proposed an antenna allocation scheme for S-ER in FD MIMO decode-and-forward (DF) relay networks and analyzed its performance. They also conducted an EE evaluation of an FD nonlinear MIMO-OFDM system with S-ER and proposed beamforming design optimization for FD S-ER relay systems using the DF protocol. The study examined both amplify-and-forward (AF) and DF transitions in S-ER supported FD networks. Other publications on S-ER include considerations of secure transmission aspects and internal S-ER NOMA systems. In particular, the authors in [6] reviewed the application of S-ER to wireless AF relay networks with multiple parallel relays and designed and analyzed forward selection algorithms to enhance the overall system reliability. This is a new and highly effective solution, so this book chooses to incorporate this method into the design of the 5G/6G network simulation toolkit.

Relay selection (RS) is a typical solution in practice to harness cooperative diversity in wireless networks with multiple relays, especially in high-density networks such as sensor networks or vehicular networks. In this book, two algorithms are examined. First, partial relay selection (PRS), where the relay with the largest channel gain from the source to the relay is selected. This algorithm is suitable for situations where channel state information (CSI) is difficult to obtain. Second, full relay selection (FRS), where the relay aiding in achieving the highest overall signal-to-interference-plus-noise ratio (SINR) is selected. This algorithm requires CSI from all channels. Additionally, the paper [6] conducts research under both general and more realistic channel conditions than previous works, i.e., independently but non-identically distributed (i.n.d.) channel gains. All subsequent analytical results are confirmed by Monte Carlo simulations.

7.2.2 System Model

The model of a multi-relay full-duplex network simulation is illustrated in Fig. 7.6. In this figure, a source S communicates with a destination node D through the assistance of M full-duplex relays denoted by R_m, where $m = 1, 2, ..., M$. Let a and b be the channel coefficients of the links $S \rightarrow R_m$ and $R_m \rightarrow D$, respectively, and f_m represents the self-interference channel gain between the transmit and receive antennas of relay R_m.

Similar to Sect. 7.1, let's assume that all channels are affected by Rayleigh fading, thus, the channel gains $\gamma_{SR_m} = |h_{SR_m}|^2$, $\gamma_{R_m D} = |h_{R_m D}|^2$ are random variables (RVs) distributed exponentially with parameters λ_{SR_m} and $\lambda_{R_m D}$, respectively. The self-interference channel power gain also follows an exponential distribution with parameter Ω_m.

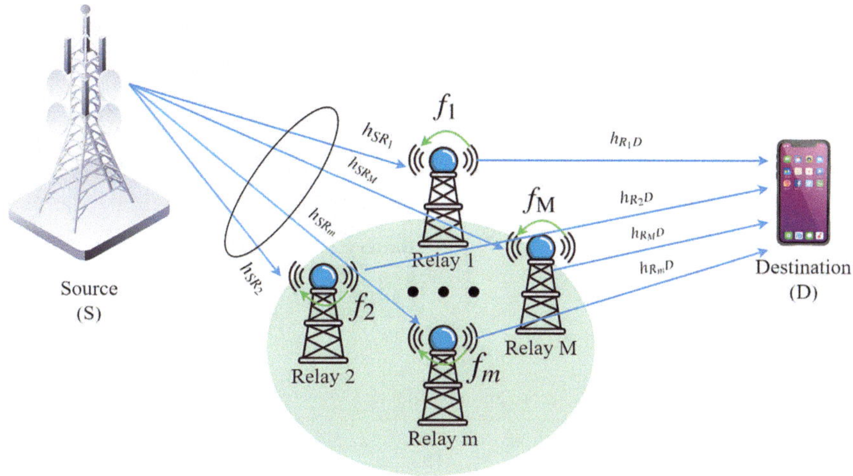

Fig. 7.6 System model of multi-relay full-duplex transmission network of interest

According to the EH power-splitting (PS) protocol, the received signal power at the relay in each transmission block is divided for EH and message decoding at a ratio of $\rho : (1-\rho)$ where $\rho \in (0, 1)$is the power-splitting coefficient, as illustrated in Fig. 7.7a. Figure 7.7 also depicts the S-ER protocol. Each transmission block is divided into two equal-length time slots (each slot has duration $\frac{T}{2}$). Two batteries alternate between two operations after each time slot: one battery is used to store harvested energy, and the other battery provides energy to the activated relay. This is referred to as the virtual harvesting protocol as summarized in Fig. 7.7b.

The received signal at relay R_m used for EH is provided by

$$y_{eh} = \sqrt{\rho} h_{SR_m} X_S + \sqrt{\rho} f_m X_{R_m} + n_{R_m}^{(eh)} \tag{7.11}$$

where $m = 1, 2, ..., M$; x_S is the signal transmitted by S, x_{R_m} is the signal generated by the transmitting antenna of R_m, and $n_{R_m}^{(eh)}$ is the AWGN noise at relay R_m. The harvested energy will be used to relay the message from the source to the destination. Therefore, the average transmission power of R_m can be determined by

$$P_{R_m} = \frac{\eta \rho P_S \gamma_{SR_m}}{1 - \eta \rho \varphi_m} \tag{7.12}$$

where $0 \leq \eta \leq 1$ is the EH efficiency, P_S is the signal transmission power, with the condition $\varphi_m \geq \dfrac{1}{\eta \rho}$.

The information part of the received signal at R_m is determined by

$$y_{R_m} = \sqrt{1-\rho} h_{SR_m} X_S + \sqrt{1-\rho} f_m X_{R_m} + n_{R_m}, \tag{7.13}$$

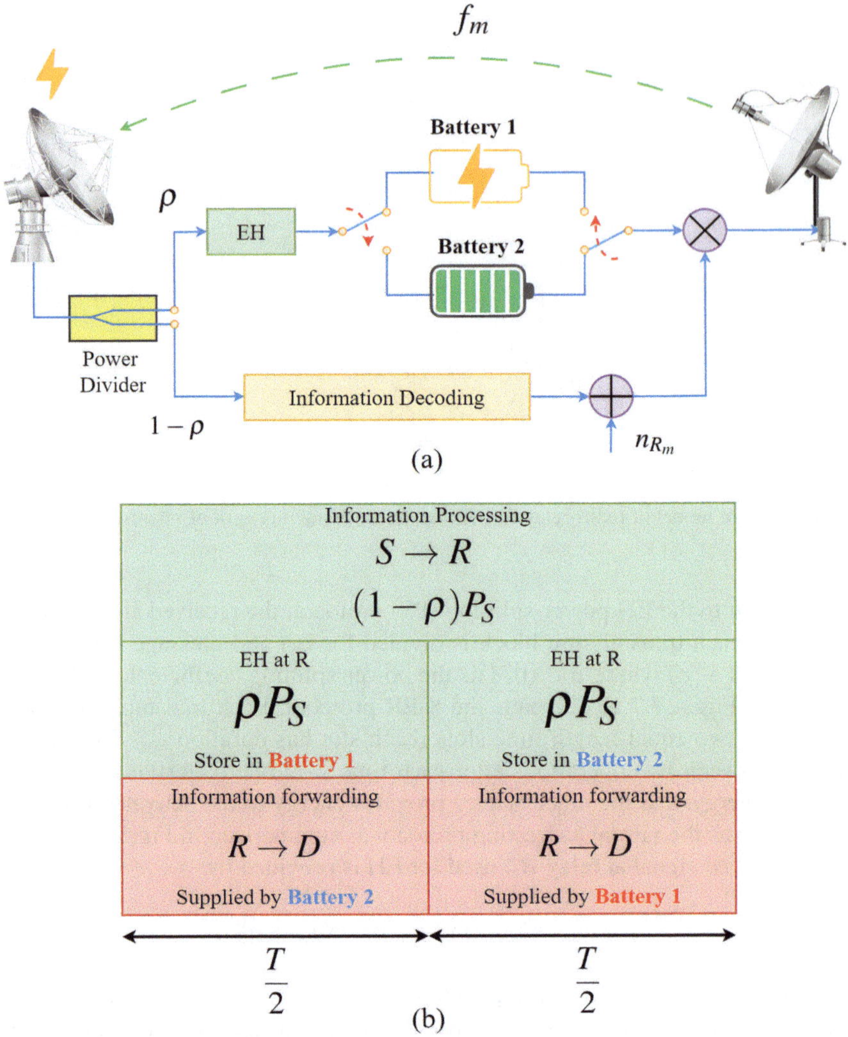

Fig. 7.7 Architecture of the S-ER protocol and hardware. **a** The EH structure at the relay. **b** Illustration of the energy reuse protocol

where $m = 1, 2, ..., M$, X_S is the signal transmitted by S, X_{R_m} is the signal emitted by R_m, and n_{R_m} is the AWGN at R_m.

If R_m is selected, it will amplify the received signal from S and relay this signal to D in the third stage (referred to as the information transmission phase). The received signal at destination D can now be expressed as

$$\begin{aligned} y_D &= h_{R_m D} X_{R_m} + n_D = h_{R_m D} G_m y_{R_m} + n_D \\ &= h_{R_m D} G_m [\sqrt{1-\rho} h_{SR_m} X_S + \sqrt{1-\rho} f_m X_{R_m} + n_{R_m}] + n_D \\ &= \underbrace{\sqrt{1-\rho} h_{R_m D} G_m h_{SR_m} X_S}_{\text{signal}} + \underbrace{\sqrt{1-\rho} f_m X_{R_m D} h_{R_m D} G_m + h_{R_m D} G_m n_{R_m} + n_D}_{\text{interference and noise}}, \end{aligned} \tag{7.14}$$

where

$$G_m = \sqrt{\frac{P_{R_m}}{(1-\rho) P_{S\gamma SR_m} + (1-\rho) P_{R_m \varphi_m} + N_0}}. \tag{7.15}$$

From there, the overall SINR expression can be calculated as follows:

$$\Psi_m = \frac{\eta\rho Q \gamma_{SR_m} \gamma_{R_m D} (1 - \eta\rho\varphi_m)}{\eta^2 \rho^2 Q \gamma_{SR_m} \gamma_{R_m D} \varphi_m - \eta\rho\varphi_m + 1}, \tag{7.16}$$

where $Q \triangleq P_s/N_0$ represents the SNR ratio at the transmitter.

The followings are the description of the operation of the two relay selection algorithms under consideration, FRS, and partial relay selection (PRS).

A. Full Relay Selection

Let R_a denote the best relay; the selection algorithm can be written as

$$a = \underset{m=1,2,\ldots,M}{argm} \{\Psi_m\} \Leftrightarrow \Psi_a = \underset{m=1,2,\ldots,M}{max} \{\Psi_m\} \tag{7.17}$$

B. Selecting individual relays

The relay achieving the highest channel gain between itself and the source will be selected. Specifically,

$$b = \underset{m=1,2,\ldots,M}{argm} \{\gamma_{SR_m}\} \Leftrightarrow \gamma_{SR_b} = \underset{m=1,2,\ldots,M}{max} \{\gamma_{SR_m}\} \tag{7.18}$$

where R_b represents the selected relay, and $b \in \{1, 2, \ldots, M\}$.

Assuming that the links from S to the relay nodes are independent, thus the CDF of γ_{SR_b} can be constructed as

$$F_{\gamma_{SR_b}}(X) = Pr(\gamma_{SR_b} \leq X) = Pr(\gamma_{SR_m} \leq X, \forall m = 1, \ldots, M) = \prod_{m=1}^{M} F_{\gamma_{SR_m}}(X) \tag{7.19}$$

In the case of i.n.d., i.e., when the parameters of γ_{SR_m} are different ($\lambda_{SR_m} \neq \lambda_{SR_n}, \forall m \neq n$): from (4.19), $F_{\gamma_{SR_b}}(X)$ can be expressed as

$$F_{\gamma_{SR_b}}(X) = \prod_{m=1}^{M}[1 - e^{-\lambda_{SR_m} X}] = 1 + \sum_{n=1}^{M}(-1)^n \sum_{1\le j_1<\ldots<j_n\le M} e^{-\sum_{i=1}^{n}\lambda_{SR_{j_i}} X} \tag{7.20}$$

With the case of γ_{SR_m} being i.i.d., i.e., $\lambda_{SR_m} = \lambda_{SR}, \forall m$. $F_{\gamma_{SR_b}(X)}$ in (4.19) becomes

$$F_{\gamma_{SR_b}(X)} = [1 - e^{-\lambda_{SR_m} X}]^M = 1 + \sum_{n=1}^{M}(-1)^n \binom{M}{n} e^{-n\lambda_{SR} X} \tag{7.21}$$

7.2.3 *Performance Evaluation*

In this section, the OP (OP) of the PRS and FRS algorithms will be established. Specifically, the exact expressions for OP in the i.n.d. case are derived first and applied to the special case of i.i.d.

7.2.3.1 Full Relay Selection

For the i.n.d. case:

$$OP_{i.n.d}^{FRS} = Pr(\Psi \le \gamma_{th}) = \prod_{m=1}^{M} \underbrace{Pr(\Psi_m < \gamma_{th})}_{OP_{m,i.n.d}^{FRS}} \tag{7.22}$$

where $\gamma_{th} = 2^R - 1$, R represents the minimum required data rate of the network and $OP_{m,i.n.d}^{FRS}$ is the probability of outage of the path $S \to R_m \to D$. The i.i.d. case can be seen as a special case of the i.n.d. case. The following results are obtained from [6].

Lemma 7.2 (OP of FRS, i.n.d. case) *The OP of the protocol under consideration using the FRS method in the i.n.d. case is given by*

$$OP_{i.n.d}^{FRS} = \prod_{m=1}^{M} OP_{m,i.n.d}^{FRS}, \tag{7.23}$$

where

$$OP_{i.n.d}^{FRS} = 1 - e^{\frac{-\Omega_m}{\eta\rho}} - \int_0^{\frac{1}{\eta\rho(1+\gamma_{th})}} \Omega_m e^{-\Omega_m X} \sqrt{\frac{4\lambda_{SR_m}\lambda_{RD_m}\gamma_{th}(1-\eta\rho X)}{\eta\rho Q[1-\eta\rho(1+\gamma_{th})X]}} K_1\left(\sqrt{\frac{4\lambda_{SR_m}\lambda_{RD_m}\gamma_{th}(1-\eta\rho X)}{\eta\rho Q[1-\eta\rho(1+\gamma_{th})X]}}\right) dx, \quad (7.24)$$

and $K_v(.)$ *represents the modified Bessel function of the second kind of order* v.

Lemma 7.3 (OP of FRS, i.i.d. case) *The OP of the protocol under consideration using the FRS algorithm in the i.i.d. case can be accurately computed as*

$$OP_{i.n.d}^{FRS} = \left(OP_{m,i.i.d}^{FRS}\right)^M, \quad (7.25)$$

where $OP_{m,i.i.d}^{FRS}$ *is computed by*

$$OP_{m,i.n.d}^{FRS} = 1 - e^{\frac{-\Omega}{\eta\rho}} - \int_0^{\frac{1}{\eta\rho(1+\gamma_{th})}} \Omega e^{-\Omega X} \sqrt{\frac{4\lambda_{SR}\lambda_{RD}\gamma_{th}(1-\eta\rho X)}{\eta\rho Q[1-\eta\rho(1+\gamma_{th})X]}} K_1\left(\sqrt{\frac{4\lambda_{SR}\lambda_{RD}\gamma_{th}(1-\eta\rho X)}{\eta\rho Q[1-\eta\rho(1+\gamma_{th})X]}}\right) dx \quad (7.26)$$

7.2.3.2 Selecting Individual Relays

For the i.n.d. case, by using the law of total probability, the OP can be computed as

$$\begin{aligned} OP_{i.n.d}^{FRS} &= \sum_{m=1}^{M} Pr(\Psi_m < \gamma_{th} | R_b \equiv R_m) \\ &= \sum_{m=1}^{M} Pr\left(\begin{array}{c} \gamma_{SR_m} \geq Z_m, \\ \dfrac{\eta\rho Q\gamma_{SR_m}\gamma_{R_mD}(1-\eta\rho\varphi_m)}{\eta^2\rho^2 Q\gamma_{SR_m}\gamma_{R_mD}\varphi_m - \eta\rho\varphi_m + 1} < \gamma_{th}, \end{array}\right) \end{aligned} \quad (7.27)$$

where $Z_m \triangleq \max_{\substack{k=1,2,\ldots,M \\ k\neq m}} \{\gamma_{SR_k}\}$.

Lemma 7.4 (OP of PRS, i.n.d. case) *The OP of the PRS algorithm for the i.n.d. case is provided by [6]*

$$OP_{i.n.d}^{PRS} = \sum_{m=1}^{M}\left(1-e^{\frac{-\Omega}{\eta\rho}}\right)\left[\sum_{n=1}^{M-1}\sum_{1\le j_1\ldots<j}\sum_{\substack{n\le M\\ j_1,\ldots,j_n\neq m}}\frac{(-1)^{n+1}\sum_{k=1}^{n}\lambda_{SR_{jk}}}{\lambda_{SR_m}+\sum_{k=1}^{n}\lambda_{SR_{jk}}}\right]$$

$$-\sum_{m=1}^{M}\int_{0}^{\frac{1}{\eta\rho(1+\gamma_{th})}}\Omega_m e^{-\Omega_m X}\sqrt{4\lambda_{R_m D}\lambda_{SR_m}\vartheta(X)}K_1\left(\sqrt{4\lambda_{R_m D}\lambda_{SR_m}\vartheta(X)}\right)dx$$

$$+\sum_{m=1}^{M}\sum_{n=1}^{M-1}\sum_{1\le j_1\ldots<j}\sum_{\substack{n\le M\\ j_1,\ldots,j_n\neq m}}\frac{(-1)^{n+1}\lambda_{SR_m}}{\lambda_{SR_m}+\sum_{k=1}^{n}\lambda_{SR_{jk}}}\int_{0}^{\frac{1}{\eta\rho(1+\gamma_{th})}}\Omega_m e^{-\Omega_m X}$$

$$*\sqrt{4\lambda_{SR_m}+\left(\lambda_{SR_m}+\sum_{k=1}^{n}\lambda_{SR_{jk}}\vartheta(X)\right)}K_1\left(\sqrt{4\lambda_{R_m D}\left(\lambda_{SR_m}+\sum_{k=1}^{n}\lambda_{SR_{jk}}\right)\vartheta(X)}\right)dx \tag{7.28}$$

Lemma 7.5 (OP of PRS, i.i.d. case) *The OP of the PRS algorithm for the i.i.d. case is provided by [6]*

$$OP_{i.i.d}^{PRS} = M\left(1-e^{\frac{-\Omega}{\eta\rho}}\right)\sum_{n=1}^{M-1}\binom{M}{n}\frac{(-1)^{n+1}n}{n+1}$$
$$-M\int_{0}^{\frac{1}{\eta\rho(1+\gamma_{th})}}\Omega e^{-\Omega x}\sqrt{4\lambda_{RD}\lambda_{SR}\vartheta}K_1\left(\sqrt{4\lambda_{RD}\lambda_{SR}\vartheta(x)}\right)dx \tag{7.29}$$
$$+M\sum_{n=1}^{M-1}\binom{M}{n}\frac{(-1)^{n+1}}{n+1}\int_{0}^{\frac{1}{\eta\rho(1+\gamma_{th})}}\Omega e^{-\Omega x}\sqrt{4(n+1)\lambda_{RD}\lambda_{SR}\vartheta}K_1$$
$$*\left(\sqrt{4(n+1)\lambda_{RD}\lambda_{SR}\vartheta(x)}\right)dx$$

The authors [6] also analyze the optimal power allocation coefficient and study the asymptotic behavior of the OP. Due to page limitations, this book does not present detailed discussions on these matters.

7.2.3.3 Spectral Efficiency

The spectral efficiency (SE) or average channel capacity per unit bandwidth of the aforementioned system is given by

$$SE = E\left[log_2(1 + SINR)\right] = \int_0^{\infty} log_2(1 + x) f_{\Psi_m}(x) dx, \tag{7.30}$$

where $SINR = \Psi_m$ is the SINR of the entire system (here, m is the index of the relay selected by the chosen RS method) and f_{Ψ_m} is the PDF of Ψ_m. By applying partial integration and then applying the trapezoidal rule by dividing the interval [0, 1] by $x_k = \frac{k}{N}$, $k = 1, 2, ..., N$, where N is a sufficiently large positive integer, we obtain:

$$\begin{aligned} SE &= \int_0^1 \frac{1 - \Xi\left(\frac{1-y}{y}\right)}{yln2} dy = \frac{g\frac{1}{N} + g(1)}{2N} + \sum_{k=1}^{N-1} \frac{g\left(\frac{k}{N}\right)}{N} \\ &= \frac{1 + N\left[1 - \Xi(N-1)\right]}{ln2} + \sum_{k=1}^{N-1} \frac{1 - \Xi\left(\frac{N-k}{k}\right)}{kln2}, \end{aligned} \tag{7.31}$$

where $\Xi(x) \triangleq Pr(\Psi_m \leqslant x)$ can be found by substituting $\gamma_{th} = x$ into the expressions of $OP^{FRS}_{i,n.d.}$,$OP^{FRS}_{i,i.d.}$,$OP^{PR}_{i,n.d.}$ and $OP^{PRS}_{i,i.d.}$ for the cases: (i) FRS and i.n.d. channels, (ii) FRS and i.i.d. channels, (iii) PRS and i.n.d. channels, (iv) PRS and i.i.d. channels.

7.2.4 MATLAB Sample Codes

In this section, sample codes for simulating the self-energy recycling technique are provided. It should be noted that only a few functions are given. Other functions can be written using a similar approach, but they are omitted here due to the page limit.

```
tic
% Input parameters
PdB          = -15:3:15;   % Transmit SNR - Q
Eta          = 0.8;        % Energy conversion efficiency - \eta
AP           = 0.3;        % Time-switching factor - \alpha
Cth          = 1.25;       % Required data rate - R
PL           = 3;          % Path loss exponent - \beta
bit_frame    = 10^7;       % Number of samples
```

```
ORR                    = 4;
ind                    = 1;

% Case 1
xR                     = 0.1 + 0.8*rand(1,3);
OP_PRS_AF_IND_SIM_2 = PRS_AF_INID_SIM(PdB,xR,
    ORR,Eta,AP,Cth,PL,bit_frame);
fprintf('Complete %d per 12 \n',ind);
ind = ind + 1;
OP_PRS_AF_IND_EXACT_2 = PRS_AF_INID_EXACT(PdB,
    xR,ORR,Eta,AP,Cth,PL);
fprintf('Complete %d per 12 \n',ind);
ind = ind + 1;
OP_FRS_AF_IND_SIM_2 = FRS_AF_INID_SIM(PdB,xR,
    ORR,Eta,AP,Cth,PL,bit_frame);
fprintf('Complete %d per 12 \n',ind);
ind = ind + 1;
OP_FRS_AF_IND_EXACT_2 = FRS_AF_INID_EXACT(PdB,
    xR,ORR,Eta,AP,Cth,PL);
fprintf('Complete %d per 12 \n',ind);
ind = ind + 1;

% Plot the output
figure
h3 = semilogy(PdB,OP_PRS_AF_IND_SIM_2,'r--','
    LineWidth',2); hold on;
h4 = semilogy(PdB,OP_PRS_AF_IND_EXACT_2,'ro','
    LineWidth',2); hold on;

h9 = semilogy(PdB,OP_FRS_AF_IND_SIM_2,'b-','
    LineWidth',2); hold on;
h10 = semilogy(PdB,OP_FRS_AF_IND_EXACT_2,'bh','
    LineWidth',2); hold on;
grid on;
title(['OP versus transmit SNR for \alpha = ',
    num2str(AP),', i.n.d. case']);
xlabel('Transmit SNR (dB)');
ylabel('Outage probability');
legend([h4,h10,h1,h7],'PRS, M = 3',...
      'FRS, M = 3','PRS Simulation','FRS
          Simulation');
toc
```

```
function OP_PRS_AF_IID = PRS_AF_IID_EXACT(PdB,
    xR,ORR,Eta,AP,Cth,PL)
OP_PRS_AF_IID    = zeros(1,length(PdB));
for aa = 1 : length(PdB)
    fprintf('Running %d per %d \n',aa,length(
        PdB));
    OP_PRS_AF_IID(aa) = ham(PdB(aa),xR,ORR,Eta,
        AP,Cth,PL);
end
end
%
function OP = ham(PdB,xR,ORR,Eta,AP,Cth,PL)
%
PP          = 10^(PdB/10);
MM          = length(xR);
kap         = Eta*AP/(1-AP);
LSR         = xR.^PL;
LRD         = (1-xR).^PL;
gth         = 2^Cth-1;
OP=1;
for j = 1 : MM
            y=@(x)(kap.*x-1)./(kap .^2.*PP.*x);
            xi=@(x)(gth.*(kap.*x-1))./(kap.*PP.*(
                kap.*x.*(1+gth)-1));
            f1=@(x)sqrt(j.*LSR(j).*LRD(j).*y(x)).*
                exp(-ORR.*x).*besselk(1,2.*sqrt(j.*
                LSR(j).*LRD(j).*y(x))));
            f2=@(x)sqrt(j.*LSR(j).*LRD(j).*xi(x)).*
                exp(-ORR.*x).*besselk(1,2.*sqrt(j.*
                LSR(j).*LRD(j).*xi(x))));
            tp1=2.*ORR.*(-1).^j.*(prod(factorial(MM
                )))./(prod(factorial([MM-j j])));
            tp2=integral(f1,1./kap,inf);
            tp3=integral(f2,1./kap,inf);
            tp4=integral(f2,0,1./kap./(1+gth));
            OP=OP+tp1.*(tp3+tp4-tp2);
end
    end
```

```
function OP_PRS_AF_IID = PRS_AF_IID_SIM(PdB, xR, ORR, Eta, AP, Cth, PL, bit_frame)
% Initialize the output
OP_PRS_AF_IID   = zeros(1, length(PdB));

for aa = 1 : length(PdB)
    PP      = 10^(PdB(aa)/10);
    MM      = length(xR);
    kap     = Eta*AP/(1-AP);
    LSR     = xR.^PL;
    LRD     = (1-xR).^PL;
    gth     = 2^Cth-1;

    for bitnum   = 1 : bit_frame
        SNR_SR = zeros(1,MM);
        SNR_RD = zeros(1,MM);
        SNR_RR = zeros(1,MM);
        for bb = 1 : MM
            hSR        = sqrt(1/2/LSR(bb))*(randn(1,1) + 1i*randn(1,1));
            SNR_SR(bb) = abs(hSR)^2;
            hRD        = sqrt(1/2/LRD(bb))*(randn(1,1) + 1i*randn(1,1));
            SNR_RD(bb) = abs(hRD)^2;
            hRR        = sqrt(1/2/ORR)*(randn(1,1) + 1i*randn(1,1));
            SNR_RR(bb) = abs(hRR)^2;
        end
         if (SNR_SR(bb)> SNR_SR)
                SNR_SR = SNR_SR(bb);
         end

        pr= (kap.*PP.*SNR_SR)./(1 - kap.*SNR_RR(bb));
        num = (PP.*SNR_SR.*SNR_RD(bb))./SNR_RR(bb);
        den = (PP.*SNR_SR)./(pr.*SNR_RR(bb))+pr.*SNR_RD(bb)+1;
        SNR_AF = num./den;
        if (SNR_AF < gth)
            OP_PRS_AF_IID(aa) = OP_PRS_AF_IID(aa) + 1;
```

Table 7.2 Parameters for simulation configuration

Symbol	Parameter name	Fixed value	Range of variation
R	Transmit data rate	1.25 bps/Hz	$0.5 \rightarrow 3$ bps/Hz
γ_{th}	SINR threshold	1.378	$0.414 \rightarrow 7$
η	EH efficiency	0.8	None
ρ	Power allocation coefficient	0.3 or arbitrary	$0 \rightarrow 1$
Ω	SI channel power gain	2	$1 \rightarrow 5$
x_R	Distance from $S \rightarrow R(i.i.d)$	0.75	None
x_{R_i}	Distance from $S \rightarrow R(i.n.d)$	[0.7, 0.405, 0.324, 0.55]	None
Q	Signal-to-noise ratio	5 dB	−5 to 30 dB
M	Number of relays	None	1, 2, 3, 4

```
            end
        end
    end

    OP_PRS_AF_IID = OP_PRS_AF_IID/bit_frame;
    end
```

7.2.5 Simulation Results

In this section, Monte Carlo simulation results are presented to validate the analysis in Sect. 7.2.3. Here, the distance between S and D is normalized to 1, and all relays available are located between the source and destination, as illustrated in Fig. 4.8. The path-loss exponent is set to $\beta = 3$. The simulation scenarios for the i.n.d. and i.i.d. cases are described respectively in Fig. 7.8a and b. Here, x_{R_i}, for $i = 1, ..., M$, are drawn from an array consisting of uniformly distributed numbers in the range from 0 to 1 (excluding 0 and 1).

The impact of various parameters, including SNR, power allocation coefficient, and data rate R, on the OP is investigated through simulation. The number of samples for the Monte Carlo simulation is 10^7. The selected values of the simulation parameters are summarized in Table 7.2.

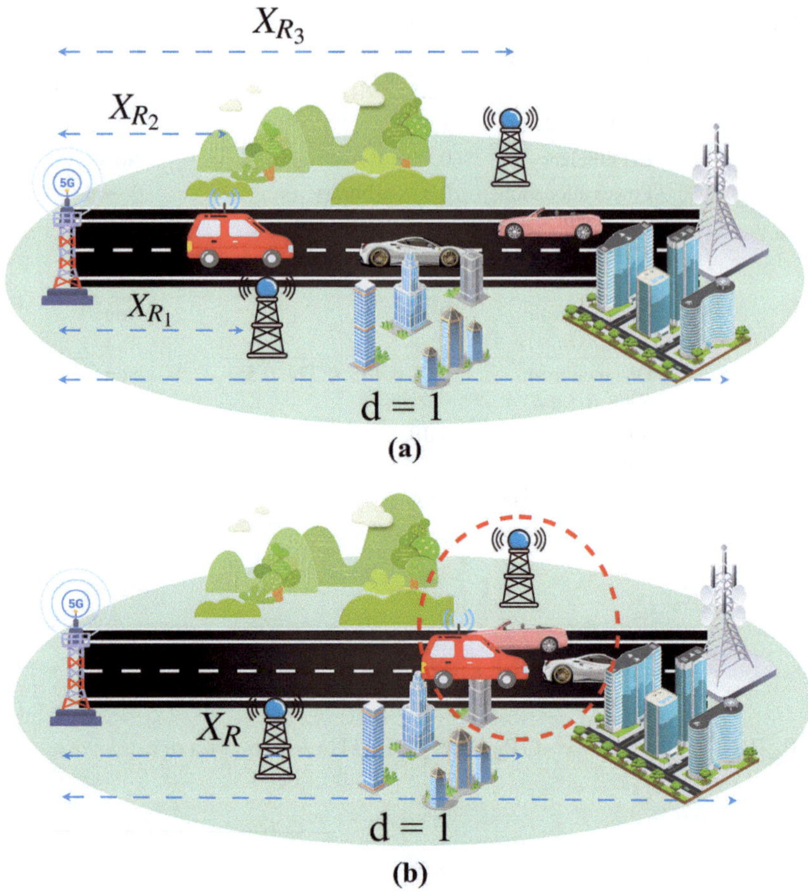

Fig. 7.8 Simulation scenarios: **a** Scenario 1: i.n.d. case. **b** Scenario 2: i.i.d. case

Figures 7.9 (i.n.d. case) and 7.10 (i.i.d. case) confirm that the OPs of the proposed system for both PRS and FRS, with varying numbers of relays, decrease as the transmit SNR increases. It can be observed that FRS significantly outperforms PRS, and both PRS and FRS outperform the single-relay scheme.

Figures 7.11 and 7.12 depict the OPs for both PRS and FRS algorithms in two cases (i.n.d. and i.i.d.) as a function of the power allocation coefficient ρ. For PRS, the variation of M does not significantly impact the OP, but for FRS, increasing M leads to a significant improvement in system quality. In Fig. 7.13, the impact of interference level on OP is investigated. As the SI level decreases, the OP also decreases. However, the OP appears to converge to a limiting value as ω becomes sufficiently large.

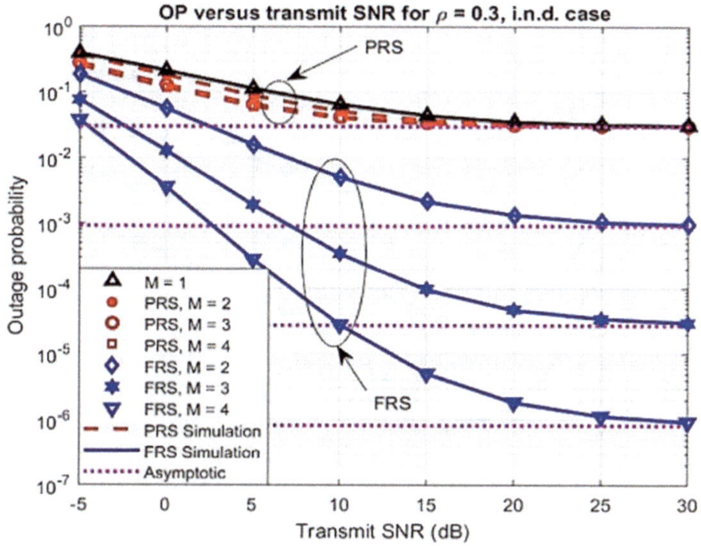

Fig. 7.9 OP plotted against the P_S/N_0 ratio (dB) for $\rho = 0.3$ and $\mathscr{R} = 1.25$ bps/Hz, i.n.d. case

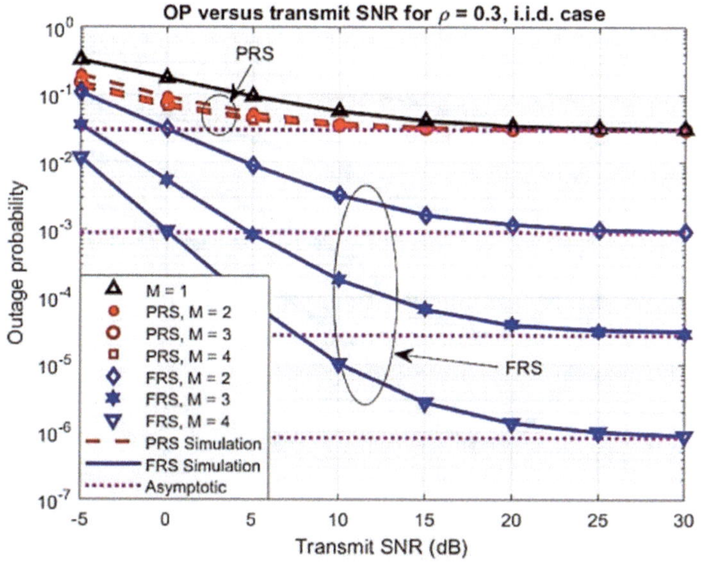

Fig. 7.10 OP plotted against transmit SNR (dB) for $\rho = 0.3$ and $\mathscr{R} = 1.25$ bps/Hz, i.i.d. case

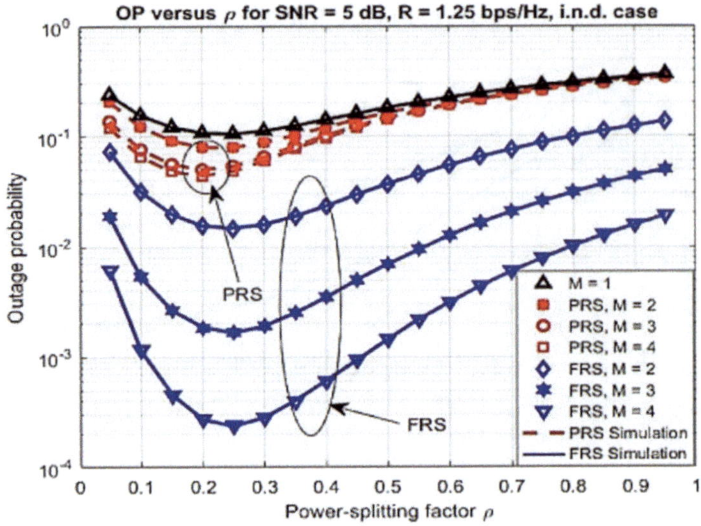

Fig. 7.11 OP plotted against ρ with SNR = 5 dB, $\mathscr{R} = 1.25$ bps/Hz, i.n.d. case

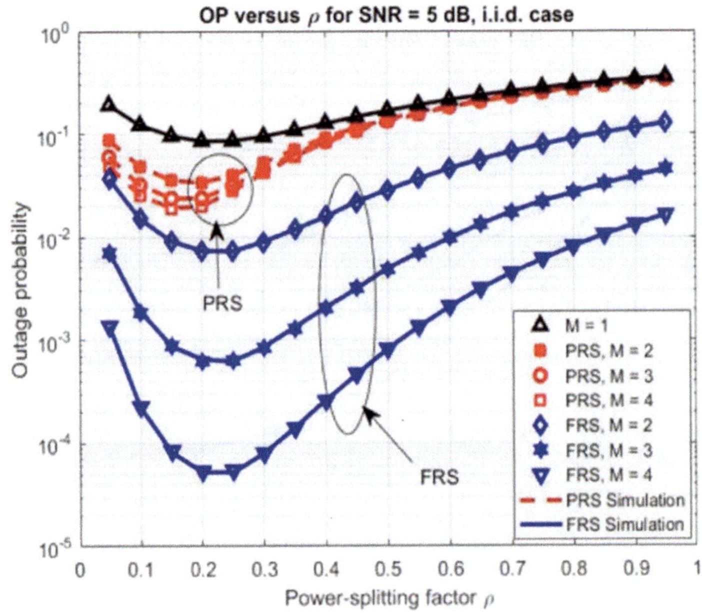

Fig. 7.12 OP plotted against ρ at SNR = 5 dB and $\mathscr{R} = 1.25$ bps/Hz, i.i.d. case

Finally, the system's SE is illustrated in Fig. 7.14. Clearly, SE improves as the transmit SNR increases, for both RS methods and for all numbers of relays. However, the SE curves seem to approach an upper limit in the high transmit SNR region.

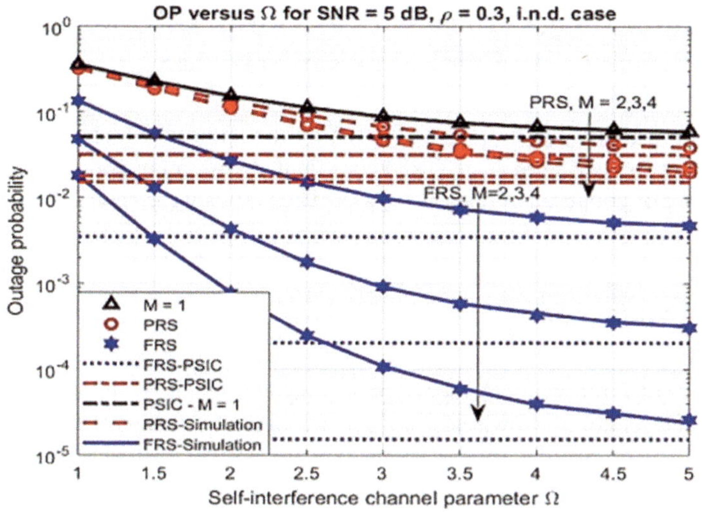

Fig. 7.13 OP plotted against interference level ω for $\mathscr{R} = 1.25$ bps/Hz, SNR = 5 dB, $\rho = 0.3$, i.n.d. case

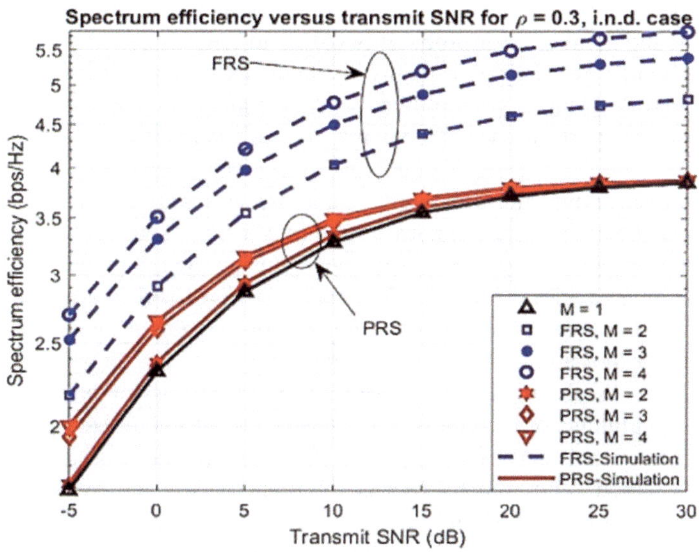

Fig. 7.14 Spectral Efficiency of the system with different RS methods for i.n.d. case, $\mathscr{R} = 1.25$ bps/Hz, and $\rho = 0.3$

7.3 Multiple-Stage Transition Model with IRS Support for Cooperative Relay Wireless Networks

This section presents the proposed solution by the authors, which is the utilization of Intelligent Reflecting Surface (IRS) coordinated with relay nodes to enhance the performance of cooperative relay networks. The results have been published in [7].

7.3.1 Introduction

IRS consists of *L* reflecting elements, presenting a promising solution for enhancing coverage and reliability in wireless communication systems, has recently been proposed for 6G. Compared to relay stations, IRS offers several advantages, such as ease of deployment, low hardware complexity, and simplicity in signal processing, as it can reflect wireless signals without the need for RF signal processing, decryption, and encryption. IRS can operate without requiring a power source and can reflect signals across any frequency.

The book contributes an analysis of system performance, demonstrated by expressions of OP and symbol error probability (SER) in cooperative and cognitive relay wireless networks over Rayleigh fading channels. IRS offers superior performance compared to traditional relay amplification. As the number of reflecting elements in IRS increases, the performance of the wireless system supported by IRS significantly improves.

As mentioned above, full-duplex (FD) communications is expected to be deployed in 5G, 6G, and beyond wireless networks to improve the spectrum utilization efficiency of current wireless systems. Specifically, using the FD mode can double the spectrum compared with the half-duplex mode. However, self-interference from the transmitter to the receiver of the same relay node remains a major challenge for full-duplex transmission.

In recent works, the integration of IRS and FD transmission has been explored in wireless communication systems. In these studies, IRS is utilized as an FD relay (FDR) to relay signals between two points. However, the transmission performance of the FD method supported by IRS is significantly affected by SI. To mitigate the substantial impacts of SI and enhance the performance of FD communication using IRS, various approaches have been proposed, such as increasing the number of RE, designing beamforming techniques, and optimizing to reduce the residual self-interference before applying FD transmission in practice.

Most of the previous publications have used IRS as a relay, but scenarios with IRS in relay networks have not been thoroughly investigated. Previous studies that explored the combination of IRS and traditional relays often focused on experimental measurements to analyze the performance of IRS systems. The major drawback of this evaluation method is the significant time and effort required to obtain results due to the lack of mathematical expressions. Additionally, only half-duplex mode is typically used in these publications. These issues drive the adoption of IRS in both HDR and FDR systems. The proposed solution in this book contributes to the following aspects:

- Using IRS in HDR and FDR systems (now referred to as HDR-IRS and FDR-IRS respectively), involves deploying IRS with L reflecting elements between the source relay and the DF relay. By leveraging IRS in HDR and FDR systems, the performance and coverage of HDR-IRS and FDR-IRS systems are significantly enhanced, while energy consumption can be notably reduced.
- The book provides accurate expressions for OP, SEP, and EC of HDR-IRS and FDR-IRS systems over Rayleigh fading channels. All expressions are validated by Monte Carlo simulations. The impact of various parameters, such as average SNR, number of reflecting elements, data transmission rate, and residual SI, on the system performance is thoroughly investigated.

7.3.2 System Model

This section presents the system models and signals of the HDR-IRS and FDR-IRS systems. The signals are transmitted from the source (S) to the destination (D) with the assistance of the IRS (I) and a relay (R). Both S and D operate in HD mode, while R can operate in HD mode (HDR-IRS system) or in FD mode (FDR-IRS system). Here, S and D are equipped with a single antenna, while I consists of L reflecting elements. Due to long-distance transmission, the S-R, IRS-D, and S-D channels are considered unavailable.

7.3.2.1 HDR-IRS System

The system model of the HDR-IRS system is illustrated in Fig. 7.15. There are two time slots required to transmit the signal from S to D, namely the odd time slot to transmit the signal from S to R and the even time slot to transmit the signal from R to D.

In the odd time slot, the signal received at R through the support of IRS is calculated by

$$y_R = \sum_{l=1}^{L} h_{SI_l} h_{I_l R} e^{j\psi_l} x_S + Z_R, \tag{7.32}$$

where h_{SI_l} and $h_{I_l R}$ represent the wireless channels from S to the lth element of I and from the lth element of I to R, respectively; ψ_l denotes the phase-shift adjustable by the lth element of IRS; x_S is the transmitted signal from S with an average transmit power of P_S ; z_R is the AWGN at R with zero mean and variance σ^2, i.e., $z_R \sim \mathcal{CN}(0, \sigma_R^2)$. We have:

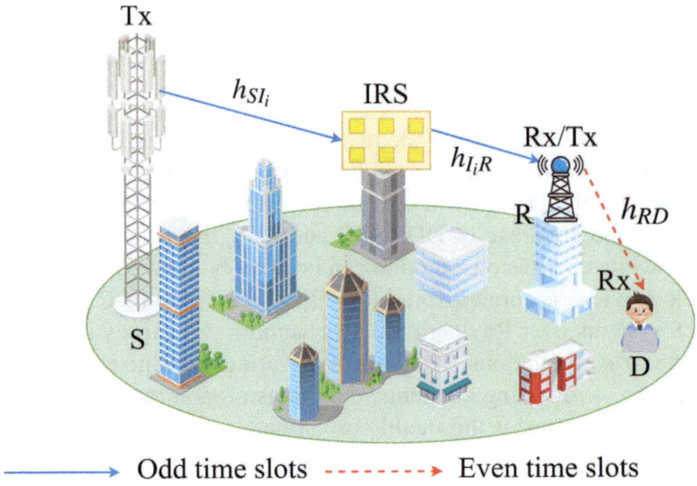

Fig. 7.15 System model of the HDR-IRS Networks

$$h_{SI_l} = a_{SI_l} e^{-j\psi_{SI_l}}, \tag{7.33}$$

$$h_{I_lR} = a_{I_lR} e^{-j\psi_{I_lR}}, \tag{7.34}$$

where a_{SI_l} and a_{I_lR} respectively represent the magnitude of h_{SI_l} and h_{I_lR}; ψ_{SI_l} and ψ_{I_lR} are the phase values of h_{SI_l} and h_{I_lR}. The phase values can be optimized to achieve maximum SNR.

$$\rho_R^{HD*} = \frac{|\sum_{l=1}^{L} a_{SI_l} a_{I_lR}|^2 P_S}{\sigma_R^2} \tag{7.35}$$

R decodes the received signal and then retransmits it to D. The received signal at D in the even time slots is represented as

$$y_D = h_{RD} x_R + z_D, \tag{7.36}$$

where h_{RD} is the channel gain from R to D; x_R is the signal transmitted at R with average power P_R; z_D is the AWGN at D with zero mean and variance σ^2, i.e., $z_D \sim \mathcal{CN}(0, \sigma_R^2)$. Therefore, the signal-to-noise ratio (SNR) at D can be calculated by

$$\rho_R^{HD} = \frac{|h_{RD}|^2 P_S}{\sigma_R^2}. \tag{7.37}$$

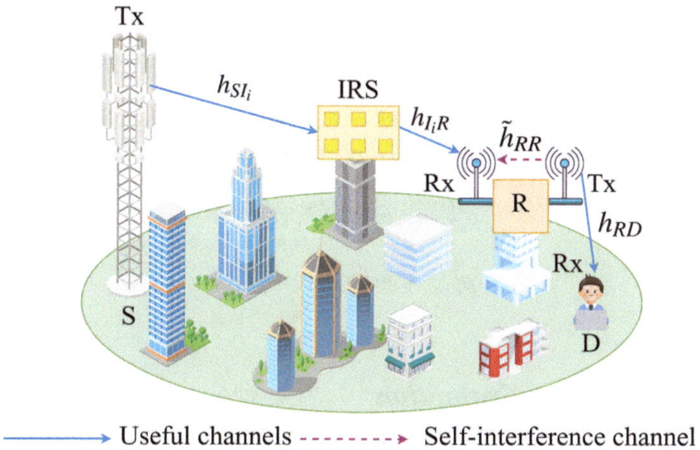

Fig. 7.16 FDR-IRS system model

7.3.2.2 FDR-IRS System

The FDR-IRS system model is illustrated in Fig. 7.16, where R is an FD device equipped with two separate antennas for transmission and reception. In each time slot, the received signal at R in the FDR-IRS systems is

$$y_R = \sum_{l=1}^{L} h_{SI_l} h_{I_l R} \exp(j\psi_l) x_S + \tilde{h}_{RR} x_R + Z_R, \tag{7.38}$$

where $\tilde{h}_{RR}$ is the self-interference (SI) channel gain from the transmit antenna to the receive antenna of R.

Assuming that after all SIC techniques are applied, residual self-interference (RSI) still exists in R and is modeled as a Gaussian random variable. Let w_R denote the RSI at R with an RSI level of σ_{SI}^2, i.e., $w_R \sim \mathcal{CN}(0, \sigma_{SI}^2 P_R)$. This assumption is widely used in previous works on FD systems. Now, the received signal at the relay can be rewritten as

$$y_R = \sum_{l=1}^{L} h_{SI_l} h_{I_l R} e^{j\psi_l} x_S + W_R + Z_R \tag{7.39}$$

The maximum SINR ratio at R of the FDR-IRS system can be calculated from (7.39) as

$$\rho_R^{FD*} = \frac{|\sum_{l=1}^{L} a_{SI_l} a_{I_l}|^2 P_S}{\sigma_{SI}^2 P_R + \sigma_R^2} \tag{7.40}$$

Similarly, the SNR at D of the FDR-IRS system is also determined by

$$\rho_R^{FD} = \frac{|h_{RD}|^2 P_R}{\sigma_D^2} \tag{7.41}$$

Because the DF protocol is applied at R, the end-to-end SNR of the HDR-IRS and FDR-IRS systems can be respectively calculated as

$$\rho_{e2e}^{HD} = m\{\rho_R^{HD*}, \rho_D^{HD}\}, \tag{7.42}$$

$$\rho_{e2e}^{FD} = m\{\rho_R^{FD*}, \rho_D^{FD}\}, \tag{7.43}$$

7.3.3 Performance Evaluation

7.3.3.1 Outage Probability

Theorem 7.1 *The OPs of HDR-IRS and FDR-IRS systems under Rayleigh fading conditions are respectively determined by*

$$\mathcal{P}_{out}^{HD} = 1 - \frac{1}{\Gamma\left(\frac{L\pi^2}{16-\pi^2}\right)} \exp\left(\frac{-\sigma_D^2 \rho_{th}^{HD}}{P_R}\right) \Gamma\left(\frac{L\pi^2}{16-\pi^2}, \frac{4\pi}{16-\pi^2}\sqrt{\frac{\sigma_R^2 \rho_{th}^{HD}}{P_S}}\right) \tag{7.44}$$

$$\mathcal{P}_{out}^{FD} = 1 - \frac{1}{\Gamma\left(\frac{L\pi^2}{16-\pi^2}\right)} \exp\left(\frac{-\sigma_D^2 \rho_{th}^{FD}}{P_R}\right) \Gamma\left(\frac{L\pi^2}{16-\pi^2}, \frac{4\pi}{16-\pi^2}\sqrt{\frac{(\sigma_{SI}^2 P_R + \sigma_R^2)\rho_{th}^{FD}}{P_S}}\right) \tag{7.45}$$

where $\Gamma(.)$ *and* $\Gamma(.,.)$ *denote the gamma function and the incomplete gamma function, respectively.*

7.3.3.2 Symbol Error Probability Analysis

Theorem 7.2 *The SEP of HDR-IRS and FDR-IRS systems (denoted as SEP_HD and SEP_FD, respectively) are determined respectively by*

$$SEP_{HD} \approx \frac{\alpha}{2} - \frac{\alpha\sqrt{\beta\pi}P_R}{4\sqrt{2}M\Gamma\left(L\frac{\pi^2}{16-\pi^2}\right)} \times \sum_{i=1}^{M}\left(\frac{1+a_i}{2}\right)^{\frac{\beta P_R}{2\sigma_D^2}} \sqrt{\frac{1-a_i^2}{P_R\sigma_D^2[\ln 2 - \ln(1+a_i)]}}\,\Gamma\left(\frac{L\pi^2}{16-\pi^2}, \frac{4\pi}{16-\pi^2}\sqrt{\frac{(P_R\sigma_R^2[\ln 2 - \ln(1+a_i)]}{P_S\sigma_D^2}}\right) \tag{7.46}$$

$$SEP_{HD} \approx \frac{\alpha}{2} - \frac{\alpha\sqrt{\beta\pi} P_R}{4\sqrt{2}M\Gamma\left(L\frac{\pi^2}{16-\pi^2}\right)}$$
$$\times \sum_{i=1}^{M} \left(\frac{1+a_i}{2}\right)^{\frac{\beta P_R}{2\sigma_D^2}} \sqrt{\frac{1-a_i^2}{P_R\sigma_D^2[\ln 2 - \ln(1+a_i)]}}\Gamma\left(\frac{L\pi^2}{16-\pi^2}, \frac{4\pi}{16-\pi^2}\sqrt{\frac{(P_R(\sigma_{SI}^2 P_R + \sigma_R^2)[\ln 2 - \ln(1+a_i)]}{P_S\sigma_D^2}}\right) \quad (7.47)$$

where $a_i \triangleq \cos\left(\frac{(2i-1)\pi}{2M}\right)$ *and M is the parameter of the Chebychev approximation.*

7.3.3.3 Ergodic Capacity

Theorem 7.3 *The EC of HDR-IRS and FDR-IRS systems are calculated, respectively, by*

$$\mathcal{C}_{HD} \approx \frac{\pi P_R}{4M\Gamma(L\frac{\pi^2}{16-\pi^2})\ln 2} \sum_{i=1}^{M} \frac{\sqrt{1-a_i^2}}{\sigma_D^2 + P_R[\ln 2 - \ln(1+a_i)]} \quad (7.48)$$

where a_i *and M are the same as in Theorem 7.2.*

7.3.4 MATLAB Coding

In this section, the sample MATLAB codes for simulating FDR-IRS are given.

```
clearvars;
clc;
close all;

% Input parameters
M = 50;
modulation = input('Please select modulation
    method (1 or 2): 1. BPSK, 2. 4-QAM ');
switch modulation
        case 1
                alpha1 = 1;
                beta1 = 2;
        case 2
                alpha1 = 2;
                beta1 = 1;
end
```

```

L = 3;
lambda_h = 1; lambda_g = 1; lambda_RD = 1;

t = 10^6;
PsN0dB = -20:5:30;
PsN0 = 10.^(PsN0dB/10);

R = 2;
gamma_HD = 2.^(2*R)-1;
gamma_FD = 2.^(R)-1;
h = sqrt(1/2/lambda_h)*(randn(L,t) + 1i*randn(L,t)); %Rayleigh fading for channel from source to IRS
g = sqrt(1/2/lambda_g)*(randn(L,t) + 1i*randn(L,t)); %Rayleigh fading for channel from IRS to relay
X = abs(h);   %Fading amplitude gain of h
Y = abs(g);  %Fading amplitude gain of g

h_RD = sqrt(1/2)*(randn(1,t) + 1i*randn(1,t)); %Rayleigh fading for channel residual interfernce in relay
Z = lambda_RD * abs(h_RD).^2;    %Fading power of h_nga_r

for p = 1: length(PsN0)
    rho_R_HD = sum(X.*Y,1).^2*PsN0(p);
    rho_D_HD = Z*PsN0(p);
    rho_R_FD = sum(X.*Y,1).^2*PsN0(p)./(1+PsN0(p)*sigma_SI^2);
    rho_D_FD = Z*PsN0(p);

    SNR_HD = min(rho_R_HD,rho_D_HD);
    SNR_FD = min(rho_R_FD,rho_D_FD);
    Pout_HD_sim(p) = sum(SNR_HD < gamma_HD) / t;
    Pout_FD_sim(p) = sum(SNR_FD < gamma_FD) / t;
    SEP_HD_sim(p) = sum(alpha1/2*erfc(sqrt(beta1*SNR_HD/2)))/t;
    SEP_FD_sim(p) = sum(alpha1/2*erfc(sqrt(beta1*SNR_FD/2)))/t;
```

```
    cap_HD_sim(p) = sum(log(1+SNR_HD)/log(2)/2)/t;
    cap_FD_sim(p) = sum(log(1+SNR_FD)/log(2))/t;


    Pout_HD_ana(p) = 1 - exp(-gamma_HD/PsN0(p))*gammainc(4*pi/(16 - pi^2)*sqrt(gamma_HD/PsN0(p)),L*pi^2/(16 - pi^2),'upper');
    Pout_FD_ana(p) = 1 - exp(-gamma_FD/PsN0(p))*gammainc(4*pi/(16 - pi^2)*sqrt(gamma_FD*(1+sigma_SI^2*PsN0(p))/PsN0(p)),L*pi^2/(16 - pi^2),'upper');
    f_0 = @(z) sqrt(PsN0(p)).*z.^(PsN0(p)*beta1/2)./sqrt(-log(z)).*gammainc(4*pi./(16 - pi^2).*sqrt(-log(z)),L*pi^2/(16 - pi^2),'upper');
    SEP_HD_ana(p) = alpha1/2 - alpha1*sqrt(beta1)/2/sqrt(2*pi)*integral(f_0,0,1);
    f_2 = @(z) sqrt(PsN0(p)).*z.^(PsN0(p)*beta1/2)./sqrt(-log(z)).*gammainc(4*pi./(16 - pi^2).*sqrt((1+PsN0(p)*sigma_SI^2).*(-log(z))),L*pi^2/(16 - pi^2),'upper');
    SEP_FD_ana(p) = alpha1/2 - alpha1*sqrt(beta1)/2/sqrt(2*pi)*integral(f_2,0,1);
    temp3 = sqrt(1-ii.^2)*PsN0(p)./(1+PsN0(p).*log(2./(1+ii))).*gammainc(4*pi/(16-pi^2)*sqrt(log(2./(1+ii))),L*pi^2/(16-pi^2),'upper');
    cap_HD_ana(p) = pi/(4*M*log(2))*sum(temp3);
    temp4 = sqrt(1-ii.^2)*PsN0(p)./(1+PsN0(p).*log(2./(1+ii))).*gammainc(4*pi/(16-pi^2)*sqrt((1+sigma_SI^2*PsN0(p))*log(2./(1+ii))),L*pi^2/(16-pi^2),'upper');
    cap_FD_ana(p) = pi/(2*M*log(2))*sum(temp4);
end
Pout_FD_ana2 = Pout_FD_ana;
Pout_FD_sim2 = Pout_FD_sim;
Pout_HD_ana2 = Pout_HD_ana;
Pout_HD_sim2 = Pout_HD_sim;
SEP_HD_ana2 = SEP_HD_ana;
SEP_FD_ana2 = SEP_FD_ana;
SEP_HD_sim2 = SEP_HD_sim;
```

```
SEP_FD_sim2 = SEP_FD_sim;
cap_HD_sim2 = cap_HD_sim;
cap_FD_sim2 = cap_FD_sim;
cap_HD_ana2 = cap_HD_ana;
cap_FD_ana2 = cap_FD_ana;

% Plot the output
figure(1)

h3 = semilogy(PsN0dB, Pout_HD_sim2,'r*','LineWidth',2.5);
hold on;
h4 = semilogy(PsN0dB, Pout_HD_ana2,'r:','LineWidth',2.5,'MarkerSize',7);
hold on;
h11 = semilogy(PsN0dB, Pout_FD_sim2,'b*','LineWidth',2.5);
hold on;
h12 = semilogy(PsN0dB, Pout_FD_ana2,'b-','LineWidth',2.5,'MarkerSize',7);
hold on;
grid on;
legend([h4,h3,h12,h11],'HDR-IRS (Analysis)',...
    'HDR-IRS (Simulation)',...
    'FDR-IRS (Analysis)',...
    'FDR-IRS (Simulation)');
xlabel('P_s/N_0 (dB)'),ylabel('Outage probability');
title('Outage probability of the IRS-assisted system versus the transmit SNR');
```

7.3.5 Simulation Results

This section presents the performance evaluation results of HDR-IRS and FDR-IRS systems using theorems and Monte Carlo simulations. Specifically, 10^6 Rayleigh-distributed channel gain samples are generated using MATLAB software, where the channels from S to IRS, from IRS to R, and from R to D are independent. Besides comparing the performance of the two proposed systems, the impact of system parameters such as the level of RSI, the number of reflecting elements, and the transmit power of S and R is also investigated. In all simulation scenarios, we

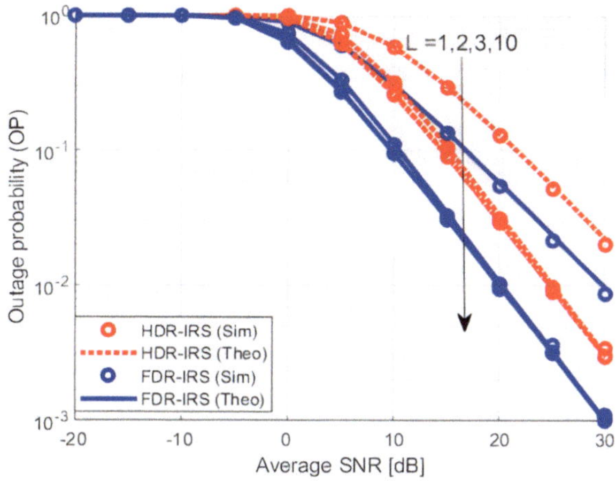

Fig. 7.17 OP of HDR-IRS and FDR-IRS systems as a function of average transmit SNR with L = 1, 2, 3, 10; $\sigma_{SI}^2 = 0.01$ and $R = 1$ bpcu

set $P_S = P_R = P$; $\sigma_R^2 = \sigma_D^2 = \sigma^2$; $M = 50$ and the average transmit SNR is defined as $SNR \triangleq \frac{P}{\sigma^2}$. The channel capacity simulation algorithm is presented in Fig. 3.19 (Sect. 3.6.3).

Figure 7.17 illustrates the OP of HDR-IRS and FDR-IRS systems as a function of average transmit SNR for $L = 1, 2, 3$ with $\Omega_{RR} = 0.01$. In the case of $L = 2, 3$ and 10, the OP of the FDR-IRS system is lower than that of the HDR-IRS system. However, in the case of $L = 1$, the OP of the FDR-IRS system is better when $SNR < 25dB$ and worse when $SNR > 25dB$ compared to the HDR-IRS system.

These results can be explained as follows. Since the IRS is used to support the S-R link, not the R-D link, increasing L only helps increase the channel gain of S-R. As a result, when $L = 1$, the channel gain of SR is lower than that of R-D. In this situation, the impact of RSI is less severe at low SNR because the power of RSI is $\sigma_{SI}^2 P_R$. Conversely, the impact of RSI will be stronger at high SNR; thus, the OP (OP) of the FDR-IRS system will be higher than that of the HDR-IRS system. When $L = 2, 3$, or 10, the OP of the FDR-IRS system is always lower than that of the HDR-IRS system because the S-R link is significantly improved and RSI is well mitigated. Furthermore, there is almost no gap between the OPs in the cases of $L = 3$ and $L = 10$, for both HDR-IRS and FDR-IRS systems. This is because in these cases, ρ_D^{HD*} and ρ_D^{FD*} are significantly higher than ρ_D^{HD} and ρ_D^{FD}. Therefore, the OPs only depend on ρ_D^{HD} and ρ_D^{FD}. As a result, for low RSI levels, for example $\sigma_{SI}^2 = 0.01$, using L=3 elements in the intelligent reflecting surfaceIRS is sufficient to completely mitigate the impact of RSI and to reduce the complexity in real-world deployment (significantly reducing the weight and size of the intelligent reflecting surfaceIRS) compared to the case of $L \geq 4$.

In Fig. 7.18, the symbol error probabilities (SEP) of HDR-IRS and FDR-IRS systems are plotted against the average transmit SNR for $\sigma_{SI}^2 = 0.01$, $L = 1, 2, 3$, using 4-QAM modulation, i.e., $(\alpha, \beta) = (2, 1)$. The SEP of HDR-IRS and FDR-IRS systems on Rician channels with a factor $K = 2$ are also provided to clearly

Fig. 7.18 SEP of HDR-IRS and FDR-IRS systems are plotted against the average transmit SNR with $\sigma_{SI}^2 = 0.01$, L = 1, 2, 3, using 4-QAM modulation

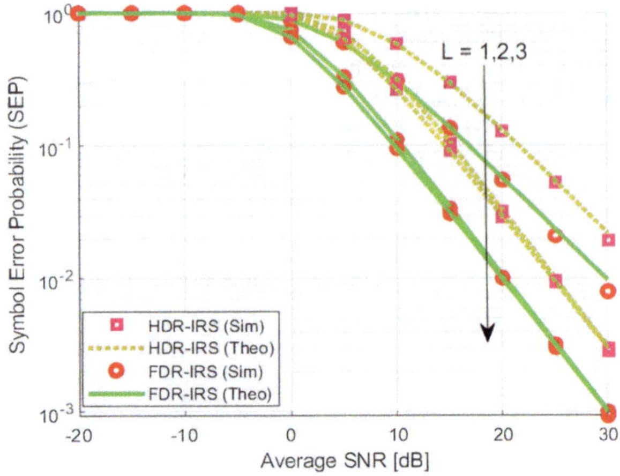

demonstrate the performance degradation due to Rayleigh channels. Unlike the Outage Probabilities (OP) in Fig. 7.17, the SEP of HDR-IRS systems is always equal to or lower than that of FDR-IRS systems. An important observation is that the difference between HDR-IRS and FDR-IRS systems is significant when $L = 1$ and $SNR > 20db$. Specifically, with $L = 1$ and $SNR = 30db$, the SEP of the HDR-IRS system is $7X10^{-3}$, and that of the FDR-IRS system is $4.4X10^{-2}$. This result demonstrates the severe impact of RSI on the SEP in the FDR-IRS system at high SNR ratios. On the other hand, the impact of RSI with $\sigma_{SI}^2 = 0.01$ on FDR-IRS can be ignored when $L \geq 3$.

The impact of RSI on the SEP of FDR-IRS systems for $L = 1, 2, 3, 4, 5, 6$ and 7 is illustrated in Fig. 7.19. Systems with higher L can withstand higher levels of RSI. Specifically, the SEP increases by 20, 10, and 1.6 times for $L = 1, 2$ and 3, respectively, when σ_{SI}^2 increases from 0 to 0.05. However, the SEP almost does not change for $L = 4, 5, 6$ and 7 when σ_{SI}^2 increases from 0 to 0.05. The SEP values in the cases of $L = 4, 5$ and 6 are only affected by RSI when $\sigma_{SI}^2 > 0.1, 0.3$ and 0.5, respectively. These results once again confirm the benefits of IRS in FDR-IRS systems. When all available techniques cannot completely eliminate SI due to FD transmission, using IRS can completely address the RSI issue. Specifically, in the case of intelligent reflecting surface IRS with $L \geq 7RE$, we can use FDR without the need for SIC instead of HDR to achieve higher spectral efficiency while maintaining the SEP at the same level.

Figure 7.20 illustrates the ergodic capacities (EC) of HDR-IRS and FDR-IRS systems plotted against the average transmit SNR with $L = 1, 2, 3, 4, 5$ and $\sigma_{SI}^2 = 0.01$. When $SNR < 0dB$, the EC of HDR-IRS and FDR-IRS systems are approximately equal. However, when $SNR > 0dB$, the EC of FDR-IRS systems is always higher than that of HDR-IRS systems. As a result, for a certain level of RSI, e.g., $\sigma_{SI}^2 = 0.01$,

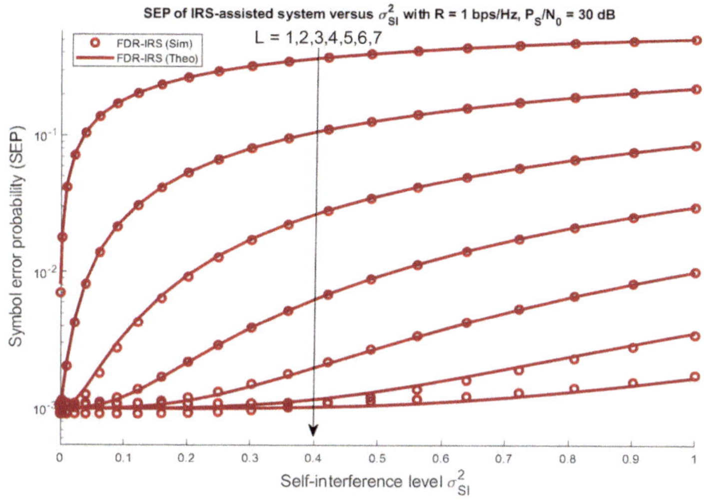

Fig. 7.19 Influence of SI on the SEP of FDR-IRS systems for L = 1, 2, 3, 4, 5, 6, and 7, using 4-QAM modulation, with SNR = 30 dB

Fig. 7.20 Ergodic capacities (EC) of HDR-IRS and FDR-IRS systems compared to the average transmit SNR for L = 1, 2, 3, 4, 5, $\sigma^2_{SI} = 0.01$

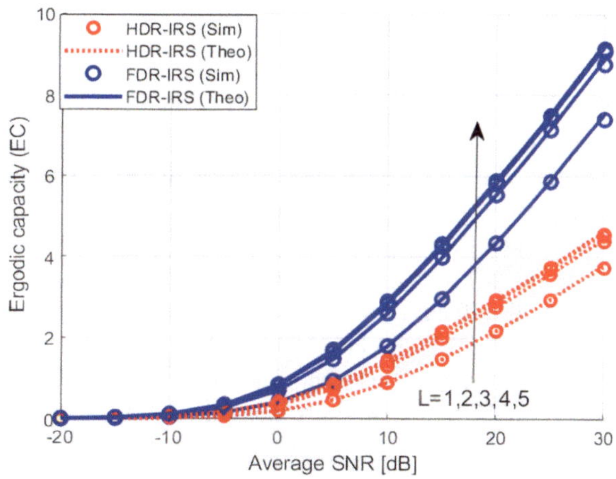

we can use $L = 3$ to achieve the lowest OP and SEP, but we need to use $L \geq 5$ to achieve the highest EC in FDR-IRS systems. This attribute indicates that RSI has different impacts on the OP, SEP, and EC metrics of FDR-IRS.

7.4 Optimal Allocation Solution for Power or Energy in EH Networks

This section presents an optimal power allocation solution in Wireless Power Transfer Networks (WPTNs) [8], where Hybrid Relay Nodes (HRNs) transmit wireless power to energy-harvesting users and relay their data to the base station using either AF or DF techniques. The authors solve the energy and time optimization problem for both AF and DF to maximize the energy utilization efficiency of HRNs. For WPTNs employing DF, the optimization problem is divided into three sub-problems: optimizing the time allocation for user data transmission, optimizing the time allocation for information transmission, and optimizing the power allocation of HRNs. For WPTNs using AF, the authors propose an iterative algorithm to optimize the transmission power and energy transmission time of HRNs. Simulation results confirm that the proposed algorithms can achieve energy efficiency improvements of approximately 300% and 150%, respectively, compared to existing AF and DF relay WPTNs.

7.4.1 Introduction

Wireless Energy Transfer (WET) is an efficient solution to extend the lifespan of energy-constrained devices, such as low-power sensors and Internet of Things (IoT) endpoints, by providing energy through radio frequency (RF) energy signals. Wireless Power Transfer Networks (WPCNs) enable the provision of energy to multiple devices and the reception of their information simultaneously using a Hybrid Access Point (HAP) to transmit RF energy signals to devices in the downlink and receive information from these devices using previously harvested energy.

Recently, WPCNs have garnered significant attention from both the academic and business communities. One limitation of traditional Wireless Power Transfer Networks (WPCNs) is the short transmission distance due to signal attenuation and poor signal quality for both uplink data transmission and downlink energy transmission. Recently, relay nodes have been introduced into WPCNs to extend coverage. There are two types of relay nodes in WPCNs: one type has no dedicated power source and only harvests energy from the Hybrid Access Point (HAP), while the other type is the Hybrid Relay Node (HRN), which integrates the functionality of HAP into the WPCN.

This research selects the WPCN solution using relays for simulation. It's worth noting that most previous results of WPCNs using relay nodes only consider a three-node system consisting of a source, a relay node, and a destination, or include an HRN and multiple source-destination pairs. However, the architecture of the WPCN supporting relay proposed by [8] includes an HRN, a BS station, and multiple users. Therefore, further research is needed to optimize the time allocation for energy and information transmission between users and HRN, as well as the information transmission between HRN and BS.

On the other hand, energy efficiency (EE) also needs to be considered in WPCNs because the use of RF energy during the energy transmission phase can significantly reduce the EE of WPCNs. Generally, network lifespan is improved for energy-constrained devices such as IoT endpoints. For WPCNs with relay support, the network continues to operate as long as the HRN has sufficient energy. Therefore, improving the energy efficiency of HRN is a further issue that needs to be addressed.

The solution proposed by [8] offers the following key contributions:

1. They proposed a WPCN model supporting relay, in which an HRN broadcasts RF energy to multiple users and forwards user data to the BS. The EE maximization problem is solved for both DF and AF relay schemes. For AF systems, optimizing the time for HRN-BS information transmission is unnecessary since the time for information transfer from users to HRN and from HRN to BS is equal.
2. For DF relay systems, the authors derive the optimal transmit power of the HRN for energy transmission and allocate optimal time for users with the given transmit power of the HRN for information transmission. Then, the EE function is transformed into an approximate function of the HRN's information transmit power with optimal time allocation and energy transmit power of the HRN. The optimal values can be quickly obtained using conventional search methods.
3. The optimal time for forwarding user information to the HRN and HRN to the BS is explicitly established for AF relay systems. Then, the optimization problem is solved by converting the original objective function into an approximate function in each dimension separately to obtain the optimal solution in each dimension. An iterative algorithm is proposed to achieve the optimal energy transmission time of the HRN and allocate optimal energy. Simulation results show that the proposed iterative algorithm converges quickly.

7.4.2 System Model

WPCN consists of one HRN and K users, each equipped with a single antenna, as shown in Fig. 7.21a. The HRN broadcasts energy signals in the initial time slot τ_0. Subsequently, all users transmit their own information to the HRN using the TDMA (Time Division Multiple Access) protocol, and the HRN forwards the received signals to the base station. Both DF and AF relay schemes are considered in the model.

During the energy transmission phase, the HRN broadcasts energy signals with a transmission power P_d. The harvested energy of user U_k is given by:

$$E_k = \eta_k P_d g_k \tau_k, k = 1, ..., K, \tag{7.49}$$

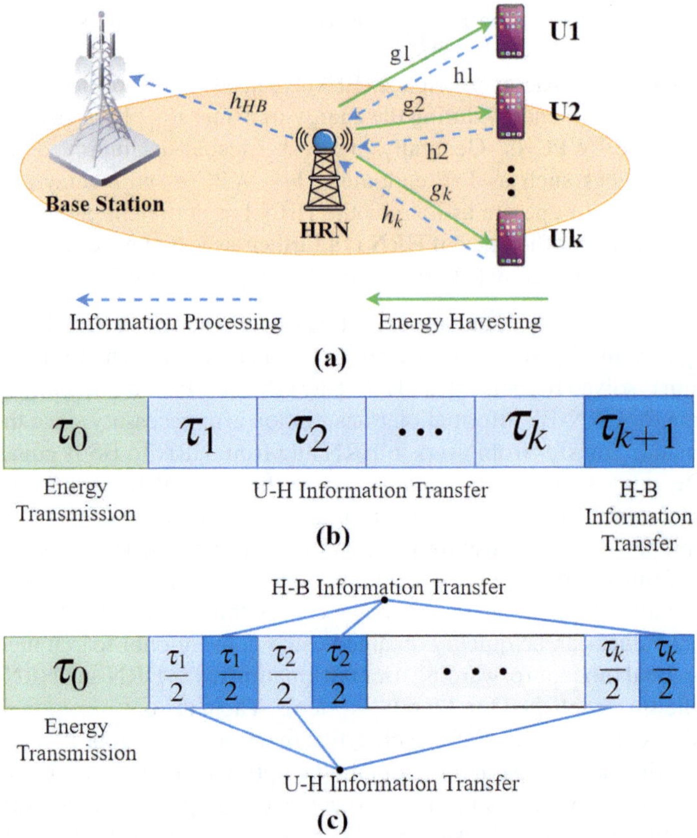

Fig. 7.21 WPCN system model. **a** System model. **b** System frame structure of DF relay. **c** System frame structure of AF relay

where g_k represents the channel gain from the HRN to the kth user (U_k), and η_k is the energy conversion efficiency at U_k.

7.4.2.1 WPCN with DF Relaying

We assume that users only use energy harvested in the current slot; thus, the maximum transmission power for U_k during the information transmission phase of WPCN with DF relaying is written as

$$P_k^{DF} = \frac{E_k}{\tau_k^{DF}} = \frac{\eta_k P_d^{DF} g_k \tau_0^{DF}}{\tau_k^{DF}}, k = 1, ..., K, \tag{7.50}$$

where τ_k^{DF} is the time U_k spends transmitting information to HRN with DF relaying, P_d^{DF} is the energy transmission power of HRN in the DF relay system, and τ_0^{DF} is the energy transmission time in the DF relay system.

The achievable capacity of U_k ($k = 1, 2, ..., K$) in the first phase (from UE to HRN) is determined by

$$R_k^{DF} = \tau_k^{DF} log_2 \left(1 + \frac{h_k P_k^{DF}}{\sigma^2}\right) = \tau_k^{DF} log_2 \left(1 + \frac{g_k \eta_k P_d^{DF} \tau_0}{\tau_k^{DF} \sigma^2}\right)$$
$$= \tau_k^{DF} log_2 \left(1 + \frac{\gamma_k P_d^{DF} \tau_0^{DF}}{\tau_k^{DF}}\right) \tag{7.51}$$

where $\gamma_k = \frac{{}_k h_k g_k}{\sigma^2}$, k = 1,...,K; h_k is the channel gain from U_k to HRN, σ^2 is the noise power.

In the last time slot τ_{K+1}^{DF}, HRN decodes the previously received signals from users and then forwards the decoded signals to the BS. The corresponding channel capacity is

$$R_{H-B}^{DF} = \tau_{K+1}^{DF} log_2 \left(1 + \frac{P_u^{DF} h_{hb}}{\sigma^2}\right), \tag{7.52}$$

where P_u^{DF} is the transmission (relay) power of HRN with DF relaying, h_{hb} is the channel gain from HRN to BS.

Therefore, the total supported user information capacity by HRN is

$$R_{sum}^{DF} = min \left(\sum_{k=1}^{K} R_k^{DF}, R_{H-B}^{DF}\right). \tag{7.53}$$

7.4.2.2 WPCN with AF Relaying

When AF is used in HRN, the transmission time from the user to the relay and from the relay to HRN must be equal. Therefore, for DF relaying, the total number of time slots allocated is 2K+1, as shown in Fig. 7.21c. All users transmit their information to HRN after HRN relays the information of the previous user within the time duration of $\frac{\tau_k^{AF}}{2}$, $\forall 1 \le k \le K$. Thus, the maximum transmission power of U_k with AF relaying is

$$P_k^{AF} = \frac{2E_k}{\tau_k^{AF}} = \frac{2_k P_d^{AF} g_k \tau_0^{AF}}{\tau_k^{AF}}, k = 1, ..., K. \tag{7.54}$$

where τ_k^{AF} is the time Uk transmits information to BS via HRN with AF relaying, τ_0^{AF} is the energy transmission time in the AF relay system, P_d^{AF} is the transmission (relay) power of HRN with AF relaying.

The signal-to-noise ratio (SNR) between U_k and BS is

$$SNR_k = \frac{\frac{P_k^{AF} h_k}{\sigma^2} \frac{P_u^{AF} h_{hb}}{\sigma^2}}{\frac{P_k^{AF} h_k}{\sigma^2} + \frac{P_u^{AF} h_{hb}}{\sigma^2} + 1}, k = 1, 2, ..., K. \tag{7.55}$$

where P_u^{AF} is the transmission (relay) power of HRN with AF relaying.

From this, we can derive the capacity of the kth user and the total user capacity for AF relaying as

$$R_k^{AF} = \frac{\tau_k^{AF}}{2} log_2 \left(1 + \frac{\frac{P_k^{AF} h_k}{\sigma^2} \frac{P_u^{AF} h_{hb}}{\sigma^2}}{\frac{P_k^{AF} h_k}{\sigma^2} + \frac{P_u^{AF} h_{hb}}{\sigma^2} + 1}\right)$$

$$= \frac{\tau_k^{AF}}{2} log_2 \left(1 + \frac{2\gamma_k P_d^{AF} \tau_0^{AF} P_u^{AF} h_{hb}}{(2\gamma_k P_d^{AF} \tau_0^{AF} \sigma^2 + P_u^{AF} h_{hb} \tau_k^{AF} + \tau_k^{AF} \sigma^2)}\right), k = 1, 2, ..., K \tag{7.56}$$

$$R_{sum}^{AF} = \sum_{k=1}^{K} R_k^{AF} \tag{7.57}$$

7.4.2.3 Optimization Problem for Energy Efficiency Utilization

For WPCN supporting DF relaying, the energy consumption level of HRN is $E_H^{DF} = \tau_0 P_d^{DF} + \tau_{K+1} P_u^{DF} + E_0^{DF}$, where E_0^{DF} is the energy consumed by the processing unit of HRN with DF relaying. The optimization problem is set up as

$$\begin{aligned} OP_1^{DF} : & \max_{\tau_k^{DF}, P_d^{DF}, P_u^{DF}} \frac{min\left(\sum_{i=1}^{K} R_k^{DF}, R_{H-B}^{DF}\right)}{\tau_0^{DF} P_d^{DF} + \tau_{K+1}^{DF} P_u^{DF} + E_0 DF} \\ s.t. \ & C_1^{DF} : \tau_k^{DF} \geq 0, k = 0, 1, ..., K, K+1, \\ & C_2^{DF} : \sum_{k=0}^{K+1} \tau_k^{DF} \leq 1, \\ & C_3^{DF} : 0 \leq P_d^{DF}, P_u^{DF} \leq P_A, \end{aligned} \tag{7.58}$$

where C_1^{DF} and C_2^{DF} are constraints on time, and C_3^{DF} represents the constraint on the maximum transmission power of HRN.

For WPCN supporting AF relaying, the energy consumption level of HRN is $E_H^{AF} = \tau_0 P_d^{DF} + \frac{1 - \tau_0}{2} P_u^{AF} + E_0^{AF}$, where E_0^{AF} is the energy consumed by the processing unit of HRN with AF relaying. The optimization problem is defined by

$$
\begin{aligned}
OP_1^{AF}: &\max_{\tau_k^{AF}, P_d^{AF}, P_u^{AF}} \frac{2\sum_{i=1}^{K} R_k^{AF}}{2\tau_0^{AF} P_d^{AF} + (1 - \tau_0^{AF} P_u^{AF} + 2E_0^{AF})} \\
s.t.\ & C_1^{AF}: \tau_k^{AF} \geq 0, k = 0, 1, ..., K, \\
& C_2^{AF}: \sum_{k=0}^{K} \tau_k^{AF} \leq 1, \\
& C_3^{AF}: 0 \leq P_d^{AF}, P_u^{AF} \leq P_A.
\end{aligned} \tag{7.59}
$$

7.4.3 Algorithm Design for WPCN with DF Relaying

The optimization problem OP_1^{DF} is decomposed into three sub-problems: optimal time allocation for user transmission, time allocation for energy and information transmission of HRN, and power allocation of HRN as follows.

Lemma 7.6 *The optimal time allocation for communication of user τ_i^{DF*}, $1 \leq i \leq K$ must satisfy the condition:*

$$
\frac{\gamma_1}{\tau_1^{DF*}} = ... = \frac{\gamma_k}{\tau_k^{DF*}} = ... = \frac{\gamma_K}{\tau_K^{DF*}}, 1 \leq k \leq K, \tag{7.60}
$$

where τ_0^{DF}, τ_{K+1}^{DF} and P_u^{DF} is given beforehand.

Lemma 7.7 *Optimal time allocation conditions τ_0^{DF*}, τ_U^{DF*}, and τ_{K+1}^{DF*}, and allocation of capacity P_d^{DF*} and P_u^{DF*} regarding the optimization problem OP must satisfy:*

$$
\tau_U^{DF*} log_2\left(1 + \frac{AP_d^{DF*}\tau_0^{DF*}}{\tau_U^{DF*}}\right) = \tau_{K+1}^{DF*} log_2\left(1 + \frac{P_u^{DF*} h_{hb}}{\sigma^2}\right). \tag{7.61}
$$

Lemma 7.8 *The optimal energy transmission power of HRN is the maximum available transmission power, i.e.,*

$$
P_d^{DF*} = P_A. \tag{7.62}
$$

7.4.4 Algorithm Design for WPCN with AF Relaying

The iterative algorithm proposed to determine the shared time slots and power allocation of WPCN supporting AF relaying for the OP_1^{AF} problem is as follows.

Lemma 7.9 *The optimal allocation of communication time τ_i^{AF*}, $1 \leq i \leq K$, must satisfy*

$$\frac{\varepsilon_1}{\tau_1^{AF*}} = ... = \frac{\varepsilon_k}{\tau_k^{AF*}} = ... = \frac{\varepsilon_K}{\tau_K^{AF*}}, 1 \leq k \leq K, \tag{7.63}$$

In this case, τ_0^{AF}, P_d^{AF}, *and* P_u^{AF} *are predetermined.*

Lemma 7.10 *Given* $\kappa = \frac{\tau_0^{AF}}{1-\tau_0^{AF}}$ *and by defining the function*

$$F(\kappa, P_d^{AF}, P_u^{AF}) \triangleq \frac{log_2\left(1 + \frac{2AP_d^{AF}\kappa P_u^{AF}h_{hb}}{(2AP_d^{AF}\kappa\sigma^2 + P_u^{AF}h_{hb} + \sigma^2)}\right)}{2\kappa P_d^{AF} + P_u^{AF} + 2E_0^{AF}(1+\kappa)}. \tag{7.64}$$

Then $F(\kappa, P_d^{AF}, P_u^{AF}$ *is a quasi-concave function in each dimension, i.e.,* κ, P_d^{AF}, P_u^{AF}, *respectively.*

7.4.5 MATLAB Coding

The following MATLAB programs are developed to implement the above optimization algorithms. Due to the large number of simulation scenarios, we only present the simulation algorithm for the proposed time duration and power allocation (PTDPA) applied for AF relaying strategy. Figure 7.22 shows the convergence of this algorithm for 4 different channel realization samples.

```
function [EE1,EE2,EE3,EE4] = EE_opt(K,P_A,
    max_ite,E0,eta0,d_ub,dmin,dmax,alpha1,
    sigma_sq)

% Input parameters
tau_0 = 0.2;
tau_k = 0.6/K;
tau_K_1 = 0.2;
kappa1 = tau_0/(1-tau_0);
epsi = 10^(-3);
P_u_DF_min = 0;
P_u_DF_max = P_A;
P_u_AF = P_A;
P_d_AF = P_A;
ite = 0;

E_0_DF = E0; %
```

```
E_0_AF = E0;
eta_k = eta0*ones(1,K);

d0 = 1;
d = dmin + (dmax-dmin)*rand(1,K); % Distance
    from HRN and users (uniformly distributed in
    [5,10] (m))
G_hb = 10^(-3)*(d_ub./d0)^(-alpha1);
G = 10^(-3)*(d./d0).^(-alpha1);

EE = zeros(1,length(sigma_sq));
for p = 1:length(sigma_sq)
    EE(p) = OTDPA_DF(K,P_A,E_0_DF,eta_k,G_hb,G,
        sigma_sq(p));
end

EE1 = EE;

for p = 1:length(sigma_sq)
    EE(p) = OTDA_DF(K,P_A,E_0_DF,eta_k,G_hb,G,
        sigma_sq(p));
end
EE2 = EE;


for p = 1:length(sigma_sq)
    EE(p) = PTDPA_AF(K,P_A,E_0_AF,eta_k,G_hb,G,
        sigma_sq(p));
end

EE3 = EE;

for p = 1:length(sigma_sq)
    EE(p) = OTDA_AF(K,P_A,E_0_AF,eta_k,G_hb,G,
        sigma_sq(p));
end

EE4 = EE;
end
```

```
function EE_out = PTDPA_AF(K,P_A,E0,etak,Ghb,Gk
    ,sigma_square)
%UNTITLED17 Summary of this function goes here
%   Detailed explanation goes here

tau_k = 0.6/K;
tau_K_1 = 0.2;
%kappa1 = tau_0/(1-tau_0);

epsi = 10^(-3);
P_u_AF = P_A;
P_d_AF = P_A;
ite = 0;
max_ite = 1000;


n = 1;
tau_0_AF = 0.2; % Initialize the value of \
    tau_0
tau_k_AF = tau_k*ones(1,K);
EE_old = 0;
EE_new = EE_AF(P_u_AF,P_d_AF,tau_0_AF,E0,
    tau_k_AF,Ghb,Gk,etak,sigma_square);
%kappa1 = tau_0_AF/(1 - tau_0_AF);
while (abs(EE_new - EE_old) >= epsi) && (n <
    max_ite)
     n = n + 1;
     gamma_k = etak.*Gk.^2./sigma_square;
     A = sum(gamma_k);

     % Update kappa(n) with Pd(n-1) and Pu(n-1)
     kappa1 = opt_kappa(A,P_d_AF,P_u_AF,Ghb,
         sigma_square,E0);
     % Update Pd(n) with kappa(n) and Pu(n-1)
     P_d_AF = opt_Pd(A,kappa1,P_u_AF,Ghb,
         sigma_square,E0,P_A);
     % Update Pu(n) with kappa(n) and Pd(n)
     P_u_AF = opt_Pu(A,kappa1,P_d_AF,Ghb,
         sigma_square,E0,P_A);
     EE_old = EE_new;
     tau_0_AF = kappa1/(kappa1 + 1);
     tau_k_AF = (1 - tau_0_AF).*gamma_k./A;
```

```
    EE_new = EE_AF(P_u_AF,P_d_AF,tau_0_AF,E0,tau_k_AF,Ghb,Gk,etak,sigma_square);
end

EE_out = EE_new;

end

function out = EE_AF(Pu,Pd,tau0,E0,tauk,G,Gk,etak,sigma1)
%UNTITLED11 Summary of this function goes here
%   Detailed explanation goes here
K = length(tauk);
if (length(Gk) ~= K) || (length(etak) ~= K)
    errormsg('Length of \tau_k, G_k, and \eta_k must be equal!');
end

gammak = etak.*Gk.*Gk./sigma1;
num = sum(tauk.*log(1 + 2*gammak.*tau0.*Pd.*Pu.*G./(2*gammak.*Pd.*tau0.*sigma1 + Pu.*G.*tauk + tauk.*sigma1))/log(2));
den = 2*tau0*Pd + (1 - tau0)*Pu + 2*E0;
out = num./den;
end

function Pu_star = opt_Pu(A0,kappa1,Pd,Ghb,sigma1,E0,PA)
max_ite = 1000;
ite = 0;
epsi = 10^(-4);
Pu_min = 0;
Pu_max = PA;
while (abs(Pu_max - Pu_min) >= epsi) && (ite < max_ite)
        Pu_l = Pu_max - 0.618*(Pu_max - Pu_min);
        Pu_r = Pu_min + 0.618*(Pu_max - Pu_min);
        F_l = log(1 + 2*Pd*Pu_l*A0*Ghb*kappa1/(2*A0*Pd*sigma1*kappa1 + Pu_l*Ghb + sigma1))/log(2)/...
    (2*Pd*kappa1 + Pu_l + 2*E0*(1 + kappa1));
```

```
        F_r = log(1 + 2*Pd*Pu_r*A0*Ghb*kappa1/(2*A0*Pd*sigma1*kappa1 + Pu_r*Ghb + sigma1))/log(2)/...
    (2*Pd*kappa1 + Pu_r + 2*E0*(1 + kappa1));
        if F_l <= F_r
            Pu_min = Pu_l;
        else
            Pu_max = Pu_r;
        end
        ite = ite + 1;
end
Pu_star = (Pu_max + Pu_min)/2;
end

function Pd_star = opt_Pd(A0,kappa1,Pu,Ghb,sigma1,E0,PA)
max_ite = 1000;
ite = 0;
epsi = 10^(-4);
Pd_min = 0;
Pd_max = PA;
while (abs(Pd_max - Pd_min) >= epsi) && (ite < max_ite)
        Pd_l = Pd_max - 0.618*(Pd_max - Pd_min);
        Pd_r = Pd_min + 0.618*(Pd_max - Pd_min);
        F_l = log(1 + 2*Pd_l*Pu*A0*Ghb*kappa1/(2*A0*Pd_l*sigma1*kappa1 + Pu*Ghb + sigma1))/log(2)/...
    (2*Pd_l*kappa1 + Pu + 2*E0*(1 + kappa1));
        F_r = log(1 + 2*Pd_r*Pu*A0*Ghb*kappa1/(2*A0*Pd_r*sigma1*kappa1 + Pu*Ghb + sigma1))/log(2)/...
    (2*Pd_r*kappa1 + Pu + 2*E0*(1 + kappa1));
        if F_l <= F_r
            Pd_min = Pd_l;
        else
            Pd_max = Pd_r;
        end
        ite = ite + 1;
end
Pd_star = (Pd_max + Pd_min)/2;
```

```
end

function kappa_star = opt_kappa(A0,Pd,Pu,Ghb,sigma1,E0)
max_ite = 1000;
ite = 0;
epsi = 10^(-4);
kappa_min = 0;
kappa_max = 10^4;
while (abs(kappa_max - kappa_min) >= epsi) && (ite < max_ite)
        kappa_l = kappa_max - 0.618*(kappa_max - kappa_min);
        kappa_r = kappa_min + 0.618*(kappa_max - kappa_min);
        F_l = log(1 + 2*Pd*Pu*A0*Ghb*kappa_l/(2*A0*Pd*sigma1*kappa_l + Pu*Ghb + sigma1))/log(2)/...
    (2*Pd*kappa_l + Pu + 2*E0*(1 + kappa_l));
        F_r = log(1 + 2*Pd*Pu*A0*Ghb*kappa_r/(2*A0*Pd*sigma1*kappa_r + Pu*Ghb + sigma1))/log(2)/...
    (2*Pd*kappa_r + Pu + 2*E0*(1 + kappa_r));
        if F_l <= F_r
            kappa_min = kappa_l;
        else
            kappa_max = kappa_r;
        end
        ite = ite + 1;
end
kappa_star = (kappa_max + kappa_min)/2;
end
```

7.4.6 Simulation Results

In this section, the book provides detailed simulation results on the analysis of average energy utilization performance. For WPCN supporting DF relay, the energy efficiency of the Optimal Time and Power Allocation (OTDPA) algorithm is compared with Optimal Time Allocation (OTDA) while maintaining fixed power and information ($P_d^{DF} = P_u^{DF} = P_A$). The energy efficiency of the Proposed Time and Energy

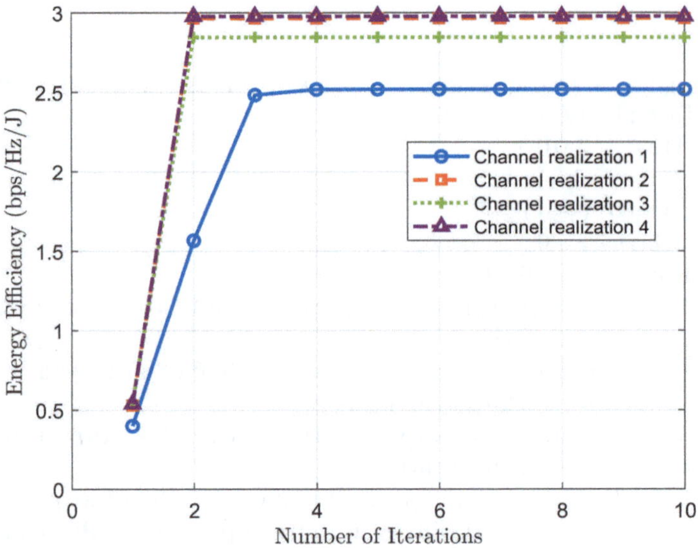

Fig. 7.22 Convergence of the iterative algorithm for WPCN supporting AF relay

Allocation (PTDPA) algorithm is also compared with OTDA with $P_d^{AF} = P_u^{AF} = P_A$ for WPCN supporting AF relay. The optimal energy transmission time T can be obtained by the Golden Section Search algorithm, and the subsequent elements are determined from Lemma 7.9, i.e., $\tau_k^{AF}, 1 \leq k \leq K$.

In the simulations, the maximum transmit power of the HRN is chosen as $P_A = 10W$. The energy consumption for signal processing at the HRN is $E_0^{DF} = 1J$ and $E_0^{AF} = 0.5J$ for the DF relay node and AF relay node, respectively, as the AF relay node does not need to decode the received signals from the UEs. All fading channels are Rayleigh, and the power gain of the channel is $G = 10^{-3}\left(\frac{d_i}{d_0}\right)^{-\alpha}$, where d represents the distance between two nodes, with the distance between HRN and UEs considered to be between 5 m to 10 m and the distance between BS and HRN set to 500 m, d_0 is the standard distance of $1m$, $\alpha = 3$ is the path-loss exponent. To maintain generality, the energy conversion efficiency is set by $\eta_k = 1, \forall k$.

Figure 7.23 compares the energy utilization efficiency corresponding to different algorithms plotted against noise power. It demonstrates that the proposed optimization algorithms exhibit a significant performance improvement compared to the reference schemes for both AF and DF relay systems. For instance, the performance improvement of the proposed algorithms is 297% and 153%, respectively, compared to OTDA schemes for AF and DF relay nodes when the noise power is –100 dBm.

Figure 7.24 illustrates the EE as a function of the maximum transmit power P_A with a noise power of –100 dBm. We observe that the OTDA schemes exhibit energy efficiency similar to the proposed optimization algorithms for both AF and DF relay systems when the maximum transmit power of the HRN is low. This is because

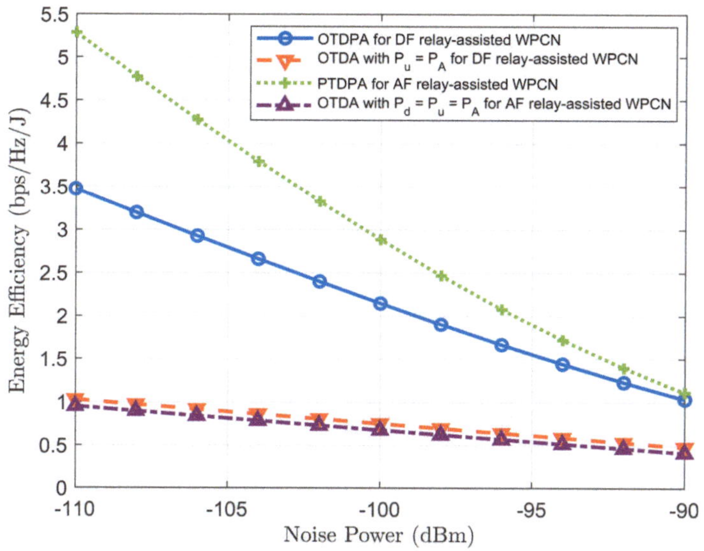

Fig. 7.23 Energy utilization efficiency plotted against noise power

when the transmit power is low, the optimal transmit power of the HRN is the maximum transmit power, i.e., $P_u^* = P_A$, and the optimal EE can be achieved by optimizing the time allocation. As P_A increases, the energy consumption of the OTDA scheme increases linearly, and the network throughput ratio increases logarithmically, resulting in decreased energy efficiency at high P_A.

7.5 Overall Evaluation of Energy-Saving Solutions

In general, 5G/6G networks require more energy due to increased network density and higher network usage and access traffic. The energy saving solutions of 5G networks are far superior to 4G, allowing for energy savings of over 90% compared to 4G. This chapter has introduced some of the most advanced solutions to improve energy efficiency and has developed a Monte Carlo simulation tool to validate the theory, serving as a guide for designing energy saving networks in practice. Specifically, for EH solutions, this chapter focuses on analyzing OP for various protocols with hardware impairment. For energy reuse solutions, this book has successfully simulated a method combining relay selection with energy reuse, applied to channels with non-uniform distribution, achieving high efficiency. This chapter also analyzes and simulates solutions to improve the performance of HD and FD relay networks using intelligent reflectors. Lastly, the optimization problem of EE for hybrid relay systems in WPCN by allocating time and power reasonably between relay nodes and users is also presented in this chapter.

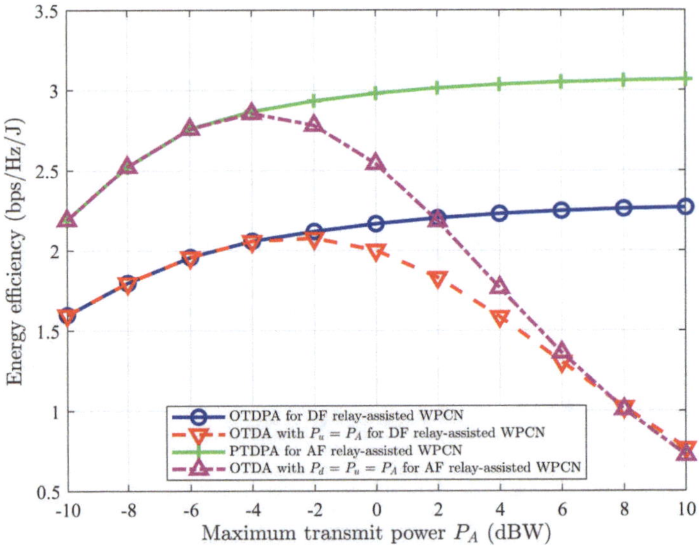

Fig. 7.24 Energy efficiency plotted against the maximum transmit power of the HRN

References

1. Varshney, L.R.: Transporting information and energy simultaneously. In: 2008 IEEE International Symposium on Information Theory, pp. 1612–1616. IEEE (2008)
2. Nasir, A.A., Zhou, X., Durrani, S., Kennedy, R.A.: Throughput and ergodic capacity of wireless energy harvesting based DF relaying network. In: 2014 IEEE International Conference on Communications (ICC), pp. 4066–4071. IEEE (2014)
3. Nguyen, T.N., Tran Tin, P., Ha, D.H., Voznak, M., Tran, P.T., Tran, M., Nguyen, T.L.: Hybrid TSR-PSR alternate energy harvesting relay network over Rician fading channels: outage probability and SER analysis. Sensors **18**(11), 3839 (2018)
4. Nguyen, T.N., Tran, P.T., Hoang, H.G., Nguyen, H.S., Voznak, M.: On the performance of decode-and-forward half-duplex relaying with time switching based energy harvesting in the condition of hardware impairment. In: Advances in Information and Communication Technology: Proceedings of the International Conference, ICTA 2016, pp. 421–430. Springer (2017)
5. Nguyen, T.N., Tran, P.T., Voznak, M., Behan, L.: Performance of time switching based energy harvesting for amplify-and-forward half-duplex relaying with hardware impairment. In: 2017 27th International Conference Radioelektronika (RADIOELEKTRONIKA), pp. 1–5. IEEE (2017)
6. Nguyen, T.N., Duy, T.T., Tran, P.T., Voznak, M., Li, X., Poor, H.V.: Partial and full relay selection algorithms for AF multi-relay full-duplex networks with self-energy recycling in non-identically distributed fading channels. IEEE Trans. Veh. Technol. **71**(6), 6173–6188 (2022). https://doi.org/10.1109/TVT.2022.3158340
7. Nguyen, B.C., Pham, T.Q., Thang, N.N., Hoang, T.M., Tran, P.T.: Improving the performance of wireless half-duplex and full-duplex relaying networks with intelligent reflecting surface. J. Frankl. Inst. **360**(4), 3095–3118 (2023)
8. Yang, F., Xu, W., Zhang, Z., Guo, L., Lin, J.: Energy efficiency maximization for relay-assisted WPCN: joint time duration and power allocation. IEEE Access **6**, 78297–78307 (2018)

Chapter 8
Physical Layer Security

8.1 Introduction

Secrecy in wireless communications remains a critical issue due to the inherent risk of information leakage associated with the broadcast nature of these systems, which leaves unprotected data vulnerable to interception. Traditional cryptographic methods have long been employed to enhance security; however, Wyner's pioneering work on wiretap channels [2], along with subsequent studies [3, 4], introduced an innovative approach known as physical layer security (PLS). This concept leverages randomness in coding or signaling at the physical layer to simultaneously achieve secrecy and reliability by confounding eavesdroppers. Over the past decade, PLS has attracted significant attention from researchers [5] due to its potential for improving communication quality and meeting the increasing demands of mobile users. The applications of PLS are diverse, spanning relay networks [6], cellular networks [7, 8], cognitive radio networks [9], IoT networks [10], and massive MIMO systems [11]. In multi-user environments, PLS can be further enhanced through node cooperation strategies such as cooperative relaying (CR) and cooperative jamming (CJ).

In CR schemes, relays not only facilitate message transmission from source to destination but also play a pivotal role in preventing the wiretap from eavesdroppers [6, 12, 13]. In particular, Wang et al. [6] proposed a generalized multi-relay selection scheme aimed at enhancing security in wireless relay networks by deriving a semi-closed-form expression for secrecy outage probability (SOP) and solving a joint optimization problem of power allocation and number of relays to minimize the SOP.

In recent years, the exploration of CJ schemes has introduced innovative methods for enhancing network security. These schemes utilize relays to forward messages while simultaneously transmitting signals that interfere with eavesdroppers. Techniques such as artificial noise (AN) and noise forwarding (NF) have been pivotal in this domain. The AN scheme, as demonstrated in studies like [14–17], involves generating Gaussian artificial noise to disrupt unauthorized listeners

P. T. Tran et al., *Developing MATLAB Comprehensive Simulation for 5G/6G Wireless Communications*, Signals and Communication Technology,
https://doi.org/10.1007/978-981-95-1117-4_8

effectively. Specifically, Ding et al.'s development of an AN-aided two-way opportunistic relay selection scheme [18] exemplifies the studies of securing two-way multi-relay networks. The NF scheme takes a different approach by employing helper nodes that transmit dummy codewords independent of source messages, yet reliably decodable at intended destinations. This introduces additional randomness into the observations made by eavesdroppers, thereby enhancing information security–a method initially proposed by [12, 19], and further optimized through covariance matrix adjustments across multiple-antenna signals from both sources and helpers to maximize secrecy rates within NF frameworks [20]. Lee and Khisti's investigation into secure degrees of freedom in Gaussian diamond-wiretap channels under rate-limited cooperation scenarios [21] highlights how interception concerns extend beyond mere transmissions to include inter-relay links themselves.

Recently, RF energy harvesting (EH) has been an emerging field due to its dual capability for supporting both energy transmission and information transfer—a concept first introduced by Varshney in [22]. Since then, advancements have occurred across three primary domains, namely wireless power transfer (WPT) [23], wireless powered communication networks (WPCN) [24], along with simultaneous wireless information and power transfer (SWIPT) [25]. Notable contributions include Nasir's work on RF EH protocols specifically designed for relay networks, including time-switching relaying (TSR) and power-splitting relaying (PSR) [26, 27]. For TSR, the receiver node switches between the information processing phase and EH phase. Differently, the PSR receiver splits the received signal into two parts using a power divider, one is used for information processing and the other is used for energy storage. In addition, a hybrid-time-switching and power-splitting (TPSR) protocol was introduced and analyzed by Tran et al. [28]. Recent studies have rigorously examined the performance of RF EH in different wireless systems, such as wireless sensor networks [29], multi-hop relay networks [30], multiple-antenna systems [25], cognitive radio networks [31], two-way relay networks [32], cellular networks [33], NOMA scheme [34], and especially in PLS scheme. In particular, Ngo et al. [35] presented a detailed analysis of a two-hop single-relay network utilizing the hybrid TPSR protocol amidst an eavesdropper's presence in proximity of the relay.

The two-way relay channel (TWRC), serving as a pivotal model for examining novel communication techniques, protocols, and algorithms—including wireless EH and PLS methods—was first introduced in Shannon's foundational work in [36]. Renewed interest in TWRC emerged during the early 2000s following a seminal work in [37], which catalyzed further research endeavors. Especially, two network coding schemes have been proposed for TWRC by Katti et al.: digital network coding (DNC) [38] and analog network coding (ANC) [39]. The former is based on the XOR-ed combination of two independently decoded packets from two sources at the relay and broadcasting it back to both sources, which effectively reduces the number of required transmission time slots by one. For the latter, the relay simultaneously receives both sources' signals and broadcasts their combined sum, which conserves one more transmission time slot than DNC.

Advancements in TWRC have spurred extensive investigations into PLS in spectrally efficient two-way relay networks, as evidenced by works such as those by

Shukla [8] and Pandey [7], among others [14–16, 40]. Specifically, Shukla [8] and Pandey [7] analyzed SOP of full-duplex multi-user two-way AF cellular relay networks where a multi-antenna BTS with a transmit antenna selection technique communicates with a selected user via relays, while countering the attack by a passive eavesdropper equipped with the maximal ratio combining technique.

The exploration of wireless EH via RF signals in two-way communication networks (TWCN) has garnered significant attention in recent years. Notably, the work by Tutuncuoglu et al. [41] in 2015 introduced EH transmission strategies for TWRC in order to optimize the system's sum-throughput. Similarly, Salem et al. [42] investigated a WPCN, where a multi-antenna two-way AF relay facilitates data exchange among single-antenna user pairs. In a related study, Zhou and Li [43] proposed a joint optimization problem of relay precoding matrices and power-splitting ratios to maximize energy efficiency for SWIPT in MIMO two-way AF relay networks, where the relay can harvest energy from both sources and use it to forward messages effectively between two sources. Furthermore, Nguyen et al.'s adaptive EH protocol tailored for two-way AF relay networks under Rician fading conditions represents another significant advancement [44].

Partial relay selection (PRS), as presented by Nguyen et al. [44], emerges as an effective strategy to bolster transmission reliability during cooperative phases of EH-based TWRNs. Nevertheless, research on PLS with RF EH-based multi-relay networks is still sparse. Recent works on the combination of RF EH and PLS, such as [35] in 2020, only considered simple configurations with a single relay.

This chapter presents one of our recent works on PLS [1], which integrates RF EH, PLS, and PRS in a TWRN. In this work, a trade-off analysis between the reliability and security performance of a TWRN employing the TPSR EH protocol alongside PRS amidst eavesdropping threats near source nodes. We illustrate the use of our developed MATLAB-based 5G/6G simulation tool to verify the main contributions of [1], which include:

- The derivation of closed-form exact and approximate expressions for critical performance metrics such as OP at the legitimate users, intercept probability (IP) by the eavesdropper, SOP, and average secrecy capacity (ASC) of the system. The complexity of this derivation is high due to the presence of numerous random variables involved.
- Analysis on the effect of various system parameters on the reliability and security, which reveals optimal values for EH parameters like time-switching factor and power-splitting factor for specific relay and jammer setups. It is also observed that the number of relays has a more important effect on the overall performance than the number of jammers, unless the transmission power is extremely high.
- Recommendation of good system configurations and parameter choices to ensure reliable and secure communication at a reasonable complexity.

This chapter is organized as follows. Section 8.2 introduces our proposed model utilizing hybrid TPSR EH protocol combined with PRS. Section 8.3 elaborates on deriving key performance metrics, including OP, IP, SOP, and ASC. Numerical results obtained by analytical expressions and Monte Carlo simulations are discussed extensively within Sect. 8.5. Finally, concluding remarks are presented throughout Sect. 8.6.

8.2 System Model

8.2.1 Energy Harvesting and Scheduling Protocols

The proposed system model, as illustrated in Fig. 8.1, examines a multi-relay TWRN consisting of M relay nodes. The relays, which have RF EH capabilities, enable communication between two sources, A and B. The relays are located in a cluster with the assumption that all of them have similar distance to each of the source nodes—an assumption particularly relevant for IoT or wireless sensor networks.

In this system, PRS is utilized. Specifically, the relay that has the best channel condition to source A is chosen.[1] All nodes operate as single-antenna devices in half-duplex mode. The hybrid TPSR EH protocol is employed here, as depicted in Fig. 8.2 and detailed in prior works such as those by Son et al. [45] and Nguyen et al. [46].

Information exchange occurs over three phases using DNC and DF relaying. In Phases 1 and 2, sources A and B respectively broadcast their messages (x_1, x_2) to both the selected relay and the other source. During Phase 3, the relay generates a combined message using an XOR operation ($x_{\oplus} = x_1 \oplus x_2$) before broadcasting it back to both sources for decoding. Each source can subtract its own transmitted message from the combined one and get the desired message from the other source.

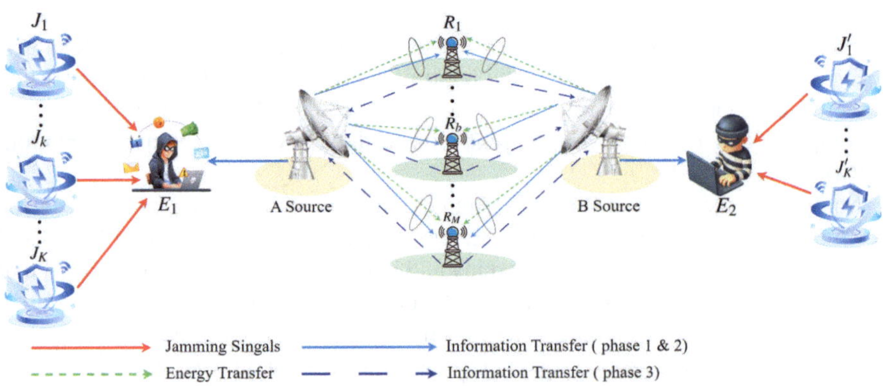

Fig. 8.1 System model of the proposed energy harvesting scheme

[1] It does not lose the generality whether we select source A or B.

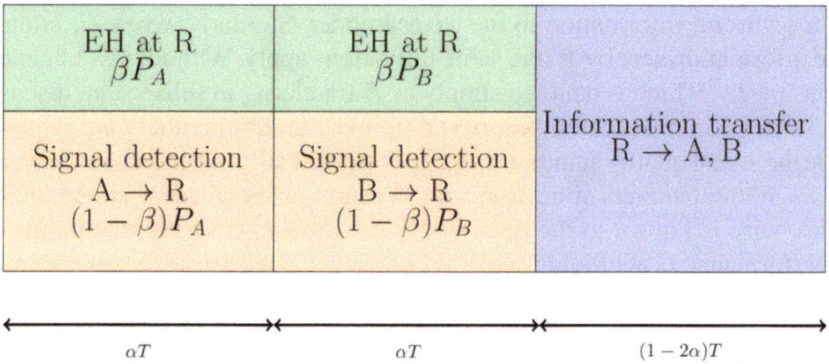

Fig. 8.2 Energy harvesting and scheduling protocol for the proposed model

It is assumed that there is no direct link between sources due to long distance or obstacles in the surrounding environment like buildings, trees, or hills. Although Physical Layer Network Coding (PNC), a.k.a. ANC, could be accomplished in two phases by enabling simultaneous transmission from both sources to a single-relay node via DF relaying, it requires either multi-user detection capability at relays [47] or specialized coding schemes at source nodes [47, 48]. Therefore, we employ DNC over three time slots here for simplicity in analysis and implementation.

Figure 8.2 shows in detail how signals transmitted from sources A and B allow EH at the selected relay using the PSR protocol during the first two phases. In particular, each received signal is split into two parts by a power divider: one part is used for EH and the other part is used for information processing at the relay. For simplicity, we assume that signals from both sources experience the same power-splitting factor, which is denoted as β, with $0 < \beta < 1$. Also, we assume that the duration of each Phase 1 and Phase 2 is equal to αT, where T is the duration of an entire transmission block, $\alpha < 0.5$ is the time-switching factor. As a result, the duration of Phase 3 is $(1 - 2\alpha)T$.

For eavesdropping threat, there are an eavesdropper E_1 near source A and an eavesdropper E_2 near source B. Each eavesdropper attempts to overhear the information sent from the source near it. Here, we consider the single eavesdropper case similar to other works like [16, 40]. The multi-eavesdropper scenario may be available in practice, but the implementation cost for eavesdropping and the risk of being detected are high. The proposed eavesdropping strategy is well-suited for practical applications, especially in military communications where eavesdroppers are strategically positioned near command centers to optimize their interception capabilities [1].

The eavesdropper E_1 aims to intercept the information transmitted from source A to other nodes. Given its proximity to A, the direct communication link from source B to E_1 is assumed to be unavailable. Consequently, even if E_1 receives signals from a relay node, it cannot effectively separate or eliminate the message sent by B from the composite received signal. As a result, the signals received from the relay node provide no useful information to the eavesdropper. Similarly, when E_1 attempts to decode information sent by B, the same limitations apply. Without loss of generality, we focus on E_1, which is denoted simply as E for clarity in subsequent discussions.

To bolster communication security, multiple friendly jammers are deployed to degrade the signal quality at the eavesdropper. Specifically, when source A broadcasts its message, the jammers simultaneously transmit artificial noise aimed at E. This artificial noise is known to the relay nodes in advance, ensuring that its impact on relay performance is negligible.

We assume that the channels between any two nodes exhibit block Rayleigh fading characteristics, i.e., channel coefficients remain constant during a single transmission block and vary independently across different blocks. Let $h_{X,Y}$, where $X, Y \in \{A, B, E, R_1, R_2, \ldots, R_M, J_1, J_2, \ldots, J_K\}$, represent the channel gain of the link between nodes X and Y. Here, R_i refers to the ith relay node, M is the total number of available relay nodes, J_k represents the kth jammer node, and K is the number of jammers.

In this work, we assume reciprocal channels, such that the channel gain from X to Y is identical to that from Y to X. Given the Rayleigh fading assumption, the squared channel amplitude, such as $|h_{A,R_i}|^2$, $|h_{B,R_i}|^2$, and $|h_{AE}|^2$, are modeled as exponential random variable. The CDF and PDF of these random variables are given by

$$F_X(x) = 1 - e^{-\lambda x} \tag{8.1}$$

$$f_X(x) = \begin{cases} \lambda e^{-\lambda x} & \text{if } x \geq 0 \\ 0 & \text{if } x < 0 \end{cases} \tag{8.2}$$

where λ is the mean of the exponential random variable X.

We will now delve into the mathematical framework that underpins the entire process. Initially, as outlined earlier, the source A utilizes the first time slot to transmit its packet to both relay R and the second source B. The signal received at the relay R is mathematically represented by

$$y_{A,R_i} = h_{A,R_i} x_A + n_{R_i}^{(1)}, \tag{8.3}$$

where $E\{|x_A|^2\} = P_A$ and P_A denotes the average transmit power at the source A. Here, $E\{.\}$ signifies the expectation operator. The term $n_{R_i}^{(1)}$ represents zero-mean AWGN with variance N_0.

By employing the PSR protocol, we can determine the energy harvested by the relay R from the source A as

$$E_{R_i} = \eta \beta \alpha T P_A |h_{AR_i}|^2, \tag{8.4}$$

where $0 < \eta \leq 1$ is the effective energy conversion efficiency, which accounts for energy losses in the harvesting circuit as well as decoding and processing circuits.

In the subsequent time slot, B transmits the signal x_2 to the relay node R_i. Consequently, the signal received at that relay is given by

$$y_{B,R_i} = h_{B,R_i} x_B + n_{R_i}^{(2)}, \tag{8.5}$$

where $E\{|x_B|^2\} = P_B$ is the average transmit power at the B, $n_{R_i}^{(2)}$ is zero-mean AWGN with variance N_0.

By aggregating this with previously harvested energy during Phase 1, we derive total harvested energy at the relay node R as

$$E_{R_i} = \eta\beta\alpha T \left(P_A |h_{AR_i}|^2 + P_B |h_{BR_i}|^2\right). \tag{8.6}$$

For simplicity, we assume the equal average transmit powers from both sources A and B, which is commonly denoted by P. Thus, (8.7) can be simplified as

$$E_{R_i} = \eta\beta\alpha T P \left(|h_{AR_i}|^2 + |h_{BR_i}|^2\right). \tag{8.7}$$

In our presented analysis, which is aligned with prior studies such as those in [24–27, 33, 34], we ignore the negligible power consumption for joint decoding/processing compared to the consumption by signal transmission, primarily because the decoding and processing energy can be estimated while RF transmission demands escalate proportionally alongside the distance traversed. Moreover, should any deduction occur regarding the harvested energy that is utilized toward decoding, it is reflected in the reduction of the overall energy efficiency η. That means our analytical approach remains the same, even if power consumption for decoding and processing is taken into account. Finally, the average transmit power of the relay node during Phase 3 can be found as

$$P_{R_i} = \frac{E_R}{T(1-2\alpha)} = \frac{\eta\beta\alpha T P \left(|h_{AR_i}|^2 + |h_{BR_i}|^2\right)}{T(1-2\alpha)} = \kappa P \left(|h_{AR_i}|^2 + |h_{BR_i}|^2\right), \tag{8.8}$$

where $\kappa \triangleq \frac{\eta\beta\alpha}{1-2\alpha}$.

The received signals received at the relay R_i from A and B undergo decoding and re-encoding following a DNC scheme, as outlined in [49]. Let $\hat{x}_A$ and $\hat{x}_B$ represent the decoded messages from A and B, respectively. In the third time slot, relay R_i, utilizing its power denoted as P_{R_i}, broadcasts the XOR of these decoded messages $x_{R_i} = \hat{x}_A \oplus \hat{x}_B$ to both sources A and B. Then the received signals at A and B can be derived, respectively, as

$$\begin{aligned} y_A &= h_{R_i A} x_{R_i} + n_A, \\ y_B &= h_{R_i B} x_{R_i} + n_B, \end{aligned} \tag{8.9}$$

where $E\{|x_{R_i}|^2\} = P_{R_i}$; n_A and n_B are i.i.d. zero-mean AWGN noise terms with variance of N_0.

From Eqs. (8.3) and (8.5), we derive the received SNRs for decoding messages x_A and x_B at the relay x_A and x_B respectively as

$$\gamma_{AR_i} = \frac{(1-\beta)|h_{AR_i}|^2 P}{N_0} = (1-\beta)|h_{AR_i}|^2\Psi, \tag{8.10}$$

$$\gamma_{BR_i} = \frac{(1-\beta)|h_{BR_i}|^2 P}{N_0} = (1-\beta)|h_{BR_i}|^2\Psi. \tag{8.11}$$

where $\Psi \triangleq \frac{P}{N_0}$ is the transmit-signal-power-to-noise-ratio. In [1], it is assumed that the channels are reciprocal, so $h_{BR_i} = h_{R_iB}$ and $h_{AR_i} = h_{R_iA}$.

In the third phase, after A and B decode the received signal from R_i successfully, they can exploit their knowledge on its own transmitted message to recover the message sent by the other source. Without loss of generality, we consider the decoding process at the source B. The SNR at this node can be derived as

$$\begin{aligned}\gamma_{R_iB} &= \frac{|h_{R_iB}|^2 P_R}{N_0} = \frac{\kappa P\left(|h_{AR_i}|^2 + |h_{BR_i}|^2\right)|h_{R_iB}|^2}{N_0}\\ &= \kappa\Psi\left(|h_{AR_i}|^2 + |h_{BR_i}|^2\right)|h_{R_iB}|^2.\end{aligned} \tag{8.12}$$

The DF relaying protocol is modified from the traditional protocol to employ the DNC technique. In particular, the transmission decision during the third time slot should depend on the outcomes of the decoding of x_1 and x_2 in the previous phases:

(1) If the relay successfully decodes both x_1 and x_2, then it will broadcast the message $x_1 \oplus x_2$ in the third time slot;
(2) If the relay decodes x_2 successfully but not x_1, it only broadcast x_2 during the third time slot without applying XOR operation;
(3) If the relay cannot decode neither x_1 nor x_2, then no message is broadcasted in the third time slot, i.e., the outage will occur.

In scenarios (1) and (2) above, ensuring robust links for the source-relay-source path is crucial for effective communication via relay. Thus, the end-to-end SNR at B can be calculated as

$$\gamma_{DF} = \min\{\gamma_{AR_i}, \gamma_{R_iB}\} = \Psi \min\left\{(1-\beta)|h_{AR_i}|^2, \kappa\left(|h_{AR_i}|^2 + |h_{BR_i}|^2\right)|h_{R_iB}|^2\right\}. \tag{8.13}$$

In analyzing the received signal at the eavesdroppers, we focus on the eavesdropper A only for two primary reasons: (i) The significant distance between A and B precludes any communication between them, so their operations are independent; (ii) Given the analogous roles of these two eavesdroppers, one can derive the IP for B by simply interchanging A and B in the analysis. The eavesdropper E_1, positioned near A, attempts to intercept signals transmitted by A during the initial phase of our protocol. However, this interception is intervened by the interference from artificial noise generated by K friendly jammers (J_k, for $k = 1, 2, ..., K$). It is crucial to note that we exclude consideration of signal interception from relay nodes during the third time slot. This exclusion is justified as follows. The relayed message is a composition of both messages from A and B, which the eavesdropper E_1 cannot have full knowledge due to the long distance between it and B.[2] Consequently, the best interception opportunity for E_1 hinges solely on the signal captured during Phase 1, which can be mathematically represented as

$$y_E = h_{AE}x_s + \sum_{k=1}^{K} h_{J_kE}x_k + nE, \tag{8.14}$$

where x_s is the message transmitted by A; x_k is the artificial noise transmitted by the jammer J_k, which satisfies $E\left(|x_k|^2\right) = P_J$, P_J is the transmit power of each jammer; and n_E is the zero-mean AWGN noise at E with variance of N_0.

Equation (8.14) helps us obtain the received SINR at the eavesdropper:

$$\gamma_E = \frac{P_s|h_{AE}|^2}{\sum_{k=0}^{K}|h_{J_k}E|^2 P_J + N_0} = \frac{\Psi|h_{AE}|^2}{\sum_{k=1}^{K}|h_{J_k}E|^2\Psi_J + 1}, \tag{8.15}$$

where $\Psi_J \triangleq \frac{P_J}{N_0}$.

As mentioned above in (8.1), the channel power gains $X_i = |h_{AR_i}|^2 = |h_{R_iA}|^2$, $Y_i = |h_{BR_i}|^2 = |h_{R_iB}|^2$, $Z = |h_{AE}|^2$, and $T = |h_{J_kE}|^2$ are modeled as independent exponential random variables (RVs). Let $\lambda_u \in \{\lambda_1, \lambda_2, \lambda_3, \lambda_4\}\}$ denote the parameters of X_i, Y_i, Z, T, respectively. To incorporate path-loss effect into our model, we express these parameters as $\lambda_1 = d_1^{-\chi}$, $\lambda_2 = d_2^{-\chi}$, $\lambda_3 = d_3^{-\chi}$, and $\lambda_4 = d_4^{-\chi}$. The terms d_1, d_2, d_3, and $d4$ denote the link distances of $A \to R_i$, $B \to R_i$, $A \to E$ and $E \to J_k$, respectively. Here, the path-loss exponent is denoted by χ with typical values ranging between 2 and 6. For simplicity of the analysis, all jammers are assumed to be equidistant from the eavesdropper. A similar assumption has been mentioned above for the relay node, i.e., they are clustered such that the source-to-relay distances are approximately equal for a specific source and all relays.

8.2.2 *Partial Relay Selection*

In our system, we employ the PRS method to improve communication quality. This method is preferred over the Optimal Relay Selection (ORS) method, which selects the relay that maximizes the secrecy capacity of the system, because the latter poses significant challenges in implementation. For PRS, a cooperative relay among M

---

[2] It may be useful for interception of the message sent by B, but this task should be done better by E_2 rather than E_1.

available ones is selected by the following condition:

$$R_b = \underset{\{R_i:i=\overline{1,M}\}}{\arg\max} |h_{AR_i}|^2. \tag{8.16}$$

In other words, the relay that possesses the highest channel gain between itself and A is selected for the cooperative communication in the third time slot. This selection is crucial for optimizing network performance while maintaining manageable computational complexity.

Remark 8.1 In the PRS algorithm, acquiring the CSI between A and the relay nodes is essential. In reality, the source A can estimate CSI through local control messages, thereby efficiently identifying the optimal candidate as demonstrated in (8.16). Importantly, we make a reasonable assumption that the relay is positioned closer to node B than to node A, which implies that relay selection should prioritize first-hop link quality to optimize overall system performance.

In PRS protocol, it will require the CSI between A and the relay nodes. In practice, A can estimate CSI via local control message and hence it can be easily determine the best candidate as in (8.16). Without loss of generality, we assume the relay is located closer to B than to A, so the selection based on the quality of the first-hop link is reasonable to enhance the system performance.

As derived in [50], both the CDF and PDF of $X = |h_{AR_b}|^2$ can be found as

$$F_X(x) = \sum_{j=0}^{M} (-1)^j \binom{M}{J} e^{-j\lambda_1 x} \tag{8.17}$$

and

$$f_X(x) = \lambda_1 \sum_{j=0}^{M-1} (-1)^j \binom{M-1}{j} e^{-(j+1)\lambda_1 x}. \tag{8.18}$$

8.3 Performance Evaluation

In this section, we rigorously derive the fundamental performance metrics of the proposed system, which include OP, IP, SOP, and ASC.

8.3.1 Outage Probability

Initially, we define the achievable data rates for both the legitimate communication link $A \to B$ and the eavesdropper link respectively as

$$C_B = (1-\alpha)\log_2(1+\gamma_{DF}), \tag{8.19}$$

$$C_E = (1-\alpha)\log_2(1+\gamma_E), \tag{8.20}$$

where γ_{DF} is given by (8.13).

Assume that the relay node R_b is selected and let $X = |h_{AR_b}|^2$, $Y = |h_{BR_i}|^2$, then we can reformulate the end-to-end SNR at the source B as

$$\gamma_{DF} = \min\{(1-\beta)X\Psi, \kappa\Psi(X+Y)Y\}. \tag{8.21}$$

The occurrence of an OP is characterized by a scenario where the system's data rate surpasses that of the achievable link rate, i.e., when $C_B < R$, with R representing the data transmission rate. This condition can be expressed mathematically as:

The OP occurs when the achievable capacity of the link is less than the required data rate, i.e., $C_B < R$, where R is the data transmission rate. Therefore, OP is calculated as

$$OP = \Pr\{C_B < R\} = \Pr\{\min\{(1-\beta)\Psi X, \kappa\Psi(X_Y)Y\} < \gamma_{th},\} \tag{8.22}$$

where $\gamma_{th} \triangleq 2^{\frac{R}{(1-\alpha)}} - 1$ is the SNR threshold of B, which is the solution of the equation $C_B = R$.

The analysis in [1] allows us to state the Theorem 8.1 on the OP of the proposed system [1].

Theorem 8.1 *The OP of the proposed TWRN with PRS and PSR energy harvesting at relay nodes can be found as*

$$OP = 1 + \sum_{j=1}^{M}(-1)^j \binom{M}{j} e^{-j\lambda_1 \frac{\gamma_{th}}{(1-\beta)\Psi} - \lambda_2\xi}$$
$$+ \sum_{n=0}^{\infty}\sum_{j=0}^{M} \frac{(-1)^j (j\lambda_1 - \lambda_2)^n (j\lambda_1\gamma_{th})^{n+1}\lambda_2}{n!(\kappa\Psi)^{n+1}} \binom{M}{j} \Gamma\left(-n-1, \frac{j\lambda_1\gamma_{th}}{\kappa\Psi\xi}\right). \tag{8.23}$$

where M is the number of available relay nodes, γ_{th} is the SNR threshold of receiver, $\kappa \triangleq \frac{\eta\beta\alpha}{1-2\alpha}$, $\Psi \triangleq \frac{P}{N_0}$, $\xi = \frac{1}{2}\left[\sqrt{\frac{\gamma_{th}^2}{(1-\beta)^2\Psi^2} + \frac{4\gamma_{th}}{\kappa\Psi}} - \frac{\gamma_{th}}{(1-\beta)\Psi}\right]$, and $\Gamma(s,x) \triangleq \int_x^\infty t^{s-1}e^{(-t)}dt$ is the incomplete Gamma function.

Proof We can rewrite (8.22) as

$$
\begin{aligned}
OP &= \Pr\{\min\{(1-\beta)\Psi X, \kappa\Psi(X+Y)Y\} < \gamma_{th}\} \\
&= 1 - \Pr\{(1-\beta)\Psi X \geq \gamma_{th}, \kappa\Psi(X+Y)Y \geq \gamma_{th}\} \\
&= 1 - \Pr\left\{X \geq \frac{\gamma_{th}}{(1-\beta)\Psi}, X \geq \frac{\gamma_{th}}{\kappa\Psi Y} - Y\right\} \\
&= 1 - \underbrace{\Pr\left(\frac{\gamma_{th}}{\kappa\Psi Y} - Y \geq \tilde{\gamma}_{th}, X \geq \frac{\gamma_{th}}{\kappa\Psi Y} - Y\right)}_{P_1} - \underbrace{\Pr\left(\frac{\gamma_{th}}{\kappa\Psi Y} - Y \leq \tilde{\gamma}_{th}, X \geq \tilde{\gamma}_{th}\right)}_{P_2}.
\end{aligned}
\tag{8.24}
$$

where $\tilde{\gamma}_{th} \triangleq \frac{\gamma_{th}}{(1-\beta)\Psi}$.

Let $\xi = \frac{1}{2}\left(\sqrt{(\tilde{\gamma}_{th})^2 + \frac{4\gamma_{th}}{\kappa\Psi}} - \tilde{\gamma}_{th}\right)$ be the positive solution of the equation $\frac{\gamma_{th}}{\kappa\Psi y} - y = \tilde{\gamma}_{th}$, then the probability terms in (8.24) can be derived as

$$
\begin{aligned}
P_1 &= \int_0^{\xi} \Pr\left(X \geq \frac{\gamma_{th}}{\kappa\Psi y} - y\right) f_Y(y)dy = \int_0^{\xi}\left[1 - F_X\left(\frac{\gamma_{th}}{\kappa\Psi y} - y\right)\right] f_Y(y)dy \\
&= -\int_0^{\xi}\left\{\sum_{j=1}^{M}(-1)^j\binom{M}{j}e^{-j\lambda_1\left(\frac{\gamma_{th}}{\kappa\Psi y}-y\right)}\right\} f_Y(y)dy = -\sum_{j=1}^{M}(-1)^j\binom{M}{j}\lambda_2\int_0^{xi} e^{y(j\lambda_1-\lambda_2)}e^{-\frac{j\lambda_1\gamma_{th}}{\kappa\Psi y}}dy,
\end{aligned}
\tag{8.25}
$$

$$
\begin{aligned}
P_2 &= \int_{\xi}^{\infty} \Pr\left(X \geq \tilde{\gamma}_{th}\right) f_Y(y)dy = \int_{\xi}^{\infty}\left[1 - F_X(\tilde{\gamma}_{th})\right] f_Y(y)dy \\
&= \sum_{j=1}^{M}(-1)^{j+1}\binom{M}{j}\cdot e^{-j\lambda_1\tilde{\gamma}_{th}}\int_{\xi}^{\infty} f_Y(y)dy \\
&= \sum_{j=1}^{M}(-1)^{j+1}\binom{M}{j}\cdot e^{-j\lambda_1\tilde{\gamma}_{th}-\lambda_2\xi}.
\end{aligned}
\tag{8.26}
$$

The integral in P_1 can be solved by changing variable $t = 1/y$ and applying Taylor's series expansion $e^x = \sum_{n=0}^{\infty}\frac{x^n}{n!}$ for $x = \frac{j\lambda_1-\lambda_2}{t}$:

$$
\begin{aligned}
P_1 &= -\sum_{j=1}^{M}(-1)^j\binom{M}{j}\lambda_2\int_{1/\xi}^{\infty} t^{-2}e^{\frac{(j\lambda_1-\lambda_2)}{t}}e^{-j\frac{\lambda_1\gamma_{th}t}{\kappa\Psi}}dt \\
&= -\sum_{n=0}^{\infty}\sum_{j=1}^{M}\frac{(-1)^j(j\lambda_1-\lambda_2)^n\lambda_2}{n!}\binom{M}{j}\int_{1/\xi}^{\infty}\frac{e^{-j\lambda_1}\gamma_{th}t}{\kappa\Psi t^{n+2}}dt \\
&= \sum_{n=0}^{\infty}\sum_{j=1}^{M}\frac{(-1)^{j+1}(j\lambda_1-\lambda_2)^n(j\lambda_1\gamma_{th})^{n+1}\lambda_2}{n!(\kappa\Psi)^{n+1}}\binom{M}{j}\times\Gamma\left(-n-1, \frac{j\lambda_1\gamma_{th}}{\kappa\Psi\xi}\right).
\end{aligned}
\tag{8.27}
$$

where the last equality comes from [51, 3.462.16].

By substituting (8.18), (8.26) and (8.27) into (8.24), we finally get (8.23). □

8.3.2 *Intercept Probability*

The IP at the eavesdropper E is defined as the probability that the capacity of the legitimate link is dominated by the capacity of the wiretap link, i.e., $C_E > C_B$. By exploiting the result from the OP analysis, we obtain the Theorem 8.2 [1].

Theorem 8.2 *The IP at the eavesdropper E in the proposed TWRN with jamming by artificial noise is given by* (8.28)*, where K is the number of jammers,* $\Psi \triangleq \frac{P}{N_0}$, $\Psi_J \triangleq \frac{P_J}{N_0}$, *and* $\tilde{\xi}_x = \frac{1}{2}\left[\sqrt{\frac{x^2}{(1-\beta)^2} + \frac{4x}{\kappa}} - \frac{x}{(1-\beta)}\right]$.

$$IP = 1 + \int_0^\infty \left\{ \sum_{n=0}^{\infty} \sum_{j=1}^{M} \frac{(-1)^j (j\lambda_1 - \lambda_2)^n [j\lambda_1 x]^{n+1} \lambda_2}{n! (\kappa)^{n+1}} \binom{M}{j} \Gamma\left(-n-1, \frac{j\lambda_1 x}{\kappa \tilde{\xi}_x}\right) \right\}$$
$$\times \frac{e^{-\lambda_3 x}\left[(\lambda_3)^2 \frac{\Psi_J}{\lambda_4} x + \lambda_3 + \frac{\lambda_3 \Psi_J K}{\lambda_4}\right]}{\left(1 + \frac{\lambda_3 \Psi_J x}{\lambda_4}\right)^{K+1}} dx$$
$$+ \int_0^\infty \left\{ \sum_{j=1}^{M} (-1)^j \binom{M}{j} e^{-\frac{j\lambda_1 x}{(1-\beta)} - \lambda_2 \tilde{\xi}_x} \right\} \frac{e^{-\lambda_3 x}\left[(\lambda_3)^2 \frac{\Psi_J}{\lambda_4} x + \lambda_3 + \frac{\lambda_3 \Psi_J K}{\lambda_4}\right]}{\left(1 + \frac{\lambda_3 \Psi_J x}{\lambda_4}\right)^{K+1}} dx. \tag{8.28}$$

Proof By definition and using (8.13) and (8.15), the IP at the eavesdropper E can be found as

$$\begin{aligned} IP &= \Pr(C_E \ge C_B) = \Pr(\gamma_E \ge \gamma_{DF}) \\ &= \Pr\left(\frac{Z}{\Psi_J T + 1} \ge \min\{(1-\beta) X, \kappa (X+Y) Y\}\right). \end{aligned} \tag{8.29}$$

where $\Psi_J \triangleq P_J/N_0$, $Z = |h_{AE}|^2$, $T = \sum_{k=1}^{K} |h_{J_k E}|^2$. □

Let's denote $V = \min\{(1-\beta) X, \kappa (X+Y) Y\}$ and $U = \frac{Z}{\Psi_J T + 1}$. Then (8.29) can be rewritten as

$$IP = \Pr(U \ge V) = \int_0^\infty f_U(u) du \int_0^u f_V(v) dv = \int_0^\infty f_U(u) \cdot F_V(u) du. \tag{8.30}$$

By using (8.15), the CDF of U can be obtained as

$$\begin{aligned} F_U(u) &= \Pr\{U \le u\} = \Pr\left\{\frac{Z}{\Psi_J T + 1} \le u\right\} \\ &= \Pr\{Z < u(T\Psi_J + 1)\} = \int_0^\infty F_Z(u(t\Psi_J + 1)) f_T(t)dt, \end{aligned} \tag{8.31}$$

where $F_Z(.)$ and $f_T(.)$ are the CDFs of Z and PDF of T, respectively.

In the previous section, we established that Z is an exponential random variable characterized by the parameter λ_3, whereas T represents the sum of K i.i.d. exponential random variables with parameter λ_4. Consequently, the CDF of Z is expressed in (8.1), and T follows a gamma distribution with parameters K and λ_4. The PDF for this gamma distribution is given by [52]

$$f_T(t) = \frac{\lambda_4^K}{(K-1)!} t^{K-1} e^{-\lambda_4 t}. \tag{8.32}$$

By substituting (8.1) and (8.36) into (8.31), we can reformulate $F_U(u)$ as

$$\begin{aligned} F_U(u) &= \int_0 \infty F_Z(u(t\Psi_J + 1)) f_T(t)dt \\ &= \int_0 \infty f_T(t)dt - \frac{\lambda_4^K e^{-\lambda_3 u}}{(K-1)!} \int_0 \infty t^{K-1} e^{-t(\lambda_4 + \lambda_3 u \Psi_J)} dt \\ &= 1 - \frac{\lambda_4^K \Gamma(K) e^{-\lambda_3 u}}{(K-1)!(\lambda_4 + \lambda_3 u \Psi_J)^K} = 1 - \frac{\lambda_4^K e^{-\lambda_3 u}}{(\lambda_4 + \lambda_3 u \Psi_J)^K}, \end{aligned} \tag{8.33}$$

where the second last equality is obtained by using [51, 3.381.4] and the last equality comes from the replacing $\Gamma(K) = (K-1)!$.

By taking the derivative of (8.37), we obtain the PDF of U as

$$f_U(u) = \frac{e^{-\lambda_3 u}\left[(\lambda_3)^2 \frac{\Psi_J}{\lambda_4} u + \lambda_3 + \frac{\lambda_3 \Psi_J K}{\lambda_4}\right]}{\left(1 + \frac{\lambda_3 u \Psi_J}{\lambda_4}\right)^{K+1}}. \tag{8.34}$$

On the other hand, the CDF of V can be derived by using Theorem 8.1 (by substituting γ_{th} by $\upsilon\Psi$ in (8.23))

$$F_v(v) = \Pr(V \le v) = \Pr\left(\frac{\gamma_{DF}}{\Psi} \le v\right)$$
$$= 1 + \sum_{j=1}^{M} (-1)^j \binom{M}{j} e^{-\frac{j\lambda_1 v}{(1-\beta)} - \lambda_2 \tilde{\xi} v}$$
$$+ \sum_{n=0}^{\infty} \sum_{j=1}^{M} \frac{(-1)^j (j\lambda_1 - \lambda_2)^n (j\lambda_1 v)^{n+1} \lambda_2}{n!\,(\kappa)^{n+1}} \binom{M}{j} \Gamma\left(-n-1, \frac{j\lambda_1 v}{\kappa \tilde{\xi} v}\right). \tag{8.35}$$

where $\tilde{\xi} v = \frac{1}{2}\left[\sqrt{\frac{v^2}{(1-\beta)^2} + \frac{4v}{\kappa}} - \frac{v}{(1-\beta)}\right]$.

By substituting (8.34) and (8.35) into (8.30), we get (8.28). The proof is complete.

Remark 8.2 It can be seen from (8.28) that the IP does not depend on Ψ.

Remark 8.3 It is worth noting that our analysis in this section can be easily extended to the general case, in which hardware noise terms exist at both relay and primary transmitters. In fact, each SINR formula for the hardware imperfection case is a one-to-one mapping of the corresponding SINR for the perfect hardware case presented above [1]. In particular, the received signal at the relay and its corresponding SINR in the hardware imperfection case can be written respectively as

$$y_{A,R_i} = h_{A,R_i}\left(x_A + \eta_A^t\right) + \eta_{R_i}^r + n_{r_i}^{(1)},$$

$$\gamma_{AR_i}^{HI} = \frac{(1-\beta)|h_{AR_i}|^2 P}{|h_{AR_i}|^2 k_{AR_i}^2 P + N_0} = \frac{(1-\beta)}{k_{AR_i}^2 + \frac{1}{|h_{AR_i}|^2 \Psi}} = \frac{1}{\frac{k_{AR_i}^2}{1-\beta} + \frac{1}{\gamma_{AR_i}}}.$$

where $\eta_A^t \sim \mathcal{CN}\left(0, \left(k_A^t\right)^2 P\right)$ and $\eta_{R_i}^r \sim \mathcal{CN}\left(0, \left(k_{R_i}^r\right)^2 P_{R_i}\right)$ are hardware impairment (HI) noise terms caused by the transmitter impairment at A and receiver impairment at R_i, respectively; $k_{AR_i}^2 = \left(k_A^t\right)^2 + \left(k_{R_i}^r\right)^2$ denotes the aggregated HI level at both transmitter A $\left(k_A^t\right)$ and receiver R_i $\left(k_{R_i}^r\right)$ [53].

Then, $P\left(\gamma_{AR_i}^{HI} < \gamma_{th}\right)$ can be derived by using

$$P\left(\gamma_{AR_i}^{HI} < \gamma_{th}\right) = P\left(\frac{1}{\frac{k_{AR_i}^2}{1-\beta} + \frac{1}{\gamma_{AR_i}}} < \gamma_{th}\right) = P\left(\gamma_{AR_i} < \frac{1-\beta-k_{AR_i}^2}{(1-\beta)\gamma_{th}}\right)$$

By replacing γ_{th} in our original analysis by $\dfrac{1-\beta-k_{AR_i}^2}{(1-\beta)\gamma_{th}}$, we can obtain the result for hardware imperfection case.

8.3.3 Secrecy Outage Probability

In the domain of PLS, as explored in works like [20], researchers focus on transmitting confidential messages at a positive rate, known as the secrecy rate, between a source and its intended destination. This is achieved while ensuring that the source-destination channel outperforms the source-eavesdropper channel. The disparity in channel gains directly correlates with an increased secrecy rate. The secrecy rate is mathematically defined by

$$C_{sec} = \max\left\{C_B - C_E, 0\right\}, \tag{8.36}$$

where $C_B = (1-\alpha)\log_2(1+\gamma_{DF})$ is the achievable date rate at the node B.

To maintain message confidentiality, it is imperative that this secrecy rate surpasses a specific threshold. A secrecy outage occurs if the secrecy rate falls below this threshold:

$$SOP = \Pr\left\{C_{sec} < R\right\} = \Pr\left\{\frac{1+\gamma_{DF}}{1+\gamma_E} < \gamma_{sc,th}\right\}, \tag{8.37}$$

where R is the threshold of the secrecy rate, and $\gamma_{sc,th} \triangleq 2^{\frac{R}{(1-\alpha)}}$.

The derivation of the closed-form expression of this SOP can be summarized in the following theorem.

Theorem 8.3 *The SOP for TWRNs employing PRS and EH at relay nodes with friendly jammers can be derived as*

$$\begin{aligned} SOP = 1 + \int_0^\infty &\left\{\sum_{n=0}^{\infty}\sum_{j=1}^{M}\frac{(-1)^j (j\lambda_1-\lambda_2)^n\left[j\lambda_1\left(\tilde{\gamma}+\gamma_{sc,th}x\right)\right]^{n+1}\lambda_2}{n!\,(\kappa\Psi)^{n+1}}\binom{M}{j}\right.\\ &\left.\times\,\Gamma\left(-n-1,\frac{j\lambda_1\left(\tilde{\gamma}+\gamma_{sc,th}x\right)}{\kappa\Psi\,\Xi_x}\right)\right\}\frac{e^{-\frac{\lambda_3 x}{\Psi}}\left[\left(\frac{\lambda_3}{\Psi}\right)^2\frac{\Psi_J}{\lambda_4}x+\frac{\lambda_3}{\Psi}+\frac{\lambda_3\Psi_J K}{\Psi\lambda_4}\right]}{\left(1+\frac{\lambda_3\Psi_J x}{\Psi\lambda_4}\right)^{K+1}}dx\\ +\int_0^\infty &\left\{\sum_{j=1}^{M}(-1)^j\binom{M}{j}e^{-\frac{j\lambda_1\left(\tilde{\gamma}+\gamma_{sc,th}x\right)}{(1-\beta)\Psi}-\lambda_2\Xi_x}\right\}\frac{e^{-\frac{\lambda_3 x}{\Psi}}\left[\left(\frac{\lambda_3}{\Psi}\right)^2\frac{\Psi_J}{\lambda_4}x+\frac{\lambda_3}{\Psi}+\frac{\lambda_3\Psi_J K}{\Psi\lambda_4}\right]}{\left(1+\frac{\lambda_3\Psi_J x}{\Psi\lambda_4}\right)^{K+1}}dx. \end{aligned} \tag{8.38}$$

where $a = \frac{j\lambda_1\gamma_{sc,th}}{(1-\beta)\Psi} + \frac{\lambda_3}{\Psi}$, $b = \frac{\lambda_3\Psi_J}{\lambda_4\Psi}$, $\tilde{\gamma} \triangleq 2^{\frac{R}{(1-\alpha)}-1}$ *is the SNR threshold at E, and* $\Gamma(s,x) \triangleq \int_x^\infty t^{s-1}e^{-t}dt$ *is the incomplete gamma function.*

Proof First, we can rewrite (8.37) as

$$
\begin{aligned}
SOP &= \Pr\left\{\frac{1+\gamma_{DF}}{1+\gamma_E} < \gamma_{sc,th}\right\} = \Pr\left\{\gamma_{DF} < (1+\gamma_E)\gamma_{sc,th} - 1\right\} \\
&= \int_0^\infty F_{\gamma_{DF}}\left((1+x)\,\gamma_{sc,th} - 1\right) f_{\gamma_E}(x)dx,
\end{aligned}
\tag{8.39}
$$

where $F_{\gamma_{DF}}(.)$ and $f_{\gamma_E}(.)$ are the CDF of γ_{DF} and PDF of γ_E, respectively. □

By applying Theorem 8.2 (by substituting u by $\frac{x}{\Psi}$ in (8.37)), the CDF and the PDF of γ_E can be found by

$$
F_{\gamma_E}(x) = \Pr\{\gamma_E < x\} = 1 - \frac{e^{-\frac{\lambda_3 x}{\Psi}}}{\left(1 + \frac{\lambda_3 \Psi_J x}{\Psi \lambda_4}\right)^K}
\tag{8.40}
$$

and

$$
f_{\gamma_e}(x) = \frac{df_{\gamma_E}}{dx} = \frac{e^{-\frac{\lambda_3 x}{\Psi}}\left[\left(\frac{\lambda_3}{\Psi}\right)^2 \frac{\Psi_J}{\lambda_4} x + \frac{\lambda_3}{\Psi} + \frac{\lambda_3 \Psi_J K}{\Psi \lambda_4}\right]}{\left(1 + \frac{\lambda_3 \Psi_J x}{\Psi \lambda_4}\right)^{K+1}}.
\tag{8.41}
$$

On the other hand, the CDF of γ_{DF} is obtained by using the result of Theorem 8.1 (substituting γ_{th} by y in (8.23))

$$
\begin{aligned}
F_{\gamma_{DF}}(y) &= 1 + \sum_{j=1}^{M} (-1)^j \binom{M}{j} e^{-j\lambda_1 \frac{y}{(1-\beta)\Psi} - \lambda_2 \xi} \\
&+ \sum_{n=0}^{\infty} \sum_{j=1}^{M} \frac{(-1)^j (j\lambda_1 - \lambda_2)^n (j\lambda_1 y)^{n+1} \lambda_2}{n!\,(\kappa\Psi)^{n+1}} \binom{M}{j} \Gamma\left(-n-1, \frac{j\lambda_1 y}{\kappa \Psi \xi_y}\right).
\end{aligned}
\tag{8.42}
$$

where $\xi_y = \frac{1}{2}\left[\sqrt{\frac{y^2}{(1-\beta)^2\Psi^2} + \frac{4y}{\kappa\Psi}} - \frac{y}{(1-\beta)\Psi}\right]$.

Let's denote $\tilde{\gamma} \triangleq \gamma_{sc,th} - 1$ and

$$
\Xi_x = \xi_{\gamma_{sc,th}}(1-x) - 1 = \sqrt{\frac{\left(x\gamma_{sc,th} + \tilde{\gamma}\right)^2}{4(1-\beta)^2\Psi^2} + \frac{\left(x\gamma_{sc,th} + \tilde{\gamma}\right)}{\kappa\Psi}} - \frac{\left(x\gamma_{sc,th} + \tilde{\gamma}\right)}{2(1-\beta)\Psi}.
$$

Now, by substituting (8.42) with $y = x\gamma_{sc,th} + \tilde{\gamma}$ into (8.39), we have (8.43). Finally, by substituting (8.41) into (8.43), we get (8.38)

$$SOP = \int_0^\infty \left\{ 1 + \sum_{n=0}^{\infty} \sum_{j=1}^{M} \frac{(-1)^j (j\lambda_1 - \lambda_2) \left[j\lambda_1 \left(\tilde{\gamma} + \gamma_{sc,th} x \right) \right]^{n+1} \lambda_2}{n! \left(\kappa \Psi \right)^{n+1}} \binom{M}{j} \right.$$
$$\left. \times \Gamma \left(-n-1, \frac{j\lambda_1 \left(\tilde{\gamma} + \gamma_{sc,th} x \right)}{\kappa \Psi \Xi_x} \right) \right\} f_{\gamma_E}(x) dx \tag{8.43}$$
$$+ \int_0^\infty \left\{ \sum_{j=1}^{M} (-1)^j \binom{M}{j} e^{-\frac{j\lambda_1 (\tilde{\gamma} + \gamma_{sc,th} x)}{(1-\beta)\Psi} - \lambda_2 \Xi_x} \right\} f_{\gamma_E}(x) dx.$$

8.3.4 Average Secrecy Capacity

The secrecy capacity of the proposed system is linked to the SNRs of both the legitimate and wiretap link, which are random variables. The average secrecy capacity (ASC) is mathematically defined as the expected value of this capacity [54]:

$$C_{avg} = E\left[C_{Sec}\right] = (1-\alpha) \iint_{\substack{x,\, y > 0 \\ y \geq x}} \log_2 \left(\frac{1+y}{1+x} \right) f_{\gamma_{DF}}(y) f_{\gamma_E}(x) dx dy$$
$$= (1-\alpha) \int_0^\infty f_{\gamma_E}(x) dx \int_x^\infty \log_2 \left(\frac{1+y}{1+x} \right) f_{\gamma_{DF}}(y) dy. \tag{8.44}$$

In this context, $f_{\gamma_E}(.)$ and $f_{\gamma_{DF}}(.)$ represent the PDFs of γ_E and γ_{DF}, respectively. We now present a theorem regarding the ASC of our proposed system:

Theorem 8.4 *The average secrecy capacity C_{avg} of the proposed TWRN with PRS, EH at relay nodes, and with friendly jammers is given by*

$$C_{avg} = (1-\alpha) \sum_{n=0}^{\infty} \sum_{j=1}^{M} \frac{(-1)^j (j\lambda_1 - \lambda_2)^n \lambda_2}{n!} \binom{M}{j} \int_0^\infty \frac{(n+1) e^{-\frac{\lambda_3 x}{\Psi}} \Lambda_x}{\left(1 + \frac{\lambda_3 \Psi_J x}{\Psi \lambda_4}\right)^{K+1}} dx \int_x^\infty \log_2 \left(\frac{1+y}{1+x} \right)$$
$$\times \left[\frac{y^n (j\lambda_1)^{n+1}}{(\kappa \Psi)^{n+1}} \Gamma \left(-n-1, \frac{j\lambda_1 y}{\kappa \Psi \xi_y} \right) \right] dy$$
$$- (1-\alpha) \sum_{n=0}^{\infty} \sum_{j=1}^{M} \frac{(-1)^j (j\lambda_1 - \lambda_2)^n \lambda_2}{n!} \binom{M}{j} \int_0^\infty \frac{e^{-\frac{\lambda_3 x}{\Psi}} \Lambda_x}{\left(1 + \frac{\lambda_3 \Psi_J x}{\Psi \lambda_4}\right)^{K+1}} dx$$
$$\times \int_0^\infty \log_2 \left(\frac{1+y}{1+x} \right) \left[e^{-\frac{j\lambda_1 y}{\kappa \Psi \xi_y}} \xi_y^{n+1} \left(\frac{1}{y} - \frac{\Upsilon_y}{\xi_y} \right) \right] dy \tag{8.45}$$
$$+ (1-\alpha) \sum_{j=1}^{M} (-1)^{j+1} \binom{M}{j} \int_0^\infty \frac{e^{-\frac{\lambda_3 x}{\Psi}} \Lambda_x}{\left(1 + \frac{\lambda_3 \Psi_J x}{\Psi \lambda_4}\right)^{K+1}} dx$$
$$\times \int_x^\infty \log_2 \left(\frac{1+y}{1+x} \right) e^{-\left[\frac{j\lambda_1 y}{(1-\beta)\Psi} + \lambda_2 \xi_y\right]} \left[\frac{j\lambda_1}{(1-\beta)\Psi} + \lambda_2 \Upsilon_y \right] dy$$

where $\Lambda_x \triangleq \left(\frac{\lambda_3}{\Psi}\right)^2 \frac{\Psi_J}{\lambda_4} x + \frac{\lambda_3}{\Psi} + \frac{\lambda_3 \Psi_J K}{\Psi \lambda_4}$ *and* $\Upsilon_y \triangleq \frac{\frac{y}{\{(1-\beta)\Psi\}^2} + \frac{2}{\kappa \Psi}}{2\sqrt{\frac{y}{(1-\beta)\Psi} + \frac{4y}{\kappa \Psi}}} - \frac{1}{2(1-\beta)\Psi}$.

Proof The CDF of γ_{DF} has been previously established in (8.42). By differentiating (8.42), we obtain its PDF γ_{DF} as shown in (8.46), where the derivation of the last equality utilizes results from Olver's work [55, 8.8.13].

$$
\begin{aligned}
f_{\gamma_{DF}}(y) &= \frac{\partial F_{\gamma_{DF}}(y)}{\partial y} \\
&= \sum_{n=0}^{\infty}\sum_{j=1}^{M} \frac{(-1)^j (j\lambda_1 - \lambda_2)^n (j\lambda_1)^{n+1}\lambda_2}{n!(\kappa\Psi)^{n+1}} \binom{M}{j} \frac{\partial}{\partial y}\left[y^{n+1}\Gamma\left(-n-1, \frac{j\lambda_1 y}{\kappa\Psi\xi_y}\right)\right] + \\
&+ \sum_{j=1}^{M}(-1)^j \binom{M}{j} \frac{\partial}{\partial y}\left[e^{\frac{-j\lambda_1 y}{(1-\beta)\Psi} - \lambda_2\xi_y}\right] \\
&= \sum_{j=1}^{M}(-1)^{j+1} + \sum_{n=0}^{\infty}\sum_{j=1}^{M} \frac{(-1)^j (j\lambda_1 - \lambda_2)^n \lambda_2}{n!} \binom{M}{j} \\
&\times \left[\frac{(n+1)y^n (j\lambda_1)^{n+1}}{(\kappa\Psi)^{n+1}} \Gamma\left(-n-1, \frac{j\lambda_1 y}{\kappa\Psi\xi_y}\right) - e^{\frac{-j\lambda_1 y}{\kappa\Psi\xi_y}} \xi_y^{n+1}\left(\frac{1}{y} - \frac{\Upsilon_y}{\xi_y}\right)\right] \\
&= \sum_{n=0}^{\infty}\sum_{j=1}^{M} \frac{(-1)^j (j\lambda_1 - \lambda_2)^n (j\lambda_1)^{n+1}\lambda_2}{n!(\kappa\Psi)^{n+1}} \binom{M}{j} e^{-\left[\frac{j\lambda_1 y}{(1-\beta)\Psi} + \lambda_2\xi_y\right]} \left[\frac{j\lambda_1}{(1-\beta)\Psi} + \lambda_2\Upsilon_y\right].
\end{aligned}
\tag{8.46}
$$

where $\Upsilon_y \triangleq \frac{\frac{y}{((1-\beta)\Psi)^2} + \frac{2}{\kappa\Psi}}{2\sqrt{\left(\frac{y}{(1-\beta)\Psi}\right)^2 + \frac{4y}{\kappa\Psi}}} - \frac{1}{2(1-\beta)\Psi}$. By substituting (8.46) and (8.41) into (8.44) we obtain (8.45). □

8.4 MATLAB Implementation

In this section, some sample MATLAB codes to implement the analytical calculation and Monte Carlo simulations for the PLS simulation module is introduced. First, we introduce the sample codes to obtain the numerical results on the OP of the proposed system as follows.

```
clearvars;
clc;
close all;

% Simulation parameters
NUM = 10^5; % number of channel realizations
max_index = 100;
```

```
chi = 2;
d1 = 0.85; d2 = 0.5; d3 = 2; d4 = 2;
lambda_1 = d1.^chi; lambda_2 = d2.^chi;
lambda_3 = d3.^chi; lambda_4 = d4.^chi;
eta = 0.8;                    %
beta = 0.5;
alpha = 0.3;
k = eta.*beta.*alpha./(1-2.*alpha);
Cth = 0.25; % Capacity threshold
gamma = 2.^(3.*Cth./(1-2.*alpha))-1;

PdB = 0:1:30;
PP = 10.^(PdB/10);

%----------------------------------------------------------------------
%----------------------------------------------
%%%% CASE: 3 RELAY NODES (M = 3)

OP1 = zeros(1,length(PP));
MM = 3;

% Mote Carlo simulation
for aa = 1:length(PP)
    % Channel generation
        har = sqrt(1/2/lambda_1)*(randn(MM,NUM) + 1i*randn(MM,NUM));
    hrb = sqrt(1/2/lambda_2)*(randn(MM,NUM) + 1i*randn(MM,NUM));
    if MM > 1
        [X,index] = max(abs(har).^2);
    else
        X = abs(har).^2;
        index = ones(1,NUM);
    end
    Y = zeros(1,NUM);
    for ii = 1:NUM
        Y(ii) = abs(hrb(index(ii),ii)).^2;
    end

        % SNR calculation
    SNR_1 = (1 - beta).*X.*PP(aa);
    SNR_2 = k.*PP(aa).*(X + Y).*Y;
    SNR_DF = min(SNR_1,SNR_2);
```

```
    OP1(aa) = sum(SNR_DF < gamma)./NUM;
end

% Analytical result
P_out_ana = zeros(1,length(PP));
for aa = 1 : length(PP)
    jj = 1:MM;
    zeta = 1./2.*(sqrt(gamma.^2./(1 - beta).^2./(PP(aa).^2) + 4.*gamma./k./PP(aa))-gamma./(1 - beta)./PP(aa));
    tong2 = sum((-1).^(jj+1).*my_nchoosek(MM,jj).*exp(-jj.*lambda_1.*gamma./(1 - beta)./PP(aa)-lambda_2.*zeta));
    tong1 = 0;
    for nn = 0:max_index
        tp1 = (-1).^(jj+1).*(jj.*lambda_1 - lambda_2).^nn.*(jj.*lambda_1.*gamma./k./PP(aa)).^(nn+1)./...
            factorial(nn).*lambda_2.*my_nchoosek(MM,jj).*incompleteGamma(-nn-1,jj.*lambda_1.*gamma./k./PP(aa)./zeta);
        tong1 = tong1 + sum(tp1);
    end
    P_out_ana(aa) = 1 - tong1 - tong2;
end
P_out_ana1 = P_out_ana;


% Plot the OP
figure(1)
h1 = semilogy(PdB,OP1,'k--','LineWidth',2);
hold on;
h2 = semilogy(PdB,P_out_ana1,'r*','LineWidth',2);
hold on;
grid on;
legend([h2,h1],'Analysis M = 3','Monte-Carlo Simulation M = 3');
xlabel('\Psi(dB)'),ylabel('Outage probability(OP)');
```

```
title('OP versus \Psi with R = 0.25 bps/Hz, \eta = 0.8, \beta = 0.5, and \alpha = 0.3');
```

Next, we provide the sample codes for analyzing the SOP. Here, we only show the main codes for calculating and simulating, and remove the lines that are used for produce the output graph.

```
clearvars;
clc;
close all;
profile on
tic
% Simulation parameters
NUM = 10^5; % number of channel realization
max_index = 50;
chi = 2;
d1 = 0.85; d2 = 0.5; d3 = 2; d4 = 2;
lambda_1 = d1.^chi; lambda_2 = d2.^chi;
lambda_3 = d3.^chi; lambda_4 = d4.^chi;
eta = 0.8;                         %
beta = 0.5;
alpha = 0.3;
k = eta.*beta.*alpha./(1-2.*alpha);

R = 0.25;
gamma_tilde = 2.^(R./(1-alpha))-1;
gamma_1 = gamma_tilde + 1;
PdB = 0:1:30;
PP   = 10.^(PdB/10);
PjdB = 15;
Pj = 10.^(PjdB/10);
SOP1 = zeros(1,length(PP));

%==========================================
%============================
SOP1    = zeros(1,length(PP));
SOP_anal    = zeros(1,length(PP));
KK = 3; % number of jammers
MM = 3; % number of relays
% Monte Carlo simulation
for aa = 1 : length(PP)
```

```
    har = sqrt(1/2/lambda_1)*(randn(MM,NUM) + 1
        i*randn(MM,NUM));
    hrb = sqrt(1/2/lambda_2)*(randn(MM,NUM) + 1
        i*randn(MM,NUM));
    if MM > 1
        [X,index] = max(abs(har).^2);
    else
        X = abs(har).^2;
        index = ones(1,NUM);
    end
    Y = zeros(1,NUM);
    for ii = 1:NUM
        Y(ii) = abs(hrb(index(ii),ii)).^2;
    end
    hae = sqrt(1/2/lambda_3)*(randn(1,NUM) + 1i
        *randn(1,NUM));
    hje = sqrt(1/2/lambda_4)*(randn(KK,NUM) + 1
        i*randn(KK,NUM));
    Z = abs(hae).^2;
    if KK == 1
        T = abs(hje).^2;
    else
        T = sum(abs(hje).^2);
    end

    SNR_1 = (1 - beta).*X.*PP(aa);
    SNR_2 = k.*PP(aa).*(X + Y).*Y;
    SNR_DF = min(SNR_1,SNR_2);
    SNR_E = PP(aa).*Z./(T.*Pj + 1);
    SOP1(aa) = sum((1 + SNR_DF)./(1 + SNR_E) <
        gamma_1)/NUM;
end

% Analytical result
for aa = 1:length(PP)
    waitbar(0.75+aa/31/4,b0);
    jj = 1:MM;
    y = SOP_int2(MM,KK,gamma_1,lambda_1,
        lambda_2,lambda_3,lambda_4,k,beta,PP(aa)
        ,Pj);
    S1 = sum((-1).^(jj+1).*y.*my_nchoosek(MM,jj
        ));
    S = 0;
```

```
        for nn = 0:max_index
            y = SOP_int1(nn,MM,KK,gamma_1,lambda_1,
                lambda_3,lambda_4,k,beta,PP(aa),Pj);
            if sum(isnan(y))>0
                break;
            end
           S = S + sum((-1).^(jj+1).*(jj.*lambda_1
              - lambda_2).^nn.*lambda_2.*
              my_nchoosek(MM,jj).*y);
        end
        SOP_anal(aa) = 1- S - S1;
    end
```

Other codes for IP and ASC can be written a similar approach. Therefore, we omit them to save the pages. To reproduce the codes, the readers can follow the flowchart given in Fig. 3.22.

8.5 Simulation Results

In this section, we present Monte Carlo simulations to substantiate the derived expressions for OP, IP, SOP, and ASC. The simulations explore how various system parameters, such as transmit-power-to-noise-ratio, time-switching factor, power-splitting factor, number of relay nodes, number of jammers, and spectrum efficiency, affect the system performance. Each simulation is based on 10^5 samples of each channel gain. Simulation parameters are detailed in Table 8.1. For channel settings, a simplified path-loss model $\lambda_i = d_i^{\chi}$ for $i \in \{1, 2, 3, 4\}$ is used; where d_1, d_2, d_3, and d_4 represent distances between A and the relay cluster, between B and the relay cluster, between A and the eavesdropper, and between the eavesdropper and the jamming cluster, respectively. The path-loss exponent χ is set to 2 with specific distance values $d_{1i} = 0.85$, $d_2 = 0.5$, $d_3 = d_4 = 2$, leading to calculated λ_i values shown in Table 8.1.

We first evaluate reliability and security through OP and IP metrics. Figures 8.3 and 8.4 illustrate OP and IP against the ratio between transmit power of the source A and the noise power density N_0, which is denoted by Ψ, for four different cluster sizes of the relay cluster and jamming clusters. These figures confirm that analytical results align precisely with simulation outcomes: OP decreases with increasing Ψ, while IP remains unaffected by changes in Ψ, consistent with Remark 8.3. Both reliability and security can be improved by augmenting relay or jammer nodes. However, beyond a certain point–evident from minimal performance gaps among cases where $M = 3$, $M = 5$, or $M = 7$–additional relays offer diminishing returns due to proximity-induced channel gain limitations within clusters. Specifically, with

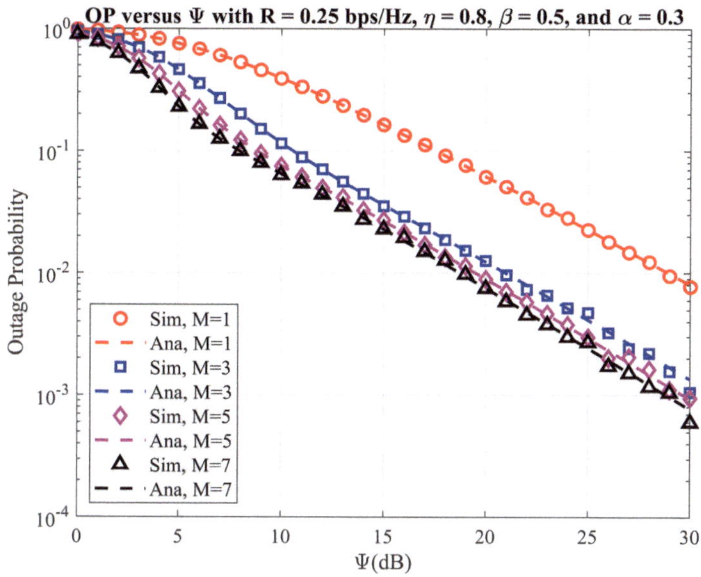

Fig. 8.3 Outage probability of $A \rightarrow B$ link versus transmit-power-to-noise-ratio for different number of relays

Table 8.1 Simulation parameters

Symbol	Parameter name	Fixed value	Varying range
R	Spectrum efficiency	0.25 bps/Hz	0.1 to 1 (bps/Hz)
η	Energy harvesting efficiency	0.8	None
α	Time division factor	0.3	0 to 0.5
β	Power splitting factor	0.5	0.1 to 1
λ_1	Mean of $\vert h_{AR_i}\vert^2$	0.7225	None
λ_2	Mean of $\vert h_{BR_i}\vert^2$	0.258	None
λ_3	Mean of $\vert h_{AE}\vert^2$	4	None
λ_4	Mean of $\vert h_{J_k E}\vert^2$	4	None
Ψ	Transmit-power-to-noise-ratio	15 dB	0 to 30 (dB)
Ψ_J	Jammer-power-to-noise-ratio	15 dB	0 to 30 (dB)
M	No. of relay nodes	1;3;5;7	None
K	No. of jamming nodes	1;3;5;7	None

PRS, if more relays are added to the cluster, the chance that the newly added relay has the best channel gain is reduced. Thus, it is advisable to employ a moderate number

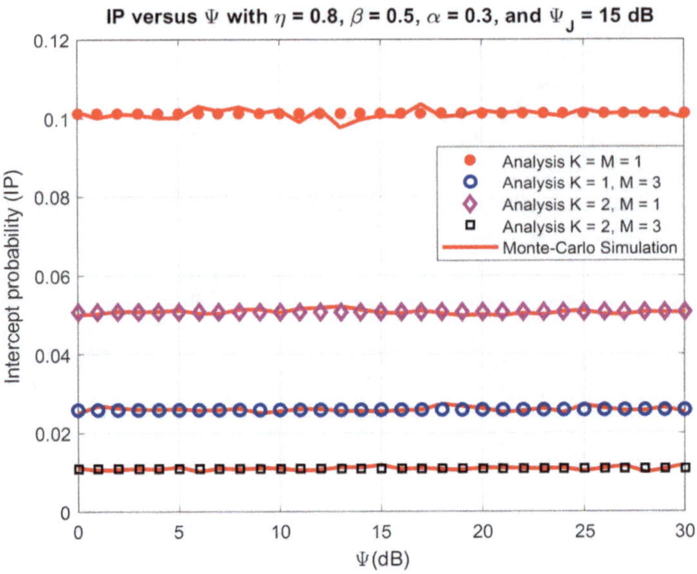

Fig. 8.4 Intercept probability versus transmit-power-to-noise-ratio for different number of jammers and jammer power levels

of relays ($M = 3$) alongside cooperative jammers ($K = 2$) for optimal performance enhancement.

The impact of the EH time-switching factor on both outage and security performance is illustrated in Figs. 8.5 and 8.6. We analyze four distinct cluster sizes for relays and jammers, specifically $M \in \{1, 3, 5, 7\}$ for relays and $K \in \{1, 2, 3\}$ for jammers. As demonstrated in Fig. 8.5, the OP decreases as the parameter α increases. This phenomenon can be logically explained: with a small α, the EH time is constrained, which limits the energy harvested by relay nodes for transmission purposes. Consequently, increasing α improves transmission quality. Again, increasing relays beyond $M = 3$ yields minimal performance gains. On the other hand, Fig. 8.6 reveals that IP also diminishes with an increase in α. This outcome is intuitive since a larger α extends the EH duration at relays, which enhances legitimate link condition without benefiting the wiretap link condition. Based on these observations from both figures, it is advisable to select a sufficiently large α, alongside an optimal number of jammers ($K = 2$) and relays ($M = 3$). Beyond these values, additional jammers or relays provide negligible advantages. Furthermore, excessive jammer transmit power isn't necessary. Maintaining it at approximately $\Psi J = 5dB$, combined with an adequately large α, ensures satisfactory performance levels for this model.

Figure 8.7 illustrates the OP as a function of the power-splitting factor β. A critical trade-off exists between allocating energy to support relay nodes and ensuring message transmission from source A. This implies that for each configuration of relay nodes, there is an optimal β that minimizes OP. As the number of relay

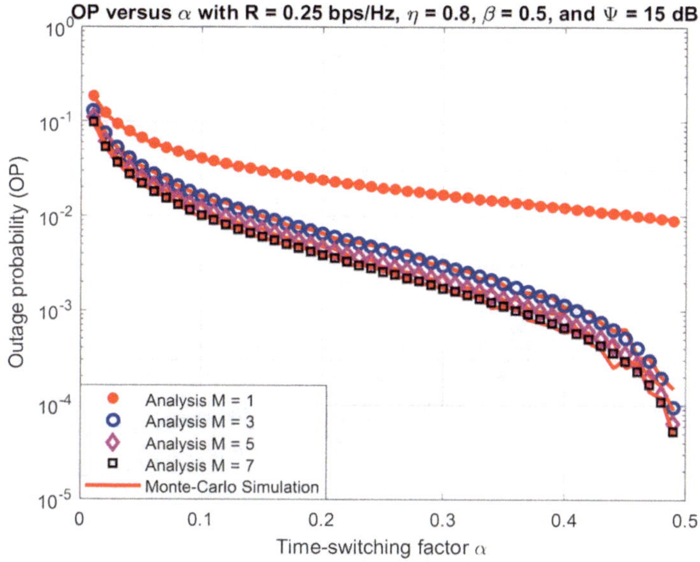

Fig. 8.5 Outage probability of $A \rightarrow B$ link versus time-switching factor different number of relays

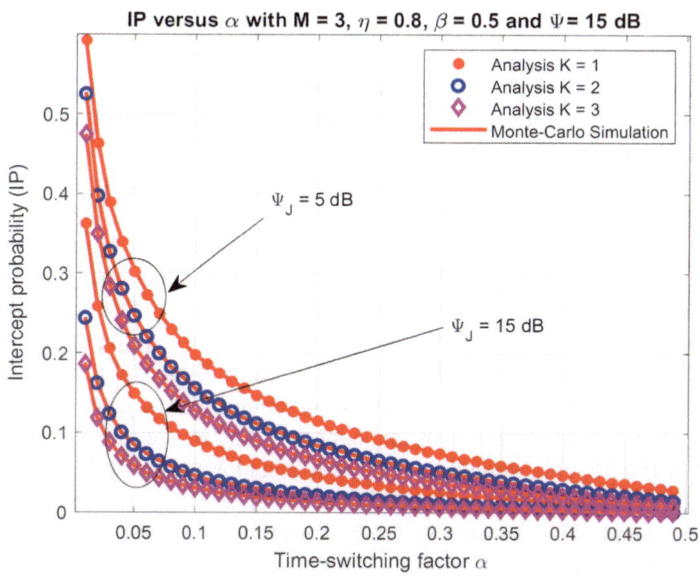

Fig. 8.6 Intercept probability of $A \rightarrow B$ versus time-switching factor for different number of jammers and jammer power levels

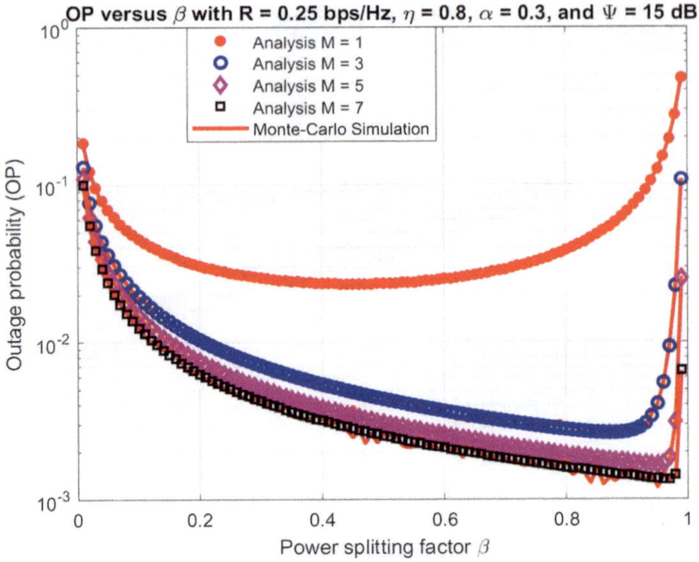

Fig. 8.7 Outage probability of $A \to B$ link versus power-splitting factor for different number of relays

nodes increases, this optimal β approaches unity ($\beta = 0.46, 0.87, 0.95$, and 0.97 for $M = 1, 3, 5$, and 7 respectively). The rationale is that with more relays, the link quality between A and the selected relay is improved significantly. Thus, less concern is needed for transmit power allocation to message signals, while more energy can be allocated to sustain the relay operation.

Furthermore, Fig. 8.8 demonstrates how β affects the IP. It is logical to deduce that an optimal β also exists for minimizing IP in each scenario since the wiretap link condition remains unaffected by β, whereas OP of the $A \to B$ link achieves a unique minimum as depicted in Fig. 8.8. However, it should be noted that the optimal value of β for minimizing IP diverges from the one that minimizes OP. Specifically, the optimal β for IP minimization is dependent on the number of jammers (it increases with more jammers present). For instance, at $\Psi_J = 5dB$, the optimal values β^* are 0.49, 0.53, and 0.56, corresponding to $K = 1, 2$, and 3, respectively. At higher jamming level ($\Psi_J = 15dB$), the optimal values β^* for $K = 1, 2, 3$ are shifted upward to 0.56, 0.66, and 0.73, respectively. Here, Ψ_J has a significant impact on IP due to an insufficiently large α.

Figures 8.9, 8.10, and 8.11 meticulously illustrate the influence of various parameters on the SOP of our proposed systems. Again, the analytical curves align perfectly with their simulation counterparts, which confirms our theoretical analysis. In Fig. 8.9, it is evident that SOP decreases as Ψ increases, similar to the OP shown in Fig. 8.3, albeit at a more rapid rate. It can be observed that an outage floor persists for each simulation scenario, indicating that beyond a certain threshold of Ψ, further increments do not significantly alter SOP. This floor tends to diminish when more

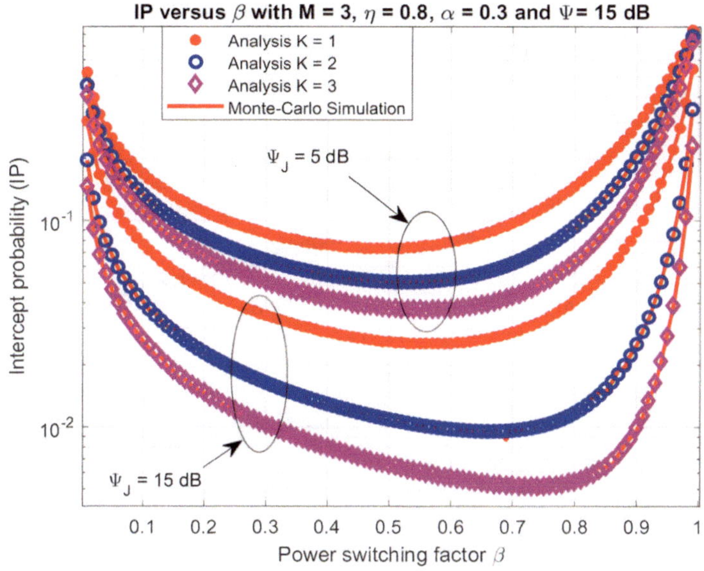

Fig. 8.8 Intercept probability of $A \rightarrow B$ link versus spectrum efficiency for different number of jammers and jammer power levels

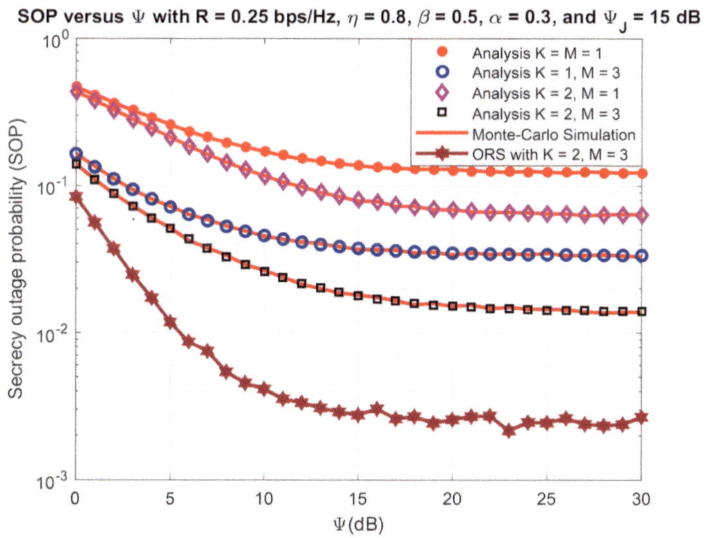

Fig. 8.9 Secrecy outage probability of the proposed system versus transmit-power-to-noise-ratio for different configurations of relay and jammer nodes

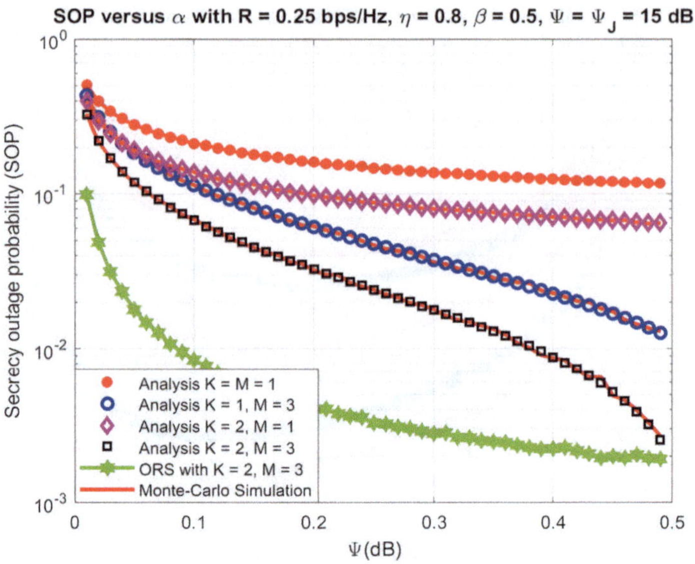

Fig. 8.10 Secrecy outage probability of the proposed system versus time-switching factor for different configurations of relay and jammer nodes

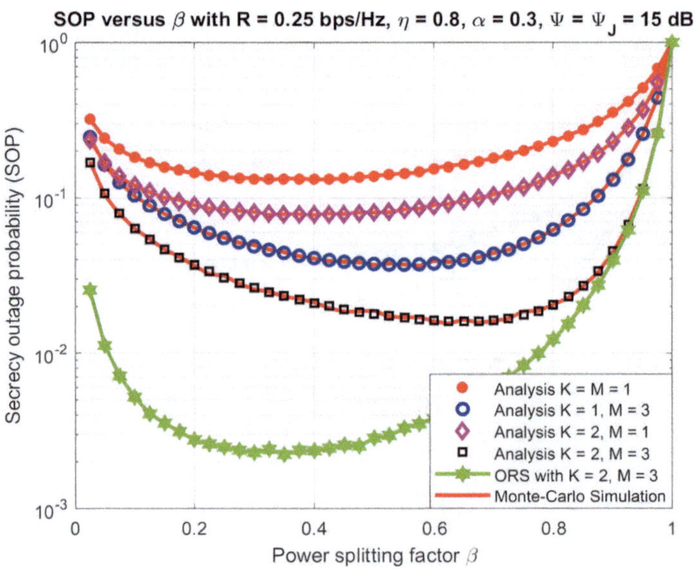

Fig. 8.11 Secrecy outage probability of the proposed system versus power-splitting factor for different configurations of relay and jammer nodes

relay and jammer nodes are incorporated. In Figs. 8.9, 8.10, and 8.11, we also present SOP for optimal relay selection (ORS) via simulation as a benchmark to evaluate the PRS strategy. Despite this comparison, PRS is favored due to its implementation simplicity.

Figure 8.10 provides a detailed analysis of how variation in the time-switching factor α can affect the secrecy performance. As α increases, we observe a decrease in the SOP. This is primarily because the secrecy rate is determined by the capacity difference between source-relay and source-eavesdropper links. With an increase in α, although there is a slight decline in source-relay link performance due to an elevated SINR threshold, there is a notable improvement in relay-destination link performance. This enhancement results from increased harvested energy (i.e., higher transmit power) and reduced transmission time. Importantly, increasing α does not enhance the capacity of the source-eavesdropper link. Thus, SOP decreases as α rises. Figure 8.10 suggests that selecting $K = 2$ relays and $M = 3$ jammers optimizes the system efficacy.

Furthermore, as depicted in Fig. 8.11, there exists an optimal value for the power-splitting factor β, which minimizes SOP. This conclusion is supported by similar reasoning as shown in Fig. 8.8. Notably, the optimal values of β for the cases $K = M = 1$, $K = 2$, $M = 1$, $K = 1$, $M = 3$, and $K = 2$, $M = 3$ are 0.375, 0.4, 0.55, and 0.65, respectively. Although deriving a closed-form solution for optimal β, which is the solution of the equation $\frac{\partial(SOP)}{\partial\beta} = 0$, is complicated due to the integral form of SOP, its single-minimum nature allows us to employ iterative numerical methods like the Golden section search algorithm [56]. The three-dimensional plot of SOP versus both parameters α and β when $K = 2$, $M = 3$ is shown in Fig. 8.12, which reveals that the global minimum of SOP is approximately 0.00168, obtained when $(\alpha, \beta) = (0.5, 0.34)$. For further development, an adaptive TPSR protocol that could dynamically update these parameters based on channel condition, can be implemented with the trade-off between cost and complexity.

Figures 8.13, 8.14, and 8.15 depict the ASC of the proposed system, demonstrating its growth alongside the transmit SNR as in Fig. 8.13. Adding more relay nodes and more jammers helps improve the ASC, in which the addition of relays provides more benefits, except in the high SINR regime. Moreover, ASC exhibits concavity with respect to the time-switching factor α due to the following reason. At first, small α values enlarge the transmission duration while decreasing SOP as we explained in Fig. 8.10, leading to the increasing of ASC. However, when α exceeds some threshold, the short transmission time adversely impacts the overall capacity, causing the decline of ASC. Figure 8.14 shows that the optimal values of α do not change much for different considered scenarios and stay around 0.17. Again, the operation of more relays or more jammers enhances the value of ASC, where the relays provide more significant benefits at $\Psi = 15dB$. On the other hand, a larger power-splitting factor β diminishes ASC since the exponential term is reduced according to (8.45). Intuitively, the increasing of β reduces the successful message decoding likelihood at the relay, which degrades the overall capacity. This is confirmed by Fig. 8.15, where ASC approaches zero as β approaches 1. In summary, the Monte Carlo simulations here verify the analysis again.

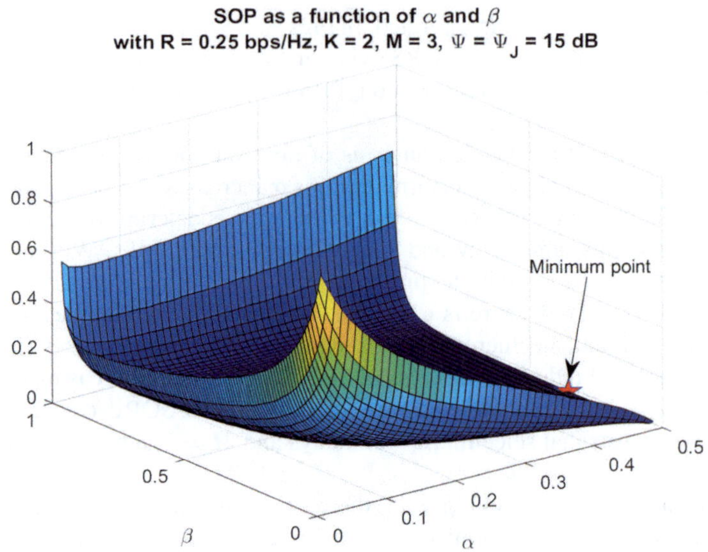

Fig. 8.12 3-D graph of SOP of $A \rightarrow B$ link versus α and β

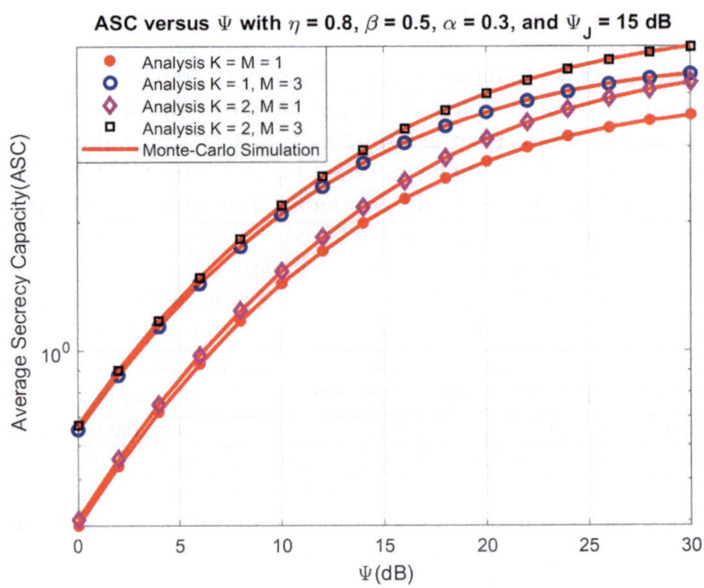

Fig. 8.13 Average secrecy capacity of A versus transmit-power-to-noise-ratio for different configurations of relays and jammers

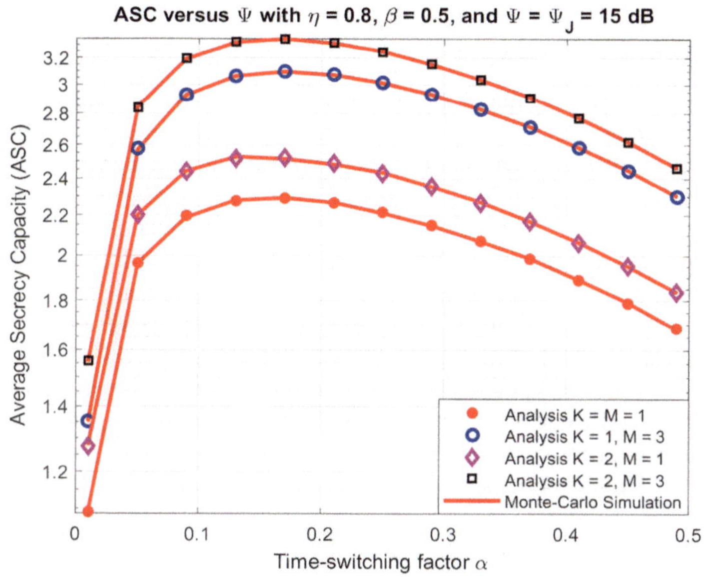

Fig. 8.14 Average secrecy capacity of A versus time-switching factor α for different configurations of relays and jammers

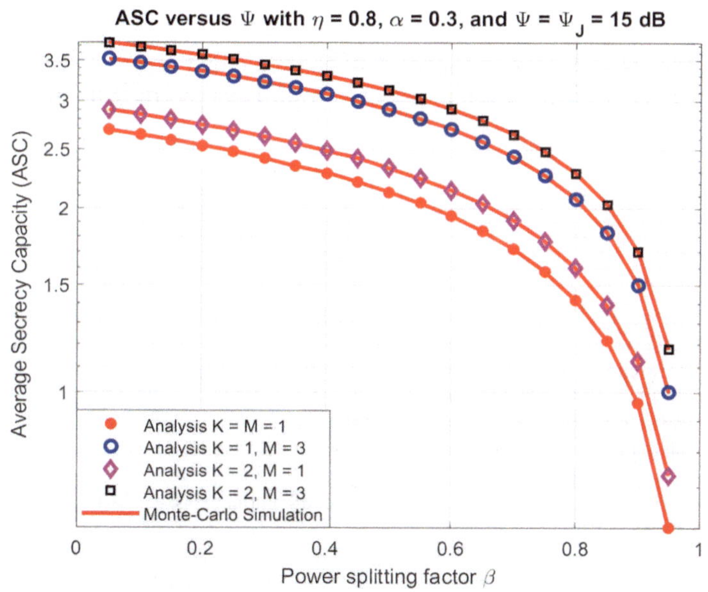

Fig. 8.15 Average secrecy capacity of A versus power-splitting factor β for different configurations of relays and jammers

8.6 Chapter Conclusion

In this chapter, we present the MATLAB simulation to verify the analysis of two-way half-duplex wireless relay networks that operate in a Rayleigh fading environment and incorporate PRS and hybrid TPSR-based EH at the relay node. A notable challenge arises from the presence of an eavesdropper near each of the source nodes. To fortify network security, friendly jammers are strategically deployed to disrupt the eavesdropper's signal reception. We rigorously derive closed-form expressions for key performance metrics, including OP at the legitimate destinations, IP at the eavesdropper, SOP, and ASC of the whole system. These analytical results are substantiated through Monte Carlo simulations. Our numerical findings reveal that increasing the number of either available-to-help relay nodes or jammers significantly enhances both reliability and security performance. Specifically, augmenting more relay nodes yields greater improvement in the average-to-high SINR regime, but for very high SINR levels, adding more jammers proves more advantageous. Furthermore, we identify a unique optimal value for either time-switching or power-splitting factors to minimize OP. The IP escalates with an increase in the time-switching factor but remains unaffected by the power-splitting factor variation. For SOP analysis, an outage floor may exist if M and K values are small. Thus, selecting α around 0.3 with $M = 3$ relay nodes and $K = 5$ or 7 jammers is recommended to achieve the desired performance without increasing the complexity. The implications of this research can be extended to military communications or IoT networks in smart cities, where message security is the main concern. Future work could explore scenarios involving multiple antennas at source nodes or eavesdroppers to further enhance system robustness against interception threats.

References

1. Ha, D.H., Nguyen, T.N., Tran, M.H.Q., Li, X., Tran, P.T., Voznak, M.: Security and reliability analysis of a two-way half-duplex wireless relaying network using partial relay selection and hybrid TPSR energy harvesting at relay nodes. IEEE Access **8**, 187165–187181 (2020). https://doi.org/10.1109/ACCESS.2020.3030794
2. Wyner, A.D.: The wire-tap channel. Bell Syst. Tech. J. **54**(8), 1355–1387 (1975)
3. Leung-Yan-Cheong, S., Hellman, M.: The gaussian wire-tap channel. IEEE Trans. Inf. Theory **24**(4), 451–456 (1978)
4. Csiszár, I., Korner, J.: Broadcast channels with confidential messages. IEEE Trans. Inf. Theory **24**(3), 339–348 (1978)
5. Mukherjee, A., Fakoorian, S.A.A., Huang, J., Swindlehurst, A.L.: Principles of physical layer security in multiuser wireless networks: a survey. IEEE Commun. Surv. & Tutor. **16**(3), 1550–1573 (2014)
6. Wang, W., Teh, K.C., Li, K.H.: Generalized relay selection for improved security in cooperative DF relay networks. IEEE Wirel. Commun. Lett. **5**(1), 28–31 (2015)
7. Pandey, A., Yadav, S.: Physical-layer security for cellular multiuser two-way relaying networks with single and multiple decode-and-forward relays. Trans. Emerg. Telecommun. Technol. **30**(12), e3639 (2019)

8. Shukla, M.K., Pandey, A., Yadav, S., Purohit, N.: Secrecy outage analysis of full duplex cellular multiuser two-way AF relay networks. In: 2019 International Conference on Wireless Communications Signal Processing and Networking (WiSPNET), pp. 458–463. IEEE (2019)
9. Liu, Y., Wang, L., Duy, T.T., Elkashlan, M., Duong, T.Q.: Relay selection for security enhancement in cognitive relay networks. IEEE Wirel. Commun. Lett. **4**(1), 46–49 (2014)
10. Li, X., Zhao, M., Gao, X.C., Li, L., Do, D.T., Rabie, K.M., Kharel, R.: Physical layer security of cooperative NOMA for IoT networks under I/Q imbalance. IEEE Access **8**, 51189–51199 (2020)
11. Kapetanovic, D., Zheng, G., Rusek, F.: Physical layer security for massive MIMO: an overview on passive eavesdropping and active attacks. IEEE Commun. Mag. **53**(6), 21–27 (2015)
12. Lai, L., El Gamal, H.: The relay-eavesdropper channel: cooperation for secrecy. IEEE Trans. Inf. Theory **54**(9), 4005–4019 (2008)
13. Li, J., Petropulu, A.P., Weber, S.: On cooperative relaying schemes for wireless physical layer security. IEEE Trans. Signal Process. **59**(10), 4985–4997 (2011)
14. Chen, J., Zhang, R., Song, L., Han, Z., Jiao, B.: Joint relay and jammer selection for secure two-way relay networks. IEEE Trans. Inf. Forensics Secur. **7**(1), 310–320 (2011)
15. Ding, X., Song, T., Zou, Y., Chen, X., Hanzo, L.: Security-reliability tradeoff analysis of artificial noise aided two-way opportunistic relay selection. IEEE Trans. Veh. Technol. **66**(5), 3930–3941 (2017). https://doi.org/10.1109/TVT.2016.2601112
16. Shukla, M.K., Yadav, S., Purohit, N.: Secure transmission in cellular multiuser two-way amplify-and-forward relay networks. IEEE Trans. Veh. Technol. **67**(12), 11886–11899 (2018)
17. Allipuram, S., Mohapatra, P., Chakrabarti, S.: Secrecy performance of an artificial noise assisted transmission scheme with active eavesdropper. IEEE Commun. Lett. **24**(5), 971–975 (2020)
18. Ding, Z., Dai, H., Poor, H.V.: Relay selection for cooperative NOMA. IEEE Wirel. Commun. Lett. **5**(4), 416–419 (2016)
19. Tang, X., Liu, R., Spasojević, P., Poor, H.V.: Interference assisted secret communication. IEEE Trans. Inf. Theory **57**(5), 3153–3167 (2011)
20. Chiang, H.T., Lehnert, J.S.: Optimal jamming with codewords. In: MILCOM 2012–2012 IEEE Military Communications Conference, pp. 1–6. IEEE (2012)
21. Lee, S.H., Khisti, A.: The gaussian diamond-wiretap channel with rate-limited relay cooperation. IEEE Commun. Lett. **21**(2), 338–341 (2016)
22. Varshney, L.R.: Transporting information and energy simultaneously. In: 2008 IEEE International Symposium on Information Theory, pp. 1612–1616. IEEE (2008)
23. Huang, K., Lau, V.K.: Enabling wireless power transfer in cellular networks: architecture, modeling and deployment. IEEE Trans. Wireless Commun. **13**(2), 902–912 (2014)
24. Ju, H., Zhang, R.: Throughput maximization in wireless powered communication networks. IEEE Trans. Wireless Commun. **13**(1), 418–428 (2013)
25. Li, X., Liu, M., Deng, C., Zhang, D., Gao, X.C., Rabie, K.M., Kharel, R.: Joint effects of residual hardware impairments and channel estimation errors on SWIPT assisted cooperative NOMA networks. IEEE Access **7**, 135499–135513 (2019)
26. Nasir, A.A., Zhou, X., Durrani, S., Kennedy, R.A.: Relaying protocols for wireless energy harvesting and information processing. IEEE Trans. Wireless Commun. **12**(7), 3622–3636 (2013)
27. Nasir, A.A., Zhou, X., Durrani, S., Kennedy, R.A.: Throughput and ergodic capacity of wireless energy harvesting based DF relaying network. In: 2014 IEEE International Conference on Communications (ICC), pp. 4066–4071. IEEE (2014)
28. Tran, P.T., Nguyen, T.N., Voznak, M.: Performance analysis of general hybrid TSR-PSR energy harvesting protocol for amplify-and-forward half-duplex relaying networks. J. Adv. Eng. Comput. **2**(2), 121–130 (2018)
29. Lu, W., Gong, Y., Liu, X., Wu, J., Peng, H.: Collaborative energy and information transfer in green wireless sensor networks for smart cities. IEEE Trans. Ind. Inform. **14**(4), 1585–1593 (2017)
30. Tin, P.T., Minh Nam, P., Trung Duy, T., Tran, P.T., Voznak, M.: Secrecy performance of TAS/SC-based multi-hop harvest-to-transmit cognitive WSNs under joint constraint of interference and hardware imperfection. Sensors **19**(5), 1160 (2019)

31. Nobar, S.K., Niya, J.M., Tazehkand, B.M.: Performance analysis of cognitive wireless powered communication networks under unsaturated traffic condition. IEEE Trans. Green Commun. Netw. **4**(3), 819–831 (2020)
32. Nguyen, T.N., Tran, P.T., Vozňák, M.: Power splitting-based energy-harvesting protocol for wireless-powered communication networks with a bidirectional relay. Int. J. Commun. Syst. **31**(13), e3721 (2018)
33. Irio, L., Oliveira, R., da Costa, D.B., Alouini, M.S.: Impact of wireless-powered communications in coexisting mobile networks. IEEE Wirel. Commun. Lett. **9**(7), 1060–1064 (2020)
34. Li, X., Li, J., Li, L.: Performance analysis of impaired SWIPT NOMA relaying networks over imperfect Weibull channels. IEEE Syst. J. **14**(1), 669–672 (2019)
35. Hoang An, N., Tran, M., Nguyen, T.N., Ha, D.H.: Physical layer security in a hybrid TPSR two-way half-duplex relaying network over a Rayleigh fading channel: outage and intercept probability analysis. Electronics **9**(3), 428 (2020)
36. Shannon, C.E.: Two-way communication channels. In: Proceedings of the Fourth Berkeley Symposium on Mathematical Statistics and Probability, Volume 1: Contributions to the Theory of Statistics, vol. 4, pp. 611–645. University of California Press (1961)
37. Rankov, B., Wittneben, A.: Achievable rate regions for the two-way relay channel. In: ISIT, vol. 6, pp. 1668–1672 (2006)
38. Katti, S., Rahul, H., Hu, W., Katabi, D., Médard, M., Crowcroft, J.: Xors in the air: practical wireless network coding. In: Proceedings of the 2006 Conference on Applications, Technologies, Architectures, and Protocols for Computer Communications, pp. 243–254 (2006)
39. Katti, S., Gollakota, S., Katabi, D.: Embracing wireless interference: analog network coding. ACM SIGCOMM Comput. Commun. Rev. **37**(4), 397–408 (2007)
40. Fu, S., Zhang, T., Colef, M.: Secrecy in two-way relay systems. In: 2010 IEEE Global Telecommunications Conference GLOBECOM 2010, pp. 1–5 (2010). https://doi.org/10.1109/GLOCOM.2010.5683210
41. Tutuncuoglu, K., Varan, B., Yener, A.: Throughput maximization for two-way relay channels with energy harvesting nodes: the impact of relaying strategies. IEEE Trans. Commun. **63**(6), 2081–2093 (2015)
42. Salem, A., Hamdi, K.A.: Wireless power transfer in multi-pair two-way AF relaying networks. IEEE Trans. Commun. **64**(11), 4578–4591 (2016)
43. Zhou, X., Li, Q.: Energy efficiency for SWIPT in MIMO two-way amplify-and-forward relay networks. IEEE Trans. Veh. Technol. **67**(6), 4910–4924 (2018)
44. Nguyen, T.N., Minh, T.H.Q., Tran, P.T., Voznak, M.: Adaptive energy harvesting relaying protocol for two-way half-duplex system network over Rician fading channels. Wirel. Commun. Mob. Comput. **2018**(1), 7693016 (2018)
45. Son, P.N., Kong, H.Y.: Improvement of the two-way decode-and-forward scheme by energy harvesting and digital network coding relay. Trans. Emerg. Telecommun. Technol. **28**(3), e2960 (2017)
46. Nguyen, T.N., Tran, P.T., Voznak, M.: Wireless energy harvesting meets receiver diversity: a successful approach for two-way half-duplex relay networks over block Rayleigh fading channel. Comput. Netw. **172**, 107176 (2020)
47. Chen, P., Xie, Z., Fang, Y., Chen, Z., Mumtaz, S., Rodrigues, J.J.: Physical-layer network coding: an efficient technique for wireless communications. IEEE Netw. **34**(2), 270–276 (2019)
48. Zhang, S., Liew, S.C.: Applying physical-layer network coding in wireless networks. EURASIP J. Wirel. Commun. Netw. **2010**, 1–12 (2010)
49. Zhong, B., Zhang, Z.: Secure full-duplex two-way relaying networks with optimal relay selection. IEEE Commun. Lett. **21**(5), 1123–1126 (2017)
50. Nguyen, T.N., Tran, M., Nguyen, T.L., Ha, D.H., Voznak, M.: Multisource power splitting energy harvesting relaying network in half-duplex system over block Rayleigh fading channel: system performance analysis. Electronics **8**(1), 67 (2019)
51. Gradshteyn, I.S., Ryzhik, I.M.: Table of integrals, series, and products. Academic Press (2014)
52. Duy, T.T., Duong, T.Q., Thanh, T.L., Bao, V.N.Q.: Secrecy performance analysis with relay selection methods under impact of co-channel interference. IET Commun. **9**(11), 1427–1435 (2015)

53. Nguyen, B.C., Nguyen-Kieu, T., Hoang, T.M., Tran, P.T., Voznak, M.: Analysis of MRT/MRC diversity techniques to enhance the detection performance for MIMO signals in full-duplex wireless relay networks with transceiver hardware impairment. Phys. Commun. **42**, 101132 (2020)
54. Liu, X.: Average secrecy capacity of the Weibull fading channel. In: 2016 13th IEEE Annual Consumer Communications & Networking Conference (CCNC), pp. 841–844. IEEE (2016)
55. Olver, F.W.: NIST handbook of mathematical functions hardback and CD-ROM. Cambridge University Press (2010)
56. Chong, E.K., Żak, S.H.: An introduction to optimization, vol. 75. John Wiley & Sons (2013)

Chapter 9
Cognitive Radio Communications

9.1 Introduction

One of the intelligent systems that has garnered significant attention from researchers both domestically and internationally in recent years is cognitive radio networks [1]. Cognitive radio emerged in the context of increasing spectrum scarcity, while current spectrum usage remains inefficient. In cognitive radio networks, two different types of users coexist and share the spectrum: primary users and secondary users. While primary users are allowed to use the spectrum at any time, secondary users can only utilize it under the condition that they do not affect the Quality of Service (QoS) of the primary network. In other words, for both primary and secondary networks to effectively use the same frequency band, efficient spectrum-sharing methods are needed to ensure the QoS for the primary network while enhancing spectrum utilization for the secondary network. One solution proposed by researchers is the underlay spectrum-sharing technique [2]. This technique allows for continuous transmission and reception of data for both primary and secondary users, as both systems can operate simultaneously on the same frequency band. However, to apply this technique, secondary users are constrained by their transmission power, which is limited to a maximum allowable threshold so that the interference caused to the primary system does not affect its QoS [3, 4]. Due to the transmission power limitation, the performance of the secondary network is significantly degraded. To improve the performance of the secondary network, diversity techniques for transmission and reception have been applied. In [5], the authors investigated transmit antenna selection (TAS) techniques for underlay cognitive radio networks. In [6], a diversity reception model using selection combining (SC) at the secondary destination node was applied. In [7], the authors proposed a combined TAS/SC model to further enhance the OP performance of the secondary network.

In this chapter, we study the OP performance of the secondary network in an underlay cognitive radio environment. In the proposed model, a secondary source node transmits data to a secondary destination node in the presence of multiple

P. T. Tran et al., *Developing MATLAB Comprehensive Simulation for 5G/6G Wireless Communications*, Signals and Communication Technology,
https://doi.org/10.1007/978-981-95-1117-4_9

primary users. The source node is equipped with a single transmit antenna, while the destination node is equipped with two antennas and employs the SC technique to combine the received signals from the source. Because the receiving antennas at the destination are in close proximity, the channels between the source and the destination are correlated. Below, we present the main contributions of the paper [8] and the points of difference compared to existing publications:

- Unlike [6], the work in [8] investigate a generalized model with multiple primary users. Moreover, while previous studies such as [5, 7, 9] have explored underlay cognitive radio models with the presence of multiple primary users, these works assume that the channels between the secondary transmit nodes and the primary users are homogeneous. However, in reality, the positions of the primary users are arbitrary, and therefore, these channels are not homogeneous. Additionally, [5, 7, 9] assume that the maximum interference level at the primary nodes is the same. However, as demonstrated in [10], this interference threshold may vary among primary nodes due to the different distances between the secondary transmit nodes and the primary users. From this observation, the proposed model in this work considers varying interference thresholds specified at the primary users.
- Unlike [6], this work investigates the impact of hardware impairments on the performance of the secondary system. Indeed, to date, most studies assume that the transceivers are ideal. However, in reality, the hardware of these transceivers is imperfect due to phase noise, I/Q imbalance, or nonlinearity from amplifiers, etc. [11, 12]. Therefore, this study will examine the combined effect of hardware impairments and channel correlation on the performance of the secondary system.
- To evaluate the performance of the investigated model, we derive closed-form expressions for the OP under Rayleigh fading channels. Subsequently, we use the Monte Carlo method to simulate the results and verify the accuracy of the analytical findings.

The remainder of this chapter is organized as follows. Section 9.2 describes the system model and the evaluation parameter, which is the signal-to-noise ratio (SNR). Next, the performance analysis with the parameter being the OP of the system is carried out in Sect. 9.3. Section 9.5 describes the analytical and simulation results. Finally, the conclusions and future research directions are presented in Sect. 9.6.

9.2 System Model

The system model under investigation is described in Fig. 9.1. In this model, a single-antenna secondary source node (SS) transmits data to a secondary destination node (SD), which is equipped with two receive antennas and employs the Selection Combining (SC) technique to enhance data decoding quality. In underlay cognitive radio, the SS node must control its transmission power such that the interference caused to the primary users satisfies the maximum interference level pre-specified by the primary users. Assume that there are N existing primary users in the primary network,

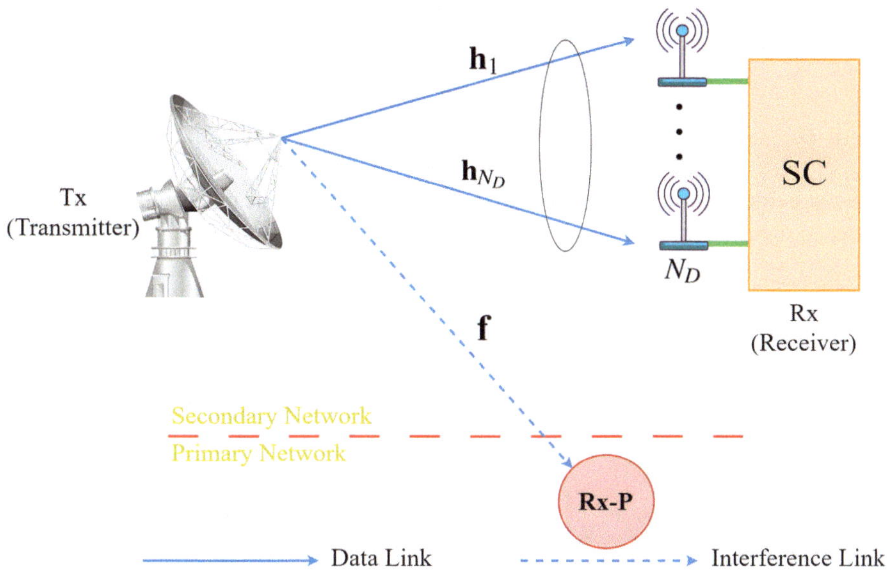

Fig. 9.1 System model

which are denoted as PU_1, PU_2, ..., PU_N. Furthermore, the maximum interference level set by each primary user PU_n is denoted as I_n, where $n = 1, 2, ..., N$.

Assume the system operates over Rayleigh fading channels, and the channel coefficients for the links between SS and SD, and SS and PUs, are denoted as h_i and g_n, respectively, where $i \in \{1, 2\}$ and $n = 1, 2, ..., N$. As mentioned in [13, 14], the channel coefficients h_i are represented as follows:

$$h_i = \sqrt{1-\rho}X_i + \sqrt{\rho}X_0 + j\left(\sqrt{1-\rho}Y_i + \sqrt{\rho}Y_0\right), \tag{9.1}$$

where $j = \sqrt{-1}$, X_0, Y_0, X_i, Y_i are random variables that follow a normal distribution with a mean of zero and a variance of $1/(2\lambda)$, ρ is the cross-correlation coefficient between h_1 and h_2, which is calculated by the following formula:

$$\rho = \frac{E\{h_1 h_2^*\}}{\sqrt{E\{|h_1|^2\}E\{|h_2|^2\}}}, 0 \le \rho \le 1, \tag{9.2}$$

Let $E\{Z\}$ represent the mean of Z, and Z^* is the complex conjugate of Z. We denote the data channel gain and interference channel gain by $\gamma_i = |h_i|^2$ and $\varphi_n = |g_n|^2$, respectively. By observing Eq. (9.1), we see that h_1 and h_2 are two correlated random variables, as they share components $\sqrt{\rho}X_0$ and $j\sqrt{\rho}Y_0$. Consequently, the channel gains γ_1 and γ_2 are also not independent, being jointly constrained by the random variable $U = X_0^2 + Y_0^2$. As presented in [6], the CDF of γ_i conditioned on U is given by

$$F_{\gamma_i|U}(z|u) = 1 - Q\left(\sqrt{\frac{2\lambda\rho u}{1-\rho}}, \sqrt{\frac{2\lambda z}{1-\rho}}\right) = 1 - Q(\sqrt{\alpha_1 u}, \sqrt{\alpha_2 z}), \tag{9.3}$$

where $Q(.,.)$ is the Marcum-Q function as defined in [15, (1)], and $\alpha_1 = \rho\alpha_2 = 2\lambda\rho/(1-\rho)$.

Before transmitting data to the destination, the source node must adjust its transmit power to meet all maximum interference threshold constraints (see [5, 7, 9]):

$$P_S \leq \frac{I_n}{(1+\kappa_p^2)\varphi_n}, \forall n \tag{9.4}$$

Here, κ_p^2 represents the total hardware impairment at the source station (SS) and PU_n. Accordingly, the maximum transmit power of SS is given by:

$$P_S = \min_{n=1,2,\ldots,N}\left(\frac{I_n}{(1+\kappa_p^2)\varphi_n}\right) = J_{\min}\min_{n=1,2,\ldots,N}\left(\frac{\mu_n}{\varphi_n}\right). \tag{9.5}$$

In Eq. (9.5), $J_{\min}$ and μ_n are defined as follows:

$$\begin{aligned} J_{\min} &= \frac{1}{1+\kappa_p^2}\min_{n=1,2,\ldots,N}(I_n), \\ \mu_n &= \frac{I_n}{\min\limits_{k=1,2,\ldots,N}(I_k)}. \end{aligned} \tag{9.6}$$

Due to the impact of hardware imperfections, the received signal-to-noise ratio (SNR) at the i-th antenna of the destination station (SD) is determined as (see (4)):

$$\Psi_i = \frac{P_S\gamma_i}{\kappa_D^2 P_S\gamma_i + N_0}, \tag{9.7}$$

Let κ_D^2 denote the total hardware impairment at both the source station (SS) and destination station (SD), and let N_0 represent the variance of Gaussian noise. By substituting P_S into ((9.7)), we can rewrite Ψ_i as follows:

$$\Psi_i = \frac{\Delta Z_{\min}\gamma_i}{\kappa_D^2\Delta Z_{\min}\gamma_i + 1}, \tag{9.8}$$

where $\Delta = J_{\min}/N_0$ and $Z_{\min} = \min\limits_{n=1,2,\ldots,N}(\mu_n/\varphi_n)$.

Since g_n experiences Rayleigh fading, φ_n follows an exponential distribution. Specifically, the CDF and PDF of φ_n are given as

$$F_{\varphi_n}(x) = 1 - e^{-\Omega_n x}, \ f_{\varphi_n}(x) = \Omega_n e^{-\Omega_n x}. \tag{9.9}$$

Assume that φ_n are independent and non-homogeneous random variables, such that $\Omega_n \neq \Omega_m$ when $n \neq m$. Therefore, the CDF of $Z_{\min}$ can be computed as

$$F_{Z_{\min}}(x) = \Pr\left(\min_{n=1,2,..,N}\left(\frac{\mu_n}{\varphi_n}\right) < x\right) = 1 - \prod_{n=1}^{N} \Pr\left(\frac{\mu_n}{\varphi_n} \geq x\right) = 1 - \prod_{n=1}^{N} F_{\varphi_n}\left(\frac{\mu_n}{x}\right). \tag{9.10}$$

By substituting the CDF of φ_n into (9.10), we have:

$$F_{Z_{\min}}(x) = 1 - \prod_{n=1}^{N}\left(1 - e^{-\frac{\Omega_n\mu_n}{x}}\right) = 1 - \prod_{n=1}^{N}\left(1 - e^{-\frac{\chi_n}{x}}\right), \tag{9.11}$$

where $\chi_n = \Omega_n \mu_n$.

For analytical convenience, we express the CDF of φ_n as

$$\begin{aligned} F_{Z_{\min}}(x) &= \sum_{k=1}^{N}(-1)^{k+1} \sum_{\substack{t_1=t_2=...=t_k=1,\\ t_1<t_2<...<t_k}}^{N} \exp\left(-\sum_{u=1}^{k}\frac{\chi_{t_u}}{x}\right) \\ &= \sum_{k=1}^{N}(-1)^{k+1} \sum_{\substack{t_1=t_2=...=t_k=1,\\ t_1<t_2<...<t_k}}^{N} \exp\left(-\frac{\omega_k}{x}\right), \end{aligned} \tag{9.12}$$

where $\omega_k = \sum_{u=1}^{k} \chi_{t_u}$.

From (9.12), we easily obtain the CDF of $Z_{\min}$ as

$$f_{Z_{\min}}(x) = \sum_{k=1}^{N}(-1)^{k+1} \sum_{\substack{t_1=t_2=...=t_k=1,\\ t_1<t_2<...<t_k}}^{N} \frac{\omega_k}{x^2} \exp\left(-\frac{\omega_k}{x}\right). \tag{9.13}$$

Now, considering the receiver SD, to enhance data decoding efficiency, the selection combining (SC) scheme can be applied. Then, the SNR for decoding the source data is given as

$$\Psi_{SC} = \max_{i=1,2}(\Psi_i). \tag{9.14}$$

9.3 Performance Analysis

In this section, we investigate the OP performance of the studied system. The OP is defined as the probability that the SNR ratio Ψ_{SC} of the secondary system falls below a predefined threshold signal-to-noise ratio, denoted by γ_{th}. From (9.8) and (9.14), the OP can be formulated as

$$
\begin{aligned}
OP &= \Pr(\Psi_{SC} \leq \gamma_{th}) = \Pr(\max(\Psi_1, \Psi_2) < \gamma_{th}) \\
&= \Pr\left(\frac{\Delta Z_{\min}\gamma_1}{\kappa_D^2 \Delta Z_{\min}\gamma_1 + 1} < \gamma_{th}, \frac{\Delta Z_{\min}\gamma_2}{\kappa_D^2 \Delta Z_{\min}\gamma_2 + 1} < \gamma_{th}\right).
\end{aligned} \tag{9.15}
$$

In (9.15)), we note that if $1 - \kappa_D^2 \gamma_{th} \leq 0$, the proposed system will always be in outage (i.e., OP = 1). In the case where $1 - \kappa_D^2 \gamma_{th} > 0$, the OP can be rewritten as

$$
OP = \Pr\left(\gamma_1 < \frac{\theta}{Z_{\min}}, \gamma_2 < \frac{\theta}{Z_{\min}}\right), \tag{9.16}
$$

where $\theta \triangleq \frac{\gamma_{th}}{(1-\kappa_D^2 \gamma_{th})\Delta}$.

By using the conditional CDF in (9.3), the OP can be expressed as

$$
\begin{aligned}
OP &= \int_0^{+\infty}\int_0^{+\infty} F_{\gamma_1|U}\left(\frac{\theta}{x}|u\right) F_{\gamma_2|U}\left(\frac{\theta}{x}|u\right) f_U(u) f_{Z_{\min}}(x) du dx \\
&= \int_0^{+\infty}\int_0^{+\infty} \left[1 - Q\left(\sqrt{\alpha_1 u}, \sqrt{\frac{\alpha_2 \theta}{x}}\right)\right]^2 f_U(u) f_{Z_{\min}}(x) du dx
\end{aligned} \tag{9.17}
$$

Furthermore, we can rewrite (9.17)) as

$$
OP = 1 - 2I_1 + I_2, \tag{9.18}
$$

where I_1 and I_2 are integrals defined respectively by

$$
I_1 = \int_0^{+\infty}\int_0^{+\infty} Q\left(\sqrt{\alpha_1 u}, \sqrt{\frac{\alpha_2 \theta}{x}}\right) f_U(u) f_{Z_{\min}}(x) du dx, \tag{9.19}
$$

$$
I_2 = \int_0^{+\infty}\int_0^{+\infty} \left[Q(\sqrt{\alpha_1 u}, \sqrt{\frac{\alpha_2 \theta}{x}})\right]^2 f_U(u) f_{Z_{\min}}(x) du dx. \tag{9.20}
$$

Next, we will calculate I_1 and I_2. By substituting the PDF $f_U(u) = \lambda \exp(-\lambda u)$ into (9.19), we obtain:

$$
I_1 = \int_0^{+\infty} A_1(x) f_{Z_{\min}}(x) dx, \tag{9.21}
$$

where

$$
A_1(x) = \int_0^{+\infty} Q\left(\sqrt{\alpha_1 u}, \sqrt{\frac{\alpha_2 \theta}{x}}\right) \lambda \exp(-\lambda u) du. \tag{9.22}
$$

By changing variable $t = \sqrt{u}$, $A_1(x)$ can be represented as

$$A_1(x) = 2\lambda \int_0^{+\infty} t \exp\left(-\lambda t^2\right) Q\left(\sqrt{\alpha_1} t, \sqrt{\frac{\alpha_2 \theta}{x}}\right) dt. \tag{9.23}$$

In this form, we can apply [14, (36)] to obtain an exact expression for $A_1(x)$ as follows:

$$\begin{aligned} A_1(x) &= \exp\left(-\frac{\lambda\alpha_2\theta}{(2\lambda+\alpha_1)x}\right) + Q\left(0, \sqrt{\frac{\alpha_2\theta}{x}}\right) - \exp\left(-\frac{\lambda\alpha_2\theta}{(2\lambda+\alpha_1)x}\right) Q\left(0, \sqrt{\frac{\alpha_1\alpha_2\theta}{(2\lambda+\alpha_1)x}}\right) \\ &= \exp\left(-\frac{\vartheta_1}{x}\right) + Q\left(0, \sqrt{\frac{\alpha_2\theta}{x}}\right) - \exp\left(-\frac{\vartheta_1}{x}\right) Q\left(0, \sqrt{\frac{\vartheta_2}{x}}\right). \end{aligned} \tag{9.24}$$

where $\vartheta_1 = \frac{\lambda\alpha_2\theta}{2\lambda+\alpha_1}$ and $\vartheta_2 = \frac{\alpha_1\alpha_2\theta}{2\lambda+\alpha_1}$.

Furthermore, by using formula [14, (2)], we can express (9.24) as a sum of exponential functions by:

$$A_1(x) = \exp\left(-\frac{\vartheta_1}{x}\right) + \exp\left(-\frac{\alpha_2\theta}{2x}\right) - \exp\left(-\frac{2\vartheta_1+\vartheta_2}{2x}\right). \tag{9.25}$$

Next, by substituting (9.13) and (9.25) into (9.22), we obtain I_1 in the following form:

$$\begin{aligned} I_1 = &\sum_{k=1}^{N} (-1)^{k+1} \sum_{\substack{t_1=t_2=\ldots=t_k=1, \\ t_1<t_2<\ldots<t_k}}^{N} \omega_k \\ &\times \int_0^{+\infty} \frac{1}{x^2} \exp\left(-\frac{\omega_k}{x}\right) \left[\exp\left(-\frac{\vartheta_1}{x}\right) + \exp\left(-\frac{\alpha_2\theta}{2x}\right) - \exp\left(-\frac{2\vartheta_1+\vartheta_2}{2x}\right)\right] dx. \end{aligned} \tag{9.26}$$

After evaluating the integrals in (9.26), we arrive at a closed-form expression for I_1 as:

$$I_1 = \sum_{k=1}^{N} (-1)^{k+1} \times \sum_{\substack{t_1=t_2=\ldots=t_k=1, \\ t_1<t_2<\ldots<t_k}}^{N} \left(\frac{\omega_k}{\omega_k+\vartheta_1} + \frac{2\omega_k}{2\omega_k+\alpha_2\theta} - \frac{2\omega_k}{2\omega_k+2\vartheta_1+\vartheta_2}\right). \tag{9.27}$$

Now, we will calculate the value of the integral I_2 in (9.20). First, we need to reformulate this integral as

$$I_2 = \int_0^{+\infty} A_2(x) f_{Z_{\min}}(x) dx, \tag{9.28}$$

where

$$A_2(x) = \int_0^{+\infty} \left[Q\left(\sqrt{\alpha_1 u}, \sqrt{\frac{\alpha_2 \theta}{x}}\right)\right]^2 \lambda \exp(-\lambda u)\, du. \tag{9.29}$$

Similar to ((9.23)), by performing the transformation $t = \sqrt{u}$, we obtain:

$$A_2(x) = 2\lambda \int_0^{+\infty} t \exp\left(-\lambda t^2\right) \left[Q\left(\sqrt{\alpha_1} t, \sqrt{\frac{\alpha_2 \theta}{x}}\right)\right]^2 dt. \tag{9.30}$$

Then, by applying [14, (59)], we achieve:

$$A_2(x) = \exp\left(-\frac{\vartheta_1}{x}\right) + \exp\left(-\frac{\vartheta_1}{x}\right) Q\left(\frac{\vartheta_3}{\sqrt{x}}, \frac{\vartheta_4}{\sqrt{x}}\right) - \exp\left(-\frac{\vartheta_1}{x}\right) Q\left(\frac{\vartheta_4}{\sqrt{x}}, \frac{\vartheta_3}{\sqrt{x}}\right), \tag{9.31}$$

where $\vartheta_3 = \frac{\alpha_1\sqrt{\alpha_2\theta}}{\sqrt{2\lambda+\alpha_1}\sqrt{2\lambda+2\alpha_1}}$ and $\vartheta_4 = \frac{\sqrt{2\lambda+\alpha_1}\sqrt{\alpha_2\theta}}{\sqrt{2\lambda+2\alpha_1}}$.

Then, by substituting (9.13) and 9.31 into (9.28), we can express I_2 in the following form:

$$I_2 = \sum_{k=1}^{N} (-1)^{k+1} \sum_{\substack{t_1=t_2=\ldots=t_k=1, \\ t_1<t_2<\ldots<t_k}}^{N} \omega_k (J_1 + J_2 - J_3), \tag{9.32}$$

where

$$J_1 = \int_0^{+\infty} \frac{1}{x^2} \exp\left(-\frac{\omega_k + \vartheta_1}{x}\right) dx, \tag{9.33}$$

$$J_2 = \int_0^{+\infty} \frac{1}{x^2} \exp\left(-\frac{\omega_k + \vartheta_1}{x}\right) Q\left(\frac{\vartheta_3}{\sqrt{x}}, \frac{\vartheta_4}{\sqrt{x}}\right) dx, \tag{9.34}$$

$$J_3 = \int_0^{+\infty} \frac{1}{x^2} \exp\left(-\frac{\omega_k + \vartheta_1}{x}\right) Q\left(\frac{\vartheta_4}{\sqrt{x}}, \frac{\vartheta_3}{\sqrt{x}}\right) dx. \tag{9.35}$$

First, we can readily compute J_1 as follows:

$$J_1 = \frac{1}{\omega_k + \vartheta_1}. \tag{9.36}$$

For the integrals J_2 and J_3, we initially perform a variable change for $w = 1/\sqrt{x}$ and rewrite these integrals as

$$J_2 = \int_0^{+\infty} 2w \exp(-(\omega_k + \vartheta_1) w^2) Q(\vartheta_3 w, \vartheta_4 w) dw, \tag{9.37}$$

$$J_3 = \int_0^{+\infty} 2w \exp(-(\omega_k + \vartheta_1)w^2) Q(\vartheta_4 w, \vartheta_3 w) dw. \tag{9.38}$$

By applying [10, Eq. (55)], we obtain:

$$J_2 = \frac{1}{2(\omega_k + \vartheta_1)} \left(1 + \frac{2(\omega_k + \vartheta_1) + \vartheta_3^2 - \vartheta_4^2}{\sqrt{(2(\omega_k + \vartheta_1) + \vartheta_3^2 + \vartheta_4^2)^2 - 4\vartheta_3^2\vartheta_4^2}} \right), \tag{9.39}$$

$$J_3 = \frac{1}{2(\omega_k + \vartheta_1)} \left(1 + \frac{2(\omega_k + \vartheta_1) + \vartheta_4^2 - \vartheta_3^2}{\sqrt{(2(\omega_k + \vartheta_1) + \vartheta_3^2 + \vartheta_4^2)^2 - 4\vartheta_3^2\vartheta_4^2}} \right). \tag{9.40}$$

By substituting the results from (9.36), (9.39), and (9.40) into (9.32), we get

$$I_2 = \sum_{k=1}^{N} (-1)^{k+1} \sum_{\substack{t_1 = t_2 = \ldots = t_k = 1, \\ t_1 < t_2 < \ldots < t_k}}^{N} \frac{\omega_k}{\omega_k + \vartheta_1} \times \left(1 - \frac{\vartheta_4^2 - \vartheta_3^2}{\sqrt{(2(\omega_k + \vartheta_1) + \vartheta_3^2 + \vartheta_4^2)^2 - 4\vartheta_3^2\vartheta_4^2}} \right) \tag{9.41}$$

9.4 MATLAB Implementation

The MATLAB coding for implementing the simulation of the cognitive radio module is presented in this section. The codes are organized into a main module and two functions to compute the OP by using an analytical expression and by using a Monte Carlo simulation. The main module is used for parameter input and producing the output graphs. A sample code for the main module is shown below.

```
clear all; clc;
% Simulation parameters
IPdB     =    0 : 5 : 30;
Muy      =    [1 1.5 1.75 2];
LD       =    0.75;
OM       =    [2 0.75 3 1.5];
kkD      =    0.1;
kkP      =    0.1;
gth      =    1.5;
NUM      =    2*10^6;
```

```

% Case 1: \rho = 0.1
Rho = 0.1;

% Call the functions to calculate theoretical and
    simulation OPs
OP_sim1 = SIMULATION(IPdB,Muy,Rho,LD,OM,kkD,kkP,gth,
    NUM);
OP_ana1 = THEORY(IPdB,Muy,Rho,LD,OM,kkD,kkP,gth);

% Case 1: \rho = 0.5
Rho = 0.5;
OP_sim2 = SIMULATION(IPdB,Muy,Rho,LD,OM,kkD,kkP,gth,
    NUM);
OP_ana2 = THEORY(IPdB,Muy,Rho,LD,OM,kkD,kkP,gth);

% Case 1: \rho = 0.9
Rho = 0.9;
OP_sim3 = SIMULATION(IPdB,Muy,Rho,LD,OM,kkD,kkP,gth,
    NUM);
OP_ana3 = THEORY(IPdB,Muy,Rho,LD,OM,kkD,kkP,gth);

% Plot the output graph
figure(1)
h1 = semilogy(IPdB, OP_sim1,'rs','LineWidth',2.5);
hold on;
h2 = semilogy(IPdB, OP_ana1,'r-','LineWidth',2.5,'
    MarkerSize',7)
hold on;
h3 = semilogy(IPdB, OP_sim2,'bo','LineWidth',2.5);
hold on;
h4 = semilogy(IPdB, OP_ana2,'b-','LineWidth',2.5,'
    MarkerSize',7)
hold on
h5 = semilogy(IPdB, OP_sim3,'kd','LineWidth',2.5);
hold on;
h6 = semilogy(IPdB, OP_ana3,'k:','LineWidth',2.5,'
    MarkerSize',7)
hold on
grid on;
legend([h1,h2,h3,h4,h5,h6],'Simulation with \rho =
    0.1','Analysis with \rho = 0.1',...
    'Simulation with \rho = 0.5','Analysis with \rho
        = 0.5',...
    'Simulation with \rho = 0.9','Analysis with \rho
        = 0.9');
```

```
xlabel(' \Delta (dB) '),ylabel('Outage probability')
    ;
title('OP versus \Delta at the rate 3 bps/Hz');
```

Next, we show the sample codes for calculating theoretical and simulation values of OP:

```
function OP = THEORY(IPdB,Muy,Rho,LD,OM,kkD,kkP,gth)
% Initialize the output array
OP    = zeros(1,length(IPdB));
% Calculation of OP for each value of interference
    level
for aa = 1 : length(IPdB)
    OP(aa) = ham(IPdB(aa),Muy,Rho,LD,OM,kkD,kkP,gth)
        ;
end
end

function out = ham(IPdB,Muy,Rho,LD,OM,kkD,kkP,gth)
%
PP       = 10.^(IPdB/10)/(1+kkP);
NN       = length(Muy);
Theta    = gth/(1 - gth*kkD)/PP;
AP1      = 2*LD*Rho/(1-Rho);
AP2      = AP1/Rho;
VThe1    = LD*AP2*Theta/(2*LD + AP1);
VThe2    = AP1*AP2*Theta/(2*LD + AP1);
VThe3    = AP1*sqrt(AP2*Theta)/sqrt(2*LD+AP1)/sqrt(2*
    LD+2*AP1);
VThe4    = sqrt(2*LD+AP1)*sqrt(AP2*Theta)/sqrt(2*LD+2*
    AP1);
%
if (1 <= gth*kkD)
    out = 1;
elseif (1 > gth*kkD && Rho == 0)
    II1      = 0;
    II2      = 0;
    SETALL = 1 : NN;
    for kk = 1 : NN
        SETS          = nchoosek(SETALL,kk);
        [hang cot] = size(SETS);
        for uu = 1 : hang
            SETU      = SETS(uu,:);
            Omegak  = 0;
```

```
                for tt = 1 : cot
                    Omegak = Omegak + OM(SETU(tt))*Muy(SETU(tt));
                end
                %
                II1     = II1 + (-1)^(kk+1)*Omegak/(Omegak + LD*Theta);
                II2     = II2 + (-1)^(kk+1)*Omegak/(Omegak + 2*LD*Theta);
            end
            out    = 1 - 2*II1 + II2;
        end
elseif (1 > gth*kkD && Rho == 1)
        out     = 1;
        SETALL = 1 : NN;
        for kk = 1 : NN
            SETS         = nchoosek(SETALL,kk);
            [hang cot] = size(SETS);
            for uu = 1 : hang
                SETU     = SETS(uu,:);
                Omegak = 0;
                for tt = 1 : cot
                    Omegak = Omegak + OM(SETU(tt))*Muy(SETU(tt));
                end
                %
                out     = out - (-1)^(kk+1)*Omegak/(Omegak + LD*Theta);
            end
        end
else
        II1     = 0;
        II2     = 0;
        SETALL = 1 : NN;
        for kk = 1 : NN
            SETS         = nchoosek(SETALL,kk);
            [hang cot] = size(SETS);
            for uu = 1 : hang
                SETU     = SETS(uu,:);
                Omegak = 0;
                for tt = 1 : cot
                    Omegak = Omegak + OM(SETU(tt))*Muy(SETU(tt));
                end
                %
                II1     = II1 + (-1)^(kk+1)*( Omegak/(Omegak + VThe1) + 2*Omegak/(2*Omegak
```

```
                    + AP2*Theta) - 2*Omegak/(2*Omegak +
                    2*VThe1 + VThe2) );
                II2     = II2 + (-1)^(kk+1)*Omegak/(
                    Omegak + VThe1)*(1 - (VThe4^2 - VThe3
                    ^2)/sqrt((2*(Omegak + VThe1) + VThe3
                    ^2 + VThe4^2)^2 - 4*VThe3^2*VThe4^2))
                    ;
            end
        end
        out     = 1 - 2*II1 + II2;
    end
end
```

```
function OP = SIMULATION(IPdB,Muy,Rho,LD,OM,kkD,kkP,
    gth,NUM)
%
PP    = 10.^(IPdB/10)/(1+kkP);
NN    = length(Muy);
OP    = zeros(1,length(PP)); % Initialize the output
    array
%
for aa = 1 : length(PP)
    aa
    for num = 1 : NUM
        Pow_min    = inf;
        for bb = 1 : NN
            hSP    = 1/sqrt(2*OM(bb))*(randn(1,1) + j
                *randn(1,1));
            power = PP(aa)*Muy(bb)/abs(hSP)^2;
            if (power < Pow_min)
                Pow_min = power;
            end
        end

                % Generate random channel gains
                X1    = 1/sqrt(2*LD)*randn(1,1);
        Y1    = 1/sqrt(2*LD)*randn(1,1);
        X2    = 1/sqrt(2*LD)*randn(1,1);
        Y2    = 1/sqrt(2*LD)*randn(1,1);
        X0    = 1/sqrt(2*LD)*randn(1,1);
        Y0    = 1/sqrt(2*LD)*randn(1,1);

```

```
            hSD1 = sqrt(1 - Rho)*X1 + sqrt(Rho)*X0 + j*(
                sqrt(1 - Rho)*Y1 + sqrt(Rho)*Y0);
            hSD2 = sqrt(1 - Rho)*X2 + sqrt(Rho)*X0 + j*(
                sqrt(1 - Rho)*Y2 + sqrt(Rho)*Y0);

                % Calculate the SNR
                GSD_max = max(abs(hSD1)^2,abs(hSD2)
                    ^2);
            SNR_SD  = Pow_min*GSD_max/(kkD*Pow_min*
                GSD_max + 1);

            if (SNR_SD < gth)
                OP(aa) = OP(aa) + 1;
            end
        end
    end
    OP = OP./NUM;
    end
```

9.5 Simulation Results

In this section, the analytical results obtained in Sect. 9.3 are verified by Monte Carlo simulation. In the simulation, we assume that the total level of hardware impairment on the links is equal $\kappa_D = \kappa_P = \kappa$, and the SNR threshold is fixed at $\gamma_{th} = 1.5$. In addition, we assume that the primary networks has four users ($N = 4$) with the values of the parameter μ are $\mu \in \{1, 1.5, 1.75, 2\}$ and the parameters of the corresponding channel gains are $\Omega_n \in \{2, 0.75, 3, 1.5\}$.

Figure 9.2 shows the effect of channel correlation coefficient ρ on the system performance. In particular, the system performance degrades significantly when the correlation is strong. In this figure, the OP is plotted with respect to Δ (dB) with $\lambda = 0.75$ and the hardware impairment level is set to $\kappa = 0.1$. When the correlation level ρ varies from 0.1 to 0.5, the OP of the secondary user does not change significantly. However, with the same value of Δ, the system performance degrades significantly as the correlation coefficient increases to $\rho = 0.99$.

Figure 9.3 depicts the OP according to the channel correlation coefficient. We fix the interference threshold power to noise ratio at $\Delta = 20dB$, and the parameter of the channel gain distribution at $\lambda = 0.75$. It can be seen that system performance degrades inversely with the channel correlation coefficient ρ. Especially, when ρ is in the range from 0.8 to 1, the OP increases rapidly. In addition, hardware impairment also has a significant impact on this performance factor. We can observe from Fig. 9.3 that the Hardware impairment level must be regulated at a permissible limit.

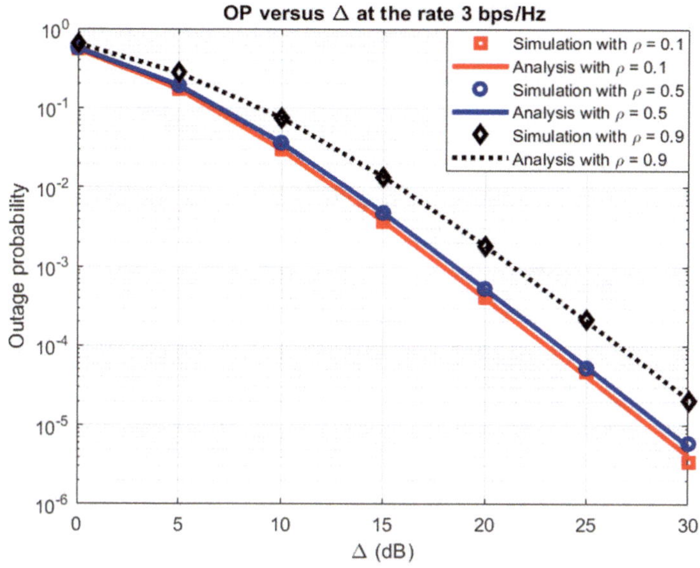

Fig. 9.2 OP versus the interference threshold power to noise ratio (Δ)

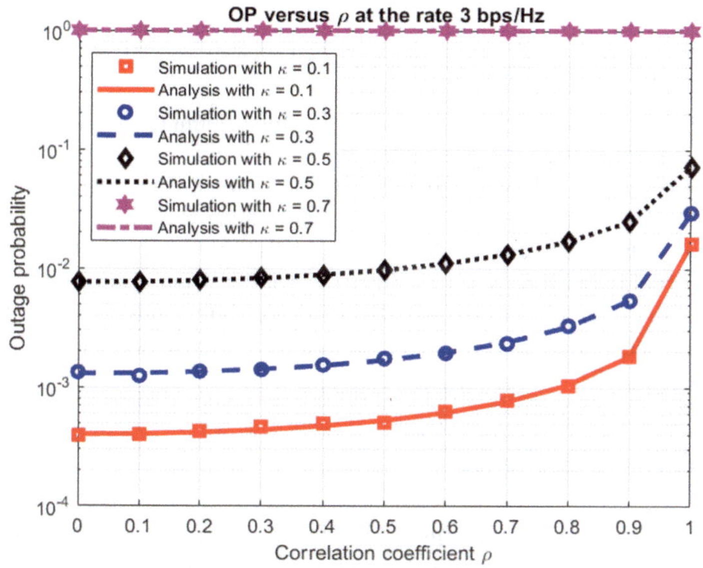

Fig. 9.3 OP versus the channel correlation coefficient ρ

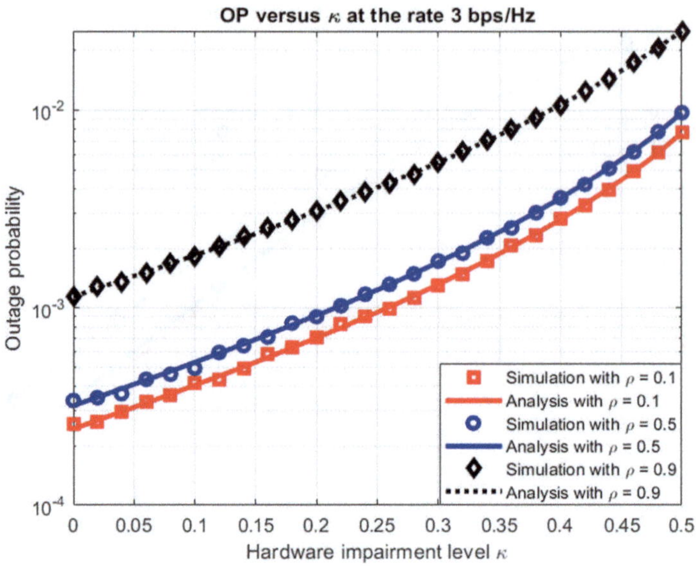

Fig. 9.4 OP versus the hardware impairment level κ

In particular, as κ goes to 0.7, outage always occurs regardless of other system parameters.

Figure 9.4 shows the OP versus the hardware impairment level κ with Δ fixed to 20 dB and $\lambda = 0.75$. For the reason that the hardware impairment strongly affects system performance, OP is drawn with respect to the hardware impairment level ranging from 0 to 0.9. This variation range indicates performance degradation according to the increase in hardware impairment level κ. Furthermore, when we adjust the channel correlation coefficient ρ as $\rho \in \{0.1, 0.5, 0.9\}$ and keep the same hardware impairment level, the outage performance also changes significantly. Again, the results show that an acceptable level for this correlation coefficient is at most 0.5.

9.6 Chapter Conclusion

In this chapter, we develop a simulation module to evaluate the system performance of cognitive radio networks. In particular, we consider an underlay cognitive radio network with a secondary transmitter and a secondary receiver, where the latter is equipped with two antennas. The OP of the secondary receiver is analyzed under the simultaneous effect of the Rayleigh fading channel correlation and hardware impairment. The analysis is verified by a Monte Carlo simulation using MATLAB. The simulation results show the joint effect of the two factors mentioned above and help recommend the best system parameters to limit the negative effects on the system.

References

1. Mitola, J., Maguire, G.Q.: Cognitive radio: making software radios more personal. IEEE Pers. Commun. **6**(4), 13–18 (1999)
2. Duy, T.T., Son, P.N.: Secrecy performances of multicast underlay cognitive protocols with partial relay selection and without eavesdropper's information. KSII Trans. Internet Inf. Syst. **9**(11), 4623–4643 (2015). https://doi.org/10.3837/tiis.2015.11.021
3. Duy, T.T., Bao, V.N.Q.: Performance analysis of cooperative-based multi-hop transmission protocols in underlay cognitive radio with hardware impairment. VNU J. Sci. Comput. Sci. Commun. Eng. **31**(2) (2015)
4. Ha, D.B., Vu, T.T., Duy, T.T., Bao, V.N.Q.: Secure cognitive reactive decode-and-forward relay networks: with and without eavesdropper. Wirel. Pers. Commun. **85**, 2619–2641 (2015)
5. Ngoc, P.T.D., Duy, T.T., Bao, V.N.Q., Ho-Van, K.: Transmit antenna selection protocols in random cognitive networks under impact of hardware impairments. In: 2016 3rd National Foundation for Science and Technology Development Conference on Information and Computer Science (NICS), pp. 38–43. IEEE (2016)
6. Thanh, T.L., Bao, V.N.Q., Ngoc, P.T.D., Duy, T.T.: On the performance of cognitive underlay simo networks over equally correlated rayleigh fading channels. REV J. Electron. Commun. **5**(1–2) (2016)
7. Ngoc, P.T.D., Duy, T.T., Bao, V.N.Q., Ho-Van, K.: Performance enhancement for underlay cognitive radio with partial relay selection methods under impact of hardware impairment. In: 2015 International Conference on Advanced Technologies for Communications (ATC), pp. 645–650. IEEE (2015)
8. Ngoc, P.T.D., Duy, T.T., Van Khuong, H.: Outage probability evaluation of underlay cognitive radio networks under joint impact of channel correlation and hardware impairments. Le Quy Don Technical University-Section on Information and Communication Technology (LQDTU-JICT) (11), 49–63 (2018)
9. Duong, T.Q., Yeoh, P.L., Bao, V.N.Q., Elkashlan, M., Yang, N.: Cognitive relay networks with multiple primary transceivers under spectrum-sharing. IEEE Signal Process. Lett. **19**(11), 741–744 (2012)
10. Duy, T.T., Thanh, T.L., Bao, V.N.Q.: A hybrid spectrum sharing approach in cognitive radio networks. In: 2014 International Conference on Computing, Management and Telecommunications (ComManTel), pp. 19–23. IEEE (2014)
11. Krone, S., Fettweis, G.: On the capacity of OFDM systems with receiver I/Q imbalance. In: 2008 IEEE International Conference on Communications, pp. 1317–1321. IEEE (2008)
12. Björnson, E., Hoydis, J., Kountouris, M., Debbah, M.: Hardware impairments in large-scale miso systems: energy efficiency, estimation, and capacity limits. In: 2013 18th International Conference on Digital Signal Processing (DSP), pp. 1–6. IEEE (2013)
13. Chen, Y., Tellambura, C.: Distribution functions of selection combiner output in equally correlated Rayleigh, Rician, and Nakagami-m fading channels. IEEE Trans. Commun. **52**(11), 1948–1956 (2004)
14. Chen, Y., Tellambura, C.: Performance of L-branch diversity combiners in equally correlated Rician fading channels. In: IEEE Global Telecommunications Conference, 2004. GLOBECOM'04., vol. 5, pp. 3379–3383. IEEE (2004)
15. Nuttall, A.H.: Some integrals involving the q-function. Tech. rep, DTIC Document (1972)

References

Chapter 10
Artificial Intelligence in 6G Wireless Communications

10.1 Introduction

10.1.1 Purpose

In the rapidly evolving landscape of wireless technology, large-scale MIMO systems have become a critical factor in shaping the future of wireless networks, especially within 5G and subsequent technologies. The objective of this study is to develop and propose novel Maximum Likelihood (ML) detection techniques, based on machine learning, for MIMO systems utilizing 1-bit Analog-to-Digital Converters (ADCs).

Although 1-bit ADCs offer benefits in terms of cost and energy efficiency, they also pose significant challenges in accurate and effective signal processing. This study aims to address these limitations by employing advanced ML-based detection methods to enhance accuracy and reliability in signal processing.

Two methods are proposed: the bias learning method and the dithering learning method. These approaches are designed to improve classification and signal detection under 1-bit quantization conditions without requiring detailed CSI. This represents a breakthrough, as CSI has traditionally been an essential component in signal detection within MIMO systems.

The ultimate goal is to optimize performance and reduce energy consumption in MIMO systems employing 1-bit ADCs. Through the application of these innovative methods, the study promises significant improvements in technical performance while also lowering operational costs and energy usage—key factors in modern wireless applications.

This research paves a new path for designing and optimizing MIMO systems, thereby contributing to the advancement of wireless communication technology. The findings and proposed methods not only offer solutions to current challenges but also aim toward a sustainable and efficient future for wireless networks.

P. T. Tran et al., *Developing MATLAB Comprehensive Simulation for 5G/6G Wireless Communications*, Signals and Communication Technology,
https://doi.org/10.1007/978-981-95-1117-4_10

10.1.2 Objectives

This chapter presents the work by [1], which aims to develop and optimize machine learning methods within the context of large-scale MIMO systems operating under 1-bit quantized signals. Such systems are particularly challenging due to information limitations imposed by 1-bit quantization and the dependency of learning efficacy on the training data length.

The focus is on minimizing reliance on extensive training data and enhancing learning efficiency in scenarios with incomplete or inaccurate data. This is achieved by introducing and advancing new learning methods to optimize detection performance in systems with 1-bit quantized signals.

The "biased-learning" method involves setting a minimum probability threshold for each probability function, preventing any probability function from reaching zero. This approach helps preserve crucial learned information, as setting a minimum probability ensures each function remains relevant and avoids information loss during training.

In addition, the "dithering-and-learning" method is introduced. Here, dithering noise is added to the signal prior to quantization, with the aim of inducing changes in the output sequence's sign, thereby allowing for a more accurate estimation of the true probability functions. This method is especially valuable for illustrating output sign variations with minimal untrained cases, thereby enhancing the learning process's quality and accuracy.

The expected outcomes of these methods include improved accuracy and reliability in detection for large-scale MIMO systems with 1-bit quantization. The proposed techniques promise enhanced detection performance in conditions of imperfect and constrained data, thus enabling applications in real-world environments where information and data resources may be limited. Notably, reducing dependency on the length of training data makes the learning process more flexible and efficient, accommodating practical situations where large, comprehensive data collection may not be feasible.

Expected simulation results will include performance evaluations. Comparisons with reference methods will show that the proposed methods closely align with optimal 1-bit ML detection performance across various SNR ranges, which significantly outperforms ZF detection and other learning-based methods. Performance gaps relative to training duration are also be evaluated, with increased training time (N_{tr}) revealing a larger performance gap between the proposed and other methods. Additionally, robustness across SNR and training durations demonstrates effective performance in low-to-high SNR regimes, suggesting near-optimal detection even with limited training duration. The dithering-and-learning method achieves performance close to perfect SNR scenarios, highlighting the method's resilience to SNR estimation errors. The effectiveness of dithering enables higher estimation accuracy by capturing a larger number of non-zero functions, significantly enhancing detection precision, especially in high SNR scenarios.

Finally, evaluation of post-processing updates shows that biased learning with post-update yields improved SER, which outperforms dithering-and-learning in high SNR conditions where most subframes are accurately decoded.

10.1.3 Concept and Methodology

This study presents an innovative approach in the field of machine learning, specifically in signal processing for large-scale MIMO systems with 1-bit quantized signals. The primary aim of this research is to develop effective machine learning methods that overcome the limitations of traditional approaches, particularly in contexts where information is often constrained due to 1-bit quantization.

The focus of this study is on reducing the dependency of the learning process on the length of training data, a common challenge in machine learning. To achieve this, two primary methods have been proposed: "biased learning" and "dithering and learning."

The "biased-learning" method has been developed to address the issue of probability functions potentially reaching zero probability, which often occurs in traditional machine learning systems and can lead to the loss of critical information. By setting a minimum probability threshold for each probability function, this method ensures each function remains relevant and prevents information loss during learning.

On the other hand, the "dithering-and-learning" method involves adding dithering noise to the signal before the quantization process. This approach improves the input signal quality for learning, thereby enhancing accuracy in signal detection and classification. The method is particularly valuable for illustrating output sign variations, allowing for more precise probability function estimations even under information constraints.

The study focuses on large-scale MIMO (Multiple-Input Multiple-Output) systems in which signals are quantized with only 1 bit. These systems involve multiple antennas at both the transmitter and receiver, significantly enhancing the capacity and reliability of wireless communication systems. However, the 1-bit quantization of signals presents a considerable challenge due to amplitude information loss, necessitating new data-processing approaches.

The study examines and analyzes various cases and scenarios within the MIMO system with 1-bit quantized signals, including performance evaluation of the "biased-learning" and "dithering-and-learning" methods under different data and environmental conditions. These scenarios may involve variations in the number of antennas, noise conditions, and other system parameters.

Investigating these cases aims to determine the performance and efficiency of the proposed methods within realistic and diverse MIMO system environments. Analysis and findings from these case studies provide evidence of the benefits and effectiveness of the proposed methods, offering insight into how these methods can be applied in practice.

10.2 System Model

In this study, we consider multi-user MIMO communication in an uplink environment, where N_u represents the number of users, each equipped with a single antenna, transmitting signals to the BTS with N_r antennas. We assume that the number of receiving antennas is much larger than the number of users, i.e., $N_r \gg N_u$. The uplink communication process consists of a pilot transmission phase and a data transmission phase: each user first transmits pilot symbols over N_ρ symbol periods, followed by data symbols over N_d symbol periods.

The total set of pilot vectors $\bar{\mathbf{s}}_k \in \mathbb{C}^{N_u}$, denoted as $\mathcal{K}$, for $k = 1, \ldots, K$, contains each pilot vector $\bar{\mathbf{s}}_k$, which is transmitted N_{tr} times during the pilot transmission phase, implying $N_\rho = K N_{\text{tr}}$.

Let $\bar{\mathbf{s}}[n] \in \mathbb{C}^{N_u}$, $n = 1, \ldots, N_d$ represent the data symbol at time n. The received signal vector at time n is given by

$$\mathbf{r}[n] = \sqrt{\rho}\,\overline{\mathbf{H}}\,\bar{\mathbf{s}}[n] + \bar{\mathbf{n}}[n], \tag{10.1}$$

where ρ denotes the average transmit power per user, $\overline{\mathbf{H}}$ is the $N_u \times N_r$ channel matrix, and $\bar{\mathbf{n}}[n]$ represents the noise vector at time n, which follows a complex Gaussian distribution with zero mean and variance N_0, i.e., $\bar{\mathbf{n}}[n] \sim \mathcal{CN}(0, N_0\mathbf{I})$. Here, $\mathbf{I}$ denotes the identity matrix of appropriate dimension. Each user symbol $\bar{s}_u[n]$ is drawn from a symbol set, $\bar{s}_u[n] \in S$, with zero mean and unit variance, such that $\mathbb{E}[\bar{s}_u[n]] = 0$ and $\mathbb{E}[|\bar{s}_u[n]|^2] = 1$, where $\bar{s}_u[n]$ denotes the u-th element of $\bar{s}[n]$.

We assume that the channel is flat and narrowband, and that $\overline{\mathbf{H}}$ remains constant over the transmission duration of $(N_\rho + N_d)$ symbol periods. We define the signal-to-noise ratio (SNR) as $\gamma = \rho/N_0$.

The received signal in (10.1) is quantized at the ADCs using 1-bit resolution. Consequently, each real and imaginary part of the received signals is quantized by 1-bit ADCs, outputting only the sign of the quantized input, i.e., either $+1$ or $- -1$. The quantized signal can be expressed as follows:

$$\bar{\mathbf{y}}[n] = Q\left(\Re\{\bar{\mathbf{r}}[n]\}\right) + jQ\left(\Im\{\bar{\mathbf{r}}[n]\}\right), \tag{10.2}$$

where $Q(.)$ represents the element-wise quantizer, and $\Re\{\bar{\mathbf{r}}[n]\}$ and $\Im\{\bar{\mathbf{r}}[n]\}$ denote the real and imaginary parts of the complex vector $\bar{\mathbf{r}}[n]$, respectively. The received signal as a complex vector $\bar{\mathbf{r}}[n]$ can be rewritten in real-valued vector form as follows:

$$\bar{\mathbf{r}}[n] = \begin{bmatrix} \Re\{\bar{\mathbf{r}}[n]\} \\ \Im\{\bar{\mathbf{r}}[n]\} \end{bmatrix} = \sqrt{\rho}\,\mathbf{H}\,s[n] + \mathbf{n}[n], \tag{10.3}$$

where

$$\mathbf{H} = \begin{bmatrix} \Re\{\overline{\mathbf{H}}\} & \Im\{\overline{\mathbf{H}}\} \\ \Im\{\overline{\mathbf{H}}\} & \Re\{\overline{\mathbf{H}}\} \end{bmatrix}, \tag{10.4}$$

$$\mathbf{s}[n] = \begin{bmatrix} \Re\{\bar{\mathbf{s}}[n]\} \\ \Im\{\bar{\mathbf{s}}[n]\} \end{bmatrix}, \mathbf{n}[n] = \begin{bmatrix} \Re\{\bar{\mathbf{n}}[n]\} \\ \Im\{\bar{\mathbf{n}}[n]\} \end{bmatrix}. \tag{10.5}$$

Thus, we can express the quantized signal in real-valued vector form as

$$\mathbf{y}[n] = \mathcal{Q}(\mathbf{r}[n]) = \mathcal{Q}(\sqrt{\rho}\,\mathbf{H}\,\mathbf{s}[n] + \mathbf{n}[n]), \tag{10.6}$$

where each element $r_i[n]$ is quantized to $y_i[n] = 1$ if $r_i[n] > 0$, or $y_i[n] = -1$ otherwise.

10.3 Problem Formulation

In this section, we formulate the overall problem and divide it into three cases as follows: efficient 1-bit maximum likelihood (ML) detection with CSI, efficient 1-bit ML detection without CSI, and updating the likelihood function. For the efficient 1-bit ML detection without CSI, we further subdivide into two cases: SNR-independent efficient 1-bit ML detection and SNR-dependent efficient 1-bit ML detection, which will be detailed in the following sections.

10.3.1 1-Bit ML Detection with CSI

Assume $\mathbf{s}_k \in \mathcal{S}$ is the k-th pilot symbol in the set of K pilot symbol vectors in real-valued form, representing the k-th element in the set of all possible symbol vectors $\mathcal{S}$. The likelihood probability of the quantized signal vector $\mathbf{y}[n]$ for a given channel $\mathbf{H}$ and transmitted symbol vector $\mathbf{s}_k$ can be estimated as follows:

$$\mathbb{P}(\mathbf{y}[n]|\mathbf{H}, \mathbf{s}_k) \approx \prod_{i=1}^{2N_r} \rho_k(y_i[n]), \tag{10.7}$$

where $\rho_k(y_i[n])$ denotes the likelihood function for the i-th element $y_i[n] \in \{1, -1\}$ when the symbol $\mathbf{s}_k$ is transmitted over a given channel, which is defined as

$$\rho_k(y_i[n]) = \mathbb{P}(y_i[n]|\mathbf{f}_i, \mathbf{s}_k) = \begin{cases} 1 - \Phi(w), & \text{if } y_i[n] = 1 \\ \Phi(w), & \text{if } y_i[n] = -1. \end{cases} \tag{10.8}$$

Here, $w = -\frac{\rho}{N_0/2}\mathbf{f}_i^{\mathbf{H}}\mathbf{s}_k$, where $\mathbf{f}_i^{\mathbf{H}}$ is the i-th row of $\mathbf{H}$. Additionally, $\Phi(x) = \int_{-\infty}^{x} \frac{1}{\sqrt{2\pi}} \exp\left(-\frac{\tau^2}{2}\right) d\tau$ represents the CDF of the standard Gaussian distribution. Note that (10.7) becomes an exact representation of $\mathbb{P}(\mathbf{y}[n]|\mathbf{H}, \mathbf{s}_k)$ when all elements of $\mathbf{y}[n]$ are independent. Based on (10.7), the ML detection rule is given by

$$k^*[n] = \arg\max_{k \in \mathcal{K}} \prod_{i=1}^{2N_r} \rho_k(y_i[n]), \tag{10.9}$$

where we define the index set of all possible symbol vectors as $\mathcal{K} = \{1, 2, \ldots, k\}$ with $k = |\mathcal{S}|^{N_u}$. Here, $|\mathcal{S}|$ denotes the cardinality of the set $\mathcal{S}$. The detection rule in (10.9) is computed using (10.8) with CSI.

10.3.2 Efficient 1-Bit ML Detection Without CSI

This section introduces a machine learning-based ML detection method with 1-bit quantization that does not require channel estimation, proposing efficient learning techniques for a training length N_{tr}. During the transmission of a wavelength with length N_ρ, each pilot symbol vector $\mathbf{s}_k$ is transmitted N_{tr} times, and the BS learns the empirical likelihood functions by measuring the frequency of $y_i[n] = 1$ and $y_i[n] = -1$ during transmission as follows:

$$\widehat{\rho}_{k,i}^{(b)} = \begin{cases} \widehat{\rho}_{k,i}^{(\mathbf{1})} = \frac{1}{N_{tr}} \sum_{t=1}^{N_{tr}} \mathbf{1}(y_i[(k-1)N_{tr} + t] = \mathbf{1}), \\ \widehat{\rho}_{k,i}^{(-1)} = 1 - \widehat{\rho}_{k,i}^{(\mathbf{1})} \end{cases}, \tag{10.10}$$

where $b \in \{1, -1\}$, and $\mathbf{1}(\mathcal{A})$ denotes an indicator function with $\mathbf{1}(\mathcal{A}) = 1$ if $\mathcal{A}$ is true and $\mathbf{1}(\mathcal{A}) = 0$ otherwise. After learning the empirical likelihood functions using the formula above, the BS has an estimate of the empirical likelihood probability for the quantized signal vector $y[n]$ during the data transmission phase as follows:

$$\mathbb{P}(y[n]|\mathbf{H}, \mathbf{s}_k) \approx \prod_{i=1}^{2N_r} \widehat{\rho}_{k,i}^{(\mathbf{1})} \mathbf{1}(y_i[n] = \mathbf{1}) + \widehat{\rho}_{k,i}^{(-1)} \mathbf{1}(y_i[n] = -1) \tag{10.11}$$

and ML detection can be performed as in (10.9).

However, with a limited training length N_{tr}, the empirical likelihood function $\widehat{\rho}_{k,i}^{(b)}$ for $b \in \{1, -1\}$ may have zero probability (i.e., $\widehat{\rho}_{k,i}^{(b)} = 0$) after learning through N_{tr} transmissions if no sign changes in the quantized sequence $y_i[n]$, with $n = (k-1)N_{tr} + 1, \ldots, kN_{tr}$, occur within the N_{tr} transmissions for symbol $\mathbf{s}_k$. Empirical likelihood functions with zero probability render the empirical likelihood probability of the observed signal $\mathbb{P}(y[n]|\mathbf{H}, \mathbf{s}_k)$ in (10.11) equal to zero for multiple candidate symbols $\mathbf{s}_k$, including potentially the desired symbol. Therefore, empirical likelihood functions with zero probability erase all acquired information from the pilot-based learning phase, significantly impairing detection capability. Note that the likelihood of zero-probability cases increases at high SNR.

10.3.3 Effective 1-Bit ML Learning Method Without SNR Information

To address the challenge of detection-based learning without SNR information γ or CSI, this study proposes a directed yet simple and highly effective ML detection method for training length N_{tr}. Subsequently, the proposed detection method will be extended to cases with available SNR information to enhance learning performance. In this method, the study constrains the minimum empirical likelihood function by $\widehat{\rho}_{k,i}^{(b)} \geq \rho_{\text{bias}}$. The polarization probability ρ_{bias} must satisfy the condition $\rho_{\text{bias}} < \frac{1}{N_{tr}}$ as no change in the sign of quantized output sequences $y_i[n]$ is observed in N_{tr} transmissions for symbol $\mathbf{s}_k$. The directed ML detection method is summarized as follows:

1. During the pilot transmission phase, the BTS computes empirical likelihood functions according to (10.10).
2. If a zero probability is observed for any empirical likelihood function $\widehat{\rho}_{k,i}^{(b)} = 0$, the BTS sets $\widehat{\rho}_{k,i}^{(b)} = \rho_{\text{bias}}$ and $\widehat{\rho}_{k,i}^{(-b)} = 1 - \rho_{\text{bias}}$.
3. During the data phase, the BS performs ML detection following (10.9) by computing (10.11).

Although this directed ML detection method prevents the loss of information acquired during the pilot transmission phase, it cannot capture the variability in probabilities among empirical likelihood functions with zero probabilities. Additionally, the number of empirical likelihood functions with zero probabilities tends to increase as the SNR increases, requiring the directed ML detection method to replace a large number of zero-probability likelihood functions with the polarization probability ρ_{bias}. Therefore, although this directed ML detection method improves detection performance, high dependence on the polarization probability in high SNR conditions may be undesirable. To address these challenges, we further develop a 1-bit ML detection method with the presence of SNR information.

10.3.4 Effective 1-Bit ML Learning Method with SNR Information

In this section, the project proposes a 1-bit ML detection method with available SNR information γ at the BS, where the noise variance N_0 and transmit power ρ are known by the BTS. As illustrated in Fig. 10.1, the BTS adds dithering signals $d[n]$ to the quantized inputs $r_i[n]$ only during the pilot transmission phase to induce changes in the sign of the output sequence $y_i[n]$. Unlike traditional polarization, this approach is used to estimate realistic empirical likelihood functions with the least number of undertrained likelihood functions. After dithering, the quantized input in the actual vector form becomes

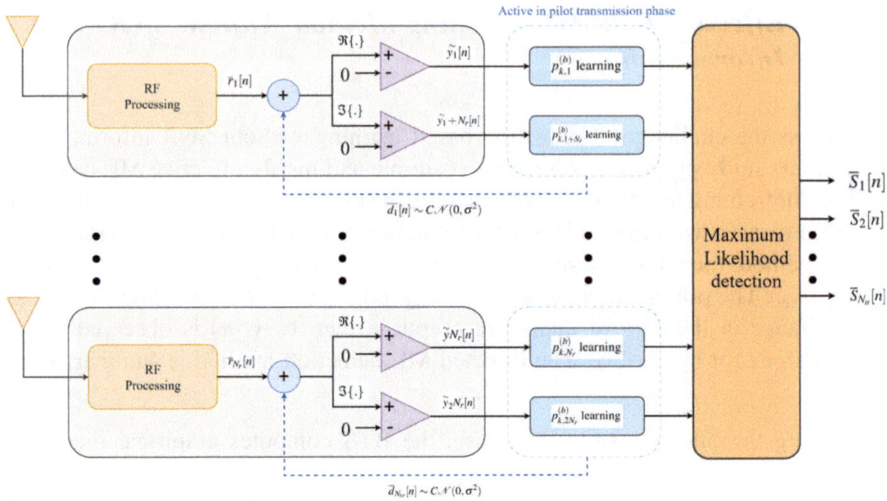

Fig. 10.1 Receiver model for pilot transmission phase with dithering signal pre-quantization

$$\tilde{r} = r[n] + d[n] = \sqrt{\rho} H s_k + n[n] + d[n]. \tag{10.12}$$

The study uses $d[n]$, which is distributed according to a Gaussian distribution with mean of 0 and variance of $\sigma^2/2$, i.e., $d[n] \sim \mathcal{N}(\mathbf{0}, \sigma^2/2\mathbf{I}_{2N_r})$, where the variance $\sigma^2/2$ is known to the BTS. The polarized and quantized signal then becomes

$$\tilde{y}[n] = Q(\sqrt{\rho} H s_k + n[n] + d[n]). \tag{10.13}$$

The BTS then calculates the predicted likelihood function for the polarized signals $\tilde{\rho}_{k,i}^{(b)}$ as in (10.10) for $b \in \{1; -1\}$. By choosing $b = 1$, as shown in (10.8), $\tilde{\rho}_{k,i}^{(1)}$ is theoretically determined as

$$\tilde{\rho}_{k,i}^{(1)} = 1 - \Phi\left(-\sqrt{\frac{2\rho}{N_0 + \sigma^2}}\mathbf{f}_i^H \mathbf{s}_k\right). \tag{10.14}$$

Since N_0 and σ^2 are known to the BTS, and $\tilde{\rho}_{k,i}^{(1)}$ is estimated from (10.10), the BTS can estimate $\psi_k \triangleq \sqrt{\rho}\mathbf{f} * \mathbf{s}_k$ using (10.14):

$$\psi_k = -\sqrt{\frac{N_0 + \sigma^2}{2}}\Phi^{-1}(1 - \overset{\sim(1)}{\rho}_{k,i}). \tag{10.15}$$

Finally, the BTS uses the estimated ψ_k and the known N_0 to estimate the true (non-polarized) likelihood function $\rho_{k,i}^{(b)}$ by using (10.8). Since the likelihood of the polarized signal $\overset{\sim(1)}{\rho}_{k,i}$ in (10.14) is less likely to have zero probability compared to the non-polarized case due to artificial SNR reduction, the BTS can learn most likelihood functions $\widehat{\rho}_{k,i}^{(b)}$ with a reasonable training length N_{tr}.

Zero-probability likelihood functions may still exist even after polarization if N_{tr} is insufficient. If the BTS observes zero-probability likelihood functions, it can also apply the polarized learning method to those functions. Unlike traditional polarization methods, the polarized signal is not discarded after quantization in the 1-bit ML detection method with polarization and learning, as polarization is solely used to induce changes in the signs of the received signal sequences within N_{tr}. Additionally, the BTS only needs to know the variance of the polarization distribution, and the dithering is not used during the data transmission phase.

10.3.5 Post-estimation Update of Likelihood Functions

The performance of the proposed algorithms can be further enhanced by adapting a post-decoding update method, which utilizes correctly decoded data symbols to update the initially estimated likelihood functions $\widehat{\rho}_{k,i}^{(b)}$. To achieve this, the BTS divides data transmission into D subframes, each of length N_d^{sub}, such that $N_d = DN_d^{sub}$, and appends multi-cycle CRC bits to each data subframe. Subsequently, when a data subframe is correctly decoded, which is determined by CRC verification, the BTS uses these correctly decoded symbols as pilot symbols to update the initial likelihood function.

By using correctly decoded symbols, the likelihood function for the polarized learning method can be updated directly: after decoding each data subframe j, the likelihood functions $\widehat{\rho}_{k,i}^{(b)}$ are updated as in (10.10) by counting the occurrences of $b \in \{1; -1\}$ over the total number $N_{tr} + d_k(j)$, where $d_k(j)$ denotes the occurrences of decoded data as $\mathbf{s}_k \in \mathcal{S}$ in the correctly decoded data subframes within the first j subframes, $0 \le d_k(j) \le jN_d^{sub}$. For the dithering-and-learning method, the likelihood functions $\widehat{\rho}_{k,i}^{(b)}$ are updated after decoding each data subframe j as shown in

$$\widehat{\rho}_{k,i}^{(b)} = \alpha_{k,i}(v_j)\widehat{\rho}_{k,i}^{(b)}(0) + (1 - \alpha_{k,i}(v_j)), \tag{10.16}$$

where v_j denotes the number of correctly decoded data frames within the first j data frames, $0 \leq \alpha_{k,i}(v_j) \leq 1$ is the update rate for $\widehat{\rho}_{k,i}^{(b)}$ after decoding the j-th subframe, $\widehat{\rho}_{k,i}^{(b)}(0)$ is the initially estimated likelihood function from the training process, and $\widehat{\rho}_{k,i}^{(b)}(v_j)$ is the likelihood function for candidate vector $\mathbf{s}_k$ at the i-th quantized signal learned from the v_j correctly decoded data frames. However, the optimal value of the parameter $\alpha_{k,i}(v_j)$ must be determined empirically. Thus, such an update method is more beneficial for the polarized learning case compared to the dithering-and-learning approach.

10.4 Problem Solution

This section addresses the core issues of the study. Large MIMO systems with 1-bit quantized signals present unique challenges. To address these, the study emphasizes reducing dependence on training length, managing likelihood functions with zero probability, developing CSI-free detection methods, and analyzing the effect of the SNR on the quantization process.

10.4.1 Reducing Dependence on Training Length

Heavy reliance on training length in machine learning systems, particularly for maximum likelihood (ML) detection in large MIMO systems with 1-bit ADCs, can lead to issues like decreased generalizability. When a machine learning model is overly dependent on specific training data, it may fail to generalize to new or unseen data, which results in poor performance in real-world applications. Furthermore, the requirement for large training datasets becomes an issue; high dependence on training length implies that substantial training data is needed to achieve the desired performance, which may be unfeasible in scenarios where data collection is costly or challenging. Increased time and computational resources for training is another drawback, as training on large datasets extends training time and demands higher computational resources, which can be inefficient for time-sensitive applications. An increase in SER is a natural consequence of reduced training length, especially at high SNR levels where the occurrence of likelihood functions with zero probability rises, highlighting the importance of training length for detection performance, particularly in high-noise environments.

To address these challenges, the study aims to develop robust learning techniques that reduce reliance on training length. These include employing biased-learning algorithms and noise-learning techniques, aiming to enhance model accuracy and stability across all SNR ranges. These efforts focus on addressing this specific issue and emphasize the importance of reducing training length dependency within the context of large MIMO systems with 1-bit signal converters. Another challenge is posed by dynamic environments, where real-world applications, especially in wireless communication systems, involve constantly changing conditions. A model must be flexible and adaptive to such variations, as heavy reliance on old training data can restrict this adaptability.

These issues highlight the importance of minimizing training length dependence in ML-based detection methods, especially within the context of large MIMO systems with 1-bit signal converters.

10.4.2 Addressing Empirical Likelihood Functions with Zero Probability

In machine learning, particularly in ML detection systems, likelihood functions with zero probability present significant challenges for machine learning models. Likelihood functions are used to model the probability of observed signals based on various states or variables. A critical issue arises when a likelihood function for a specific signal has zero probability, implying that the signal never occurs under any condition.

One major consequence is the exclusion of useful information from the training data. The model cannot consider such signals as potential events, thereby limiting the effectiveness of machine learning. If a model cannot account for a certain signal, its ability to learn effectively is restricted, preventing the utilization of critical information within the data.

A second issue is the effect on model generalizability. When a likelihood function assigns a probability of zero to a signal, it eliminates the possibility of that signal being included in future predictions, irrespective of new input data. This reduces the model's ability to generalize from training data to real-world scenarios, especially in complex MIMO systems where signal variation is extensive.

Moreover, handling likelihood functions with zero probability can complicate training, as models must accurately determine which situations should be entirely excluded—an often impractical or undesirable process in real-world environments.

Efficiently managing likelihood functions with zero probability is therefore crucial to improving detection performance and ensuring that machine learning models can generalize effectively from training data to real-world situations.

The occurrence of zero-probability likelihood functions in large MIMO systems with 1-bit ADCs often originates from multiple factors. One-bit quantization, where continuous signals are converted to two states (−1 and 1), poses a significant challenge. In this context, small amplitude signals may be quantized into the same value, leading to loss of critical information regarding the original signal. This phenomenon increases the likelihood of indistinguishable signals, which affects the performance of the likelihood function.

High noise levels in the signal are also a key factor. When the received signal is highly noisy, the ability to distinguish between different signal states decreases substantially, which can lead to certain signal states being unobserved during training, thus assigning zero probability to these states in the likelihood function.

Limited training data contributes to this challenge. If the data lacks diversity or does not adequately represent the range of possible signal states, the machine learning model may be unaware of the existence of certain states. Consequently, the model may assign zero probability to these states in the likelihood function, affecting generalizability and model performance in real-world scenarios.

To tackle these challenges, future studies must focus on developing new methods for handling likelihood functions to enhance model performance in 1-bit quantized MIMO systems.

10.4.3 The Impact of CSI on Quantization

CSI provides information not only about signal attenuation but also about noise levels within the transmission channel. During the quantization process, understanding and adapting to these conditions can improve quantization performance, particularly in high-noise or complex channel conditions. Accurate CSI can also be used to optimize 1-bit converters. When complete CSI information is available, this optimization can be performed to minimize information loss during quantization, thus enhancing the system's performance and reliability. CSI also plays a key role in determining the most effective approach to detect and decode the signal post-quantization. Especially in high-noise situations or complex channel conditions, utilizing accurate CSI helps improve detection performance and ensure system stability.

Although CSI is crucial, developing detection methods that do not require CSI estimation in wireless communication systems, particularly in MIMO systems with 1-bit converters, is important because CSI estimation often requires collecting and processing large amounts of data, which can increase system complexity and costs. By developing CSI-free detection methods, we can reduce computational and energy resource requirements while optimizing cost efficiency. In dynamic wireless communication environments where channel conditions change rapidly, continuously estimating CSI may be impractical or inaccurate. CSI-free detection methods offer more stable performance in such dynamic environments, where flexibility is essential. Lack of CSI dependency also makes the system more resilient to CSI estimation errors or deficiencies in data collection, which is particularly important in systems with limited accuracy or stability of channel estimation. In MIMO systems with 1-bit converters, accurate CSI estimation becomes especially challenging due to information loss during quantization. Developing CSI-free methods helps overcome this limitation and provides more effective detection solutions in 1-bit quantized systems.

10.4.4 The Effect of SNR on Quantization

SNR has significant impact on the quantization process in large MIMO systems with 1-bit signal converters because as SNR increases, the number of likelihood functions with zero probability in non-dithering scenarios grows rapidly. At high SNR levels, more than 80% (around 50 out of 64) of the likelihood functions have zero probability, illustrating a strong dependence of the quantization process on SNR levels. In high SNR scenarios, 1-bit quantization easily leads to information loss, increasing the likelihood of unobservable likelihood functions. In dithering scenarios, the number of zero-probability likelihood functions significantly decreases, even as SNR increases. This shows that dithering techniques enhance the quantization process at high SNR levels by minimizing information loss.

ML detection in large MIMO systems with 1-bit signal converters may vary depending on whether SNR knowledge is available. Machine learning-based ML detection without SNR knowledge (biased-learning) approach does not require knowledge of SNR or CSI. It utilizes a biased-learning algorithm that assigns a minimum probability to each likelihood function to prevent zero-probability likelihood functions from erasing learned information. This approach is simple yet highly robust to training length and can provide stable detection performance across a wide SNR range.

Machine learning-based ML detection with SNR knowledge (dithering-and-learning) method uses SNR knowledge at the BTS, including noise variations and transmit power. During the pilot data transmission phase, the BS adds noise to the quantization input to estimate likelihood functions more accurately. This approach can achieve near-optimal detection performance across low-to-high SNR ranges and provides performance robustness relative to training length. It also demonstrates equivalence to the perfect SNR case, underscoring the effectiveness of the proposed detection method's offline learning capabilities.

10.4.5 Problem Solution

To address this problem, we tackle the following key tasks: channel generation, training phase, testing the 1-bit ML system, performance evaluation, and visualization. Below is the pseudocode for the proposed approach:

Algorithm 4: Proposed algorithm–Main program

```
Main_ML() % System parameters SP.N_u =; ;        // Number of users
SP.M = ...; ;                          // Number of QAM levels
SP.Nd = ...; ;               // Number of data transmissions
SP.CRC1 = ...; ;                             // CRC1 parameter
SP.CRC2 = ...; ;                             // CRC2 parameter
SP.SNR_dB = ...; ;   // Signal-to-Noise Ratio (SNR) in dB
SP.Num_h = ...; ; // Number of iterations for channel simulation
SP.Num_case = ...; ;       // Number of iterations for performance evaluation
% Initialize variables for storing results
% Iterate over test cases
for i = 1 to SP.Num_case do
    % Iterate over channels
    for h = 1 to SP.Num_h do
        [H] = Channel_Gen_ML(SP); ; // Channel generation
        [TR] = Train_ML(i, SP, H); ;        // Model training
        [DT] = Data_ML(i, SP, H, TR); ;              // Data transmission
        % Update results
    end
    % Calculate the average % Display results
end
```

10.5 Channel Generation

In this section, we introduce the function Channel_Gen_ML, where the input parameters include the number of transmit antennas, the number of users, the channel type (Rayleigh or mmWave), the number of paths for primary users, and the number of paths for virtual users. The function outputs the channel matrices $\mathbf{H}$ and $\mathbf{H}_{\mathbf{ici}}$, representing the main and interference channels, respectively, according to the specified communication model.

The pseudocode is presented as follows:

Algorithm 5: Proposed algorithm for channel generation module

```
Channel_Gen_ML ((Num_Users)) Channel.Type =' Rayleigh';;
 // Channel type: 'Rayleigh' or 'mmWave'
Channel.Nr = ...; ;        // Number of antennas per user
Channel.Nu = Num_Users; ;              // Number of users
Channel.H_ici = eye(Channel.Nr); ;                  // ICI
 (Inter-Carrier Interference) matrix
switch Channel.Type case 'Rayleigh' ;  // Rayleigh channel
Channel.H = ...; case 'mmWave' ;         // mmWave channel
Channel.H = zeros(Channel.Nr, Channel.Nu); for u =
 1:Channel.Nu do
  | UE(u).L = ...; ;                     // Number of paths
  | UE(u).g = ...; UE(u).theta = ...;
  | A = SteeringGen(UE(u).theta, Channel.Nr);
  | Channel.H(:, u) = ...;
end
end
```

Here, the steering matrix is frequently used in antenna system simulations to characterize the directional effects of signals across different antennas. This matrix includes columns, each of which represents the response of an individual antenna in the system when a signal arrives from a particular direction. The formula for the steering matrix is

$$A = \frac{1}{\sqrt{N}} \exp(-j\pi\theta.vector), \tag{10.17}$$

where N is the number of antennas, $vector$ is an array from 0 to $N-1, \theta$ is the angle of arrival, and j is the imaginary unit. The output of this function is a column vector a, where each element represents the response of the corresponding antenna when a signal arrives from a specified angle θ. The steering matrix is commonly utilized in applications involving antenna movement, such as in MIMO systems in wireless communications.

10.5.1 Training Phase

In this phase, we conduct the training process for the 1-bit ML detection system in a wireless communication environment. The function "`Train_ML`" is designed to perform the training of the 1-bit ML detection system. During this process, it generates symbols, applies 1-bit quantization on the received signal, and computes the nonflip count to evaluate quantization performance.

The term "nonflip" refers to the number of times a bit does not undergo flipping from –1 to 1 or from 1 to –1 during the 1-bit quantization process. In a 1-bit ML detection system, this quantization maps a continuous signal to a binary form (either –1 or 1). Multiple flips can introduce potential issues affecting signal quality and resilience. Each flip can lose signal detail, especially for small variations and low-amplitude signals, decreasing the distinction between original symbols and adding noise to the quantized signal. Irreversible information loss is significant; each flip might lead to permanent loss during quantization, affecting signal recovery and quality. Furthermore, excessive flipping may increase noise in the signal, particularly with dithering. This noise increase may degrade system performance and predictability. Lastly, post-processing algorithms may struggle with flipped signals, reducing processing efficiency. Therefore, in applications requiring high quality and resilience, minimizing bit flips during quantization is essential for preserving information and enhancing system performance.

To better assess the 1-bit ML detection system's performance, the nonflip count serves as a key metric. A lower bit flip count during quantization is desirable, as excessive flipping may result in information loss and impact signal resilience.

Let "TR.NonFlip" represent the average nonflip count for 1-bit quantization without dithering, and "TR.NonFlip_dither" represent the average nonflip count with dithering. These values evaluate the 1-bit ML detection system's performance with and without dithering during training. The objective is to reduce the bit flip count to enhance the quantized signal quality, thereby improving system performance. The pseudocode for the training phase is as follows:

Algorithm 6: Training algorithm

```
Function [TrainingResult] =Train_ML(i, SP, h)
    /* Generate symbol combinations from the codebook                */
    symbols = createSymbolCombinations(SP.Nu, SP.M); /* Initialize
        training results                                             */
    TrainingResult.codebook = generateCodebook(SP.Nu, SP.M);
        TrainingResult.Nr = SP.Nr; TrainingResult.Nu = SP.Nu;
        TrainingResult.M = SP.M; TrainingResult.Ntr = SP.Ntr;
        TrainingResult.p = SP.p; TrainingResult.p_dither = SP.p_dither;
        TrainingResult.p_ici = SP.p_ici; /* Conduct training          */
    [w1, w1d] = performTraining(H, TrainingResult.codebook, SP);
        /* Compute NonFlip and NonFlip_dither                        */
    calculateNonFlip(TrainingResult, w1);
        calculateNonFlipDither(TrainingResult, w1d); /* Assign results
        to TrainingResult                                            */
    /* ...                                                           */
end
```

```
Algorithm 7: Algorithm for creating symbol combinations
Function symbols =createSymbolCombiniations(Nu, M)
  /* Create all symbol combinations from the
     codebook                                              */
  symbols = 0 : M − 1; for u = 1:Nu-1 do
    symbols = symbols:, 0 : M − 1;
  end
  combinations = cell(1, numel(symbols));
  [combinations:] = ndgrid(symbols:); combinations = ...;
  symbols = [combinations:]';
end
```

```
Algorithm 8: Pseudocode for codebook generating
Function symbols =generateCodebook(Nu, M)
  /* Generate codebook from symbol combinations
     */
  symbols = createSymbolCombinations(Nu, M); codebook =
  qammod(0 : M^N u − 1, M,' UnitAveragePower', true);
end
```

```
Algorithm 9: Pseudocode for performing the training
Function [w1,w1d] =performTraining(H, codebook, SP)
  /* Conduct training and calculate w1, w1d               */
  w1 = zeros(2 * SP.Nr, SP.M^S P.Nu);
  w1d = zeros(2 * SP.Nr, SP.M^S P.Nu); for sym = 1:SP.MŜP.Nu do
    yq = zeros(2 * SP.Nr, SP.Ntr); ydq = zeros(2 * SP.Nr, SP.Ntr);
    for j = 1:SP.Ntr do
      /* Generate channel and signals y, yd               */
      [H_i ci] = Channel_Gen_ML(SP); /* Compute w and wd
         */
      /* ...                                              */
      /* Assign values to w1 and w1d                      */
      /* ...                                              */
    end
  end
end
```

Algorithm 10: Pseudocode for calculating the non-clip training results
Function `calculateNonFlip` (*TrainingResult,w1*)
  Calculate NonFlip and assign value to TrainingResult /* ... */
end

Algorithm 11: Pseudocode for calculating the non-clip dithering
Function `calculateNonFlipDither` (*TrainingResult,w1d*)
  Calculate NonFlip_dither and assign value to TrainingResult
  /* ... */
end

In this context, the codebook represents a set of signals that can be used to represent input data. In digital communication, the codebook is often a set of symbols, each representing a message or part of a message. Here, the codebook is generated from all possible combinations of symbols from the symbol set.

10.5.2 Implementation of the 1-Bit ML System Testing Process

To execute the 1-bit ML system testing process, we define the function `Data_ML`. This function is responsible for data transmission within the system, encompassing tasks from signal generation to decoding and CRC verification.

CRC (cyclic redundancy check) is an error-detection technique used to ensure the integrity of data in communication and storage systems. The CRC value is appended to the data before transmission, and upon reception, it is verified to detect any transmission errors. Fundamentally, the CRC algorithm works by performing a division on the transmitted data, with the remainder representing the CRC. A specific polynomial, called the CRC polynomial, is used for division. The resulting CRC value is appended to the data, and when the data is received, the same division is performed. If the remainder is zero, the data is considered valid. CRC can detect various error types, including random modifications and transmission noise; however, it does not correct errors. In this project, CRC is used to verify data integrity during transmission in the system.

The pseudocode of the function is presented below:

Algorithm 12: Pseudocode for data transmission

```
Function ML_result = Data_ML(SP, H, TR)
    /* ...                                                    */
    for d = 1 : D do
        /* Each data transmission frame cycle                 */
        [x_crc, s_crc] = Transmitter_ML(SP);  // Generate
         transmission signal
        for t = 1 : Nd do
            x = x_crc(:, t);  s = s_crc(:, t); /* Decode with
                Receiver_ML                                    */
            [DC] = Receiver_ML(SP, H, TR, x) /* Note:
                The results of the decoding process are
                stored in the DC variable                      */
            /* Other operations can be performed here,
                such as performance evaluation                 */
        end
        Check CRC and perform post-update if necessary; CRC is
            checked for each symbol in each data transmission cycle;
            Perform post-update if CRC is valid; /* ...        */
    end
    /* The results of the testing process are
        stored in the ML_result structure                      */
    /* Important parameters or other results can
        be returned from this function                         */
end
```

In Step 1, we generate the transmission signal x_{crc} and actual symbol s_{crc} by calling the `Transmitter_ML` function, which prepares data prior to channel transmission. The `Transmitter_ML` function supports `Data_ML` by performing data generation, QAM symbol preparation, and transmission signal setup, thus simplifying code and improving reusability.

Step 2 iterates over each moment within a transmission cycle (**for** $t = 1 : Nd$). Each iteration selects a specific transmission signal (x) from x_{crc} and the corresponding actual symbol (s). Step 3 calls `Receiver_ML` for decoding and verification. This function receives parameters such as the system configuration (SP), the channel matrix (H), and training information (TR). The function then estimates the transmitted symbols using decoding algorithms.

In Step 4, results of the decoding process are stored in `DC`. Performance checks or error logging can be done here. Step 5 performs CRC validation and, if necessary, a post-update, to ensure data integrity. Step 6 repeats the process for each transmission frame cycle (**for** $t = 1 : Nd$), with final testing results saved in `ML_result`.

10.5.3 Performance Evaluation and Visualization

The communication system's performance is primarily assessed through metrics such as SER and CRC. SER measures the error rate in recovering transmitted symbols versus the original symbols. The proportion of correctly decoded frames, based on CRC checks, is computed and displayed. Additionally, Estimated Signal-to-Noise Ratio (`Est.SNR`), which approximates the system's SNR, and NonFlip, the average count of non-flipped symbols during transmission, are computed for both biased and dithered learning methods.

The SER metric is compared across different methods, including biased learning, dithered learning, empirical MLD, minimum mean distance, and minimum center distance. Optimal 1-bit ML, assumed to have accurate CSI knowledge, is used as a reference for comparison. Other methods include 1-bit ZF (zero forcing decoding), conventional learning-based 1-bit ML (without dithering or bias), biased-learning 1-bit ML, dithering-and-learning 1-bit ML, empirical MLD, minimum mean distance, and minimum center distance.

These methods are evaluated based on SER, estimated SNR, non-flipped symbol count, and CRC checks, providing insights into communication system performance under varied conditions, including channel interference and SNR levels.

10.6 Simulation Results

10.6.1 System Model and Simulation Parameters

This study focuses on evaluating the performance of a multi-user wireless communication system under diverse and noisy conditions. The system is simulated with the number of receivers (N_r) set to 32, the number of users (N_u) set to 4, and the QAM symbol count (M) set to 4, allowing for flexibility and diversity testing within the system. Training iterations for each symbol vector are set to 30, 50, and 100 (N_{tr}) to ensure reliable simulation results. The SNR varies from –10 to 14 dB to create different noise scenarios, simulating the effects of Rayleigh fading channels. The diversity of the communication environment is represented by the number of propagation paths (L), which is set to 4. A dither value (p_{dither}) ranging from 0.05 to 2.37 is used to simulate noise influence and maintain system stability.

10.6.2 SNR Estimation Variance with Different N_{tr} Values

The following chart illustrates the SNR estimation variance across different N_{tr} values (30, 50, 100). The x-axis represents SNR, showing values that range from –8 dB to 8 dB, a measure of signal quality relative to noise. The y-axis (`Est.SNR`) displays

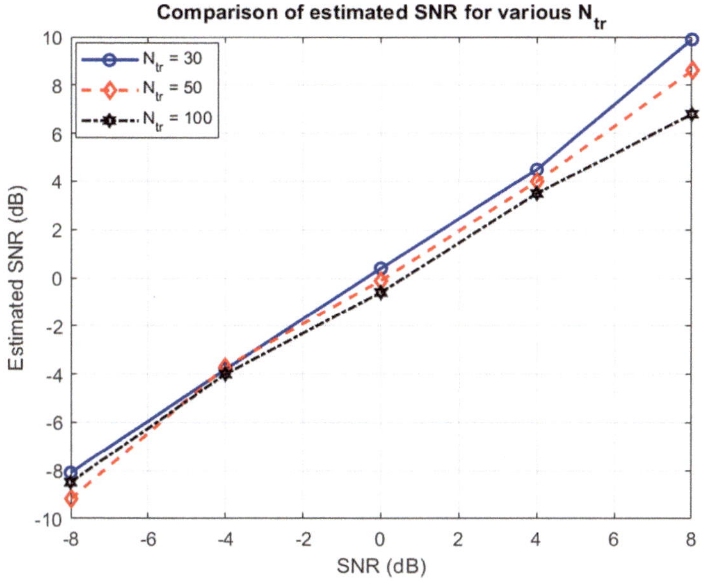

Fig. 10.2 Comparison of SNR estimation (`Est.SNR`) across different N_{tr} values for both bias and dithering methods

the estimated SNR values for each corresponding SNR input. `Est.SNR` represents the model or algorithm's estimated SNR. Data points or lines ($N_{tr} = 30$, $N_{tr} = 50$, $N_{tr} = 100$) represent `Est.SNR` for each SNR level based on the number of simulated transmissions (N_{tr}). In this case, three curves or data series are shown for ($N_{tr} = 30$, $N_{tr} = 50$, $N_{tr} = 100$).

The chart in Fig. 10.2 compares the SNR estimation performance of the communication system under varying N_{tr} conditions. If the lines or data points are close and follow an increasing trend, it may indicate that the SNR estimation is not strongly dependent on the number of transmissions. Conversely, significant deviations may highlight the impact of N_{tr} on SNR estimation.

Examining the graph reveals the stability of the SNR estimation. The data lines or points for $N_{tr} = 30$, $N_{tr} = 50$, and $N_{tr} = 100$ lie closely together with no significant deviation, suggesting that the SNR estimation is not strongly dependent on the number of transmissions. This indicates a positive stability in the proposed SNR estimation algorithm. Additionally, there are notable differences between the values of N_{tr}, suggesting a potential impact of the transmission count on the SNR estimation performance.

At low input SNR levels (e.g., $-10\ dB$), the differences in estimated SNR (`Est.SNR`) for various N_{tr} values are minimal. However, as the input SNR increases, the discrepancy between `Est.SNR` for $N_{tr} = 100$ and $N_{tr} = 30$ becomes more pronounced. Certain points where the curves nearly overlap, particularly in the low SNR

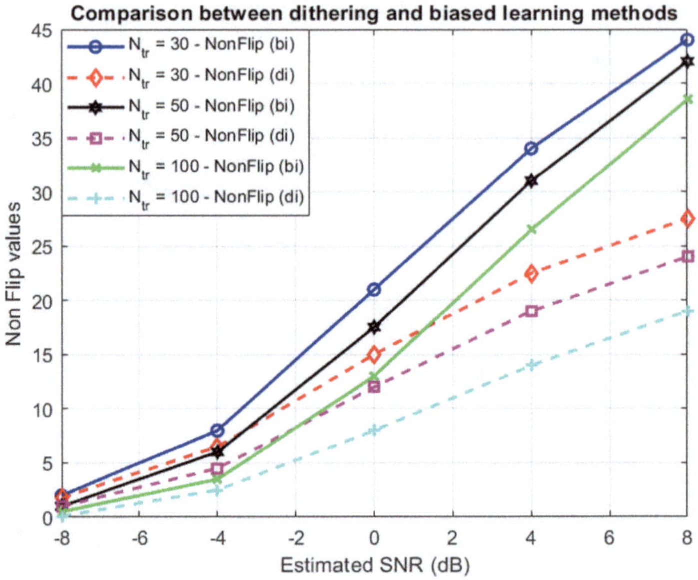

Fig. 10.3 Comparison of SNR estimates (`Est.SNR`) with different values of N_{tr} for two bias and dithering methods

region, indicate that the influence of transmission count N_{tr} on the SNR estimation may not be significant at low SNR levels.

In summary, the SNR estimated values (`Est.SNR`) for the three different transmission counts, $N_{tr} = 30$, $N_{tr} = 50$, and $N_{tr} = 100$, exhibit a consistent upward trend as the input SNR increases. This implies that `Est.SNR` is better estimated with improving input SNR, and this increase is approximately linear. However, the degree of divergence between the curves for different N_{tr} values is minimal, especially in the SNR range from –2 dB to 8 dB, indicating that the SNR estimation is not strongly dependent on transmission count. The near-parallel alignment of the curves demonstrates that increasing the transmission count N_{tr} does not significantly alter the relationship between input SNR and the estimated `Est.SNR`.

10.6.3 Comparison of SNR Estimation (`Est.SNR`) with Different N_{tr} Values for Bias and Dithering Methods

The plot in Fig. 10.3 provides a comparative analysis of SNR estimation (`Est.SNR`) for varying values of N_{tr} (number of iterations) across two methods: bias and dithering. The curves represent the relationship between `Est.SNR`, measured in dB, and the iteration counts N_{tr}. Six distinct curves are displayed, differentiated by color and marker shape, and grouped according to the two methods (bias and dithering) for each value of N_{tr} (30, 50, 100). Each method has three corresponding curves for these N_{tr} values, showing distinct results for each SNR level.

For both methods, as `Est.SNR` increases from –8 dB to 8 dB, the NonFlip value rises correspondingly, indicating a positive correlation between `Est.SNR` and NonFlip values. Within each method, NonFlip values increase as N_{tr} grows. The curve for $N_{tr} = 100$ consistently appears at the top, followed by $N_{tr} = 50$, with $N_{tr} = 30$ yielding the lowest NonFlip values. Curves associated with higher N_{tr} values (e.g., $N_{tr} = 100$) are positioned above those for lower N_{tr} values (e.g., $N_{tr} = 30$ and $N_{tr} = 50$), suggesting that higher iteration counts enhance the estimation or output (NonFlip values). Increasing the number of iterations seems to smooth out noise, improve the stability of the estimation, and produce more accurate results. A higher N_{tr} reduces iteration errors, resulting in a smoother curve and potentially diminishing dependence on random fluctuations or noise within the data.

When comparing the two methods for the same N_{tr} level, the dithering method yields consistently higher NonFlip values than the bias method across all `Est.SNR` levels. This suggests that the dithering method may be more optimal than the bias method in achieving higher NonFlip values. Data points along each curve appear to align closely with the reference line, demonstrating the precision and consistency of the `Est.SNR` estimation process.

In summary, the plot indicates a clear correlation between NonFlip values and `Est.SNR`, with this correlation strengthening as N_{tr} increases. The plot highlights significant differences in estimation or output between the two methods, as well as the impact of iteration count N_{tr} on these results.

10.6.4 Performance of Algorithms Based on SER

In Fig. 10.4, we present the SER plot over a range of SNR values from –8 dB to 8 dB. Different data points represent the performance of various algorithms in signal detection. Notably, there are distinct differences between detection methods: at lower SNR levels (around –8 dB to –2 dB), the ability to differentiate between methods is less clear compared with higher SNR levels. This may indicate that all methods are similarly affected by high noise, with performance not significantly improved by using different algorithms. As SNR increases (0 dB and above), distinctions between methods become clearer. Some methods, such as "Optimal 1-bit ML" and certain machine learning approaches, show significantly lower error rates compared to others like "1-bit ZF" or "Minimum Center Distance (MCD)."

The machine learning methods ("Learning 1-bit ML," "Biased-Learning 1-bit ML," and "Dithered Learning 1-bit ML") demonstrate strong performance and tend to approach "Optimal 1-bit ML" as SNR increases. This indicates the capability of machine learning to optimize signal classification and detection under various conditions, potentially leveraging data-driven approaches to better distinguish signal from noise. "1-bit ZF" (zero forcing), a traditional method in signal processing,

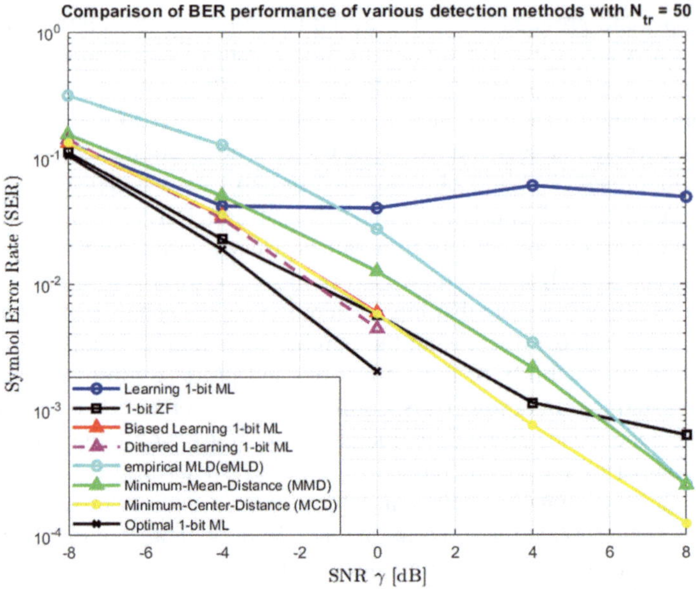

Fig. 10.4 Comparison of the performance of algorithms based on SER ($N_{tr} = 50$)

appears to perform worse than both machine learning methods and Optimal 1-bit ML, particularly at high SNR levels. This may reflect the limitations of traditional approaches when dealing with complex noise models that machine learning can more effectively address. "Minimum Mean Distance (MMD)" and "Minimum Center Distance (MCD)" are distance-based methods that likely use specific distance functions to assess discrepancies between the detected signal and the actual signal. These methods show intermediate performance, not as effective as machine learning but potentially outperforming zero forcing.

Although not directly shown in the simulation figures, computational complexity is also a critical consideration for each method. Machine learning methods may demand higher computational resources but yield superior performance, while traditional approaches like ZF may be less computationally intensive but are less effective in challenging scenarios.

Machine learning algorithms can adapt better to network condition changes, such as SNR variations, by tuning their models through learning. This adaptability helps maintain strong performance even under shifting conditions. Machine learning approaches can offer greater flexibility, making them applicable to diverse types of signals and noise, while traditional algorithms may be constrained to specific scenarios they were designed to handle. For machine learning methods, access to high-quality representative training data is essential for building effective models, whereas fixed-rule methods like ZF may not require this data dependency.

The following simulation results provide a performance comparison of algorithms based on SER with $N_{tr} = 30$ and $N_{tr} = 100$ (Figs. 10.5 and 10.6).

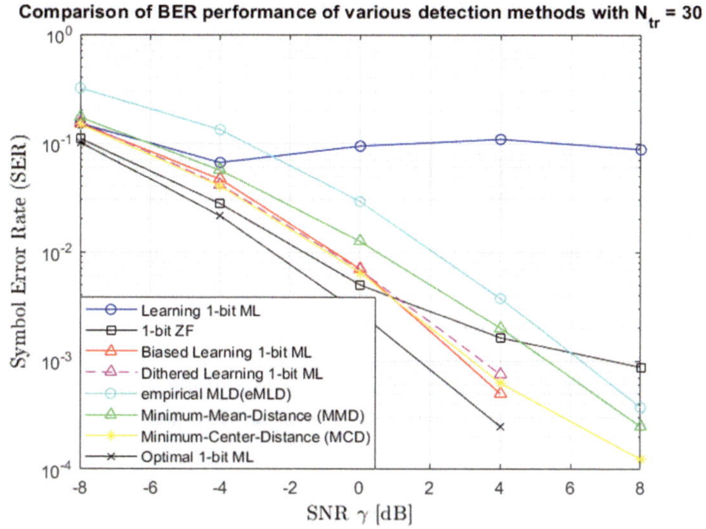

Fig. 10.5 Comparison of algorithm performance based on SER ($N_{tr} = 30$)

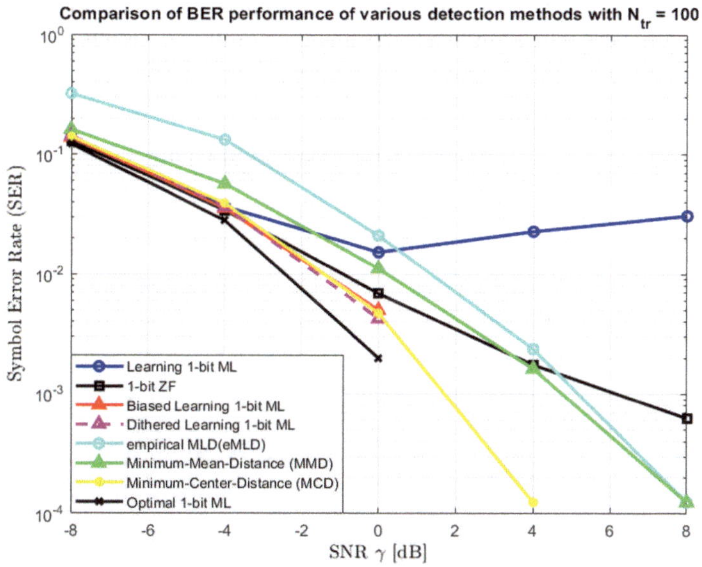

Fig. 10.6 Comparison of algorithm performance based on SER ($N_{tr} = 100$)

10.6.5 Program Runtime Across SNR Conditions

The runtime versus SNR graph in Fig. 10.7 evaluates the performance of a wireless communication system under various noise conditions.

The graph depicts runtime (measured in seconds) for three different values of N_{tr} across SNR levels ranging from –10 dB to 10 dB. The runtime for each N_{tr} value remains relatively stable across different SNR levels, showing no substantial variation. This stability indicates that runtime is not significantly affected by the noise conditions of the signal.

There is a clear difference in runtime between the different values of N_{tr}. $N_{tr} = 30$ has the shortest runtime, followed by $N_{tr} = 50$, while $N_{tr} = 100$ has the highest runtime. This reflects the fact that increasing the number of iterations or transmissions can increase the time required to complete a process. The nearly flat curve for $N_{tr} = 100$ suggests consistent performance even as the number of transmissions increases, which is important for ensuring predictable runtime in different conditions. The notably higher runtime for $N_{tr} = 100$ may indicate added computational complexity, thus requiring more processing time.

The slight variations in runtime for $N_{tr} = 30$ and $N_{tr} = 50$ could be due to external factors or noise in computation, but these are not significant enough to alter the overall results. In summary, the graph illustrates that the number of iterations (N_{tr}) impacts runtime, but it is unaffected by input signal noise level (SNR).

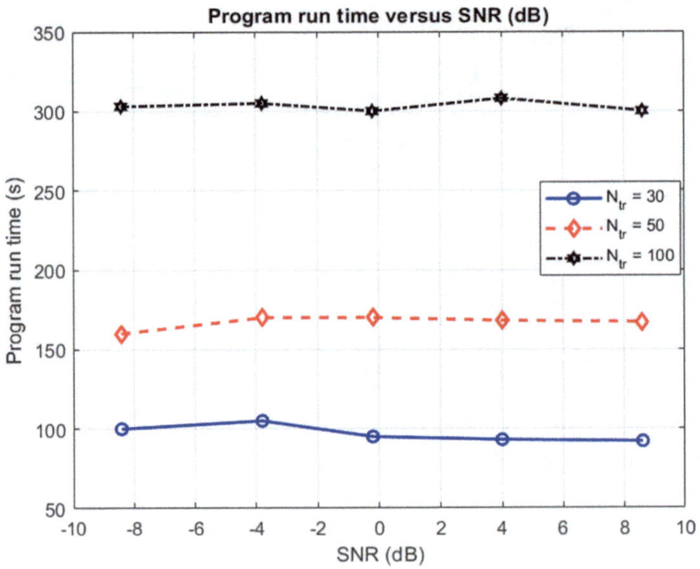

Fig. 10.7 Program runtime by SNR

10.7 Conclusion

In this work, we proposed robust ML-based 1-bit detection methods for large MIMO uplink communications. Given that the performance of ML-based 1-bit detection degrades significantly with insufficient training data, our proposed methods address this issue by introducing bias probability and dithering. First, estimating SNR in a communication system is crucial, and our results demonstrate a high degree of accuracy in this estimation. The small discrepancy between estimated and actual SNR values highlights the system's precision, which is essential for maintaining consistent and reliable communication.

When examining the decoding methods, we observe the enhanced efficiency of optimized techniques such as optimal 1-bit ML, 1-bit ZF detection, empirical ML detection (eMLD), and minimum mean distance (MMD) over traditional 1-bit decoding. Among these, optimal 1-bit ML and 1-bit ZF detection show the lowest bit error rates, although it should be noted that empirical ML detection (eMLD) and minimum mean distance (MMD) have higher bit error rates and may require further parameter adjustments or complexity enhancements.

Program runtime is also an important aspect. As SNR increases, runtime decreases, indicating improved performance as the signal becomes stronger relative to noise. However, achieving a balance between communication quality and processing time may require additional optimization strategies.

Future research directions could focus on developing and implementing new decoding methods to enhance their performance. Expanding the simulation model to cover various environmental conditions and parameters is also a potential research avenue. Furthermore, integrating optimization techniques to improve runtime and communication efficiency will be critical to ensuring the stability and effectiveness of wireless communication systems.

Reference

1. Choi, J., Cho, Y., Evans, B.L., Gatherer, A.: Robust learning-based ML detection for massive MIMO systems with one-bit quantized signals. In: 2019 IEEE Global Communications Conference (GLOBECOM), pp. 1–6 (2019). https://doi.org/10.1109/GLOBECOM38437.2019.9013332

Chapter 11
Experimental Results and Evaluation

11.1 Graphical User Interface Design

Software simulation with a main interface as shown in Fig. 11.1. The interface structure consists of the following components:

1. The header and footer: At the top of the interface is the name of the simulation tool and the logo of Ton Duc Thang University. At the bottom of the page are additional details such as the author's name and the update date.
2. The directory tree for selecting simulation modules: Arranged in the leftmost column of the screen, includes main modules as presented in Chap. 3.
3. System model: Positioned to the right of the directory tree and below the header. This section provides a visual diagram of the system model under investigation. Below the model image are a few concise lines of description to help users understand the model and its relevant parameters.
4. Data input panel: The program consists of four data input panels located below the system model section, including three permanent panels: simulation, communication protocols, and channel parameters. The remaining panel is dedicated to specific parameters of each module (for example, multiplexing panel, energy harvesting panel, cognitive radio panel, MIMO panel).
5. Result graph: Placed in the top right corner, featuring dropdown menus for users to select scanning parameters and output graph types.
6. Output data table: Positioned below the result graph to list detailed values of the output variables.
7. Control panel: Located in the bottom left corner, comprising program control buttons. Additionally, there is a slider to display the execution progress (used for programs with long execution times).

The user interface demonstration for each module is presented in the following sections.

P. T. Tran et al., *Developing MATLAB Comprehensive Simulation for 5G/6G Wireless Communications*, Signals and Communication Technology,
https://doi.org/10.1007/978-981-95-1117-4_11

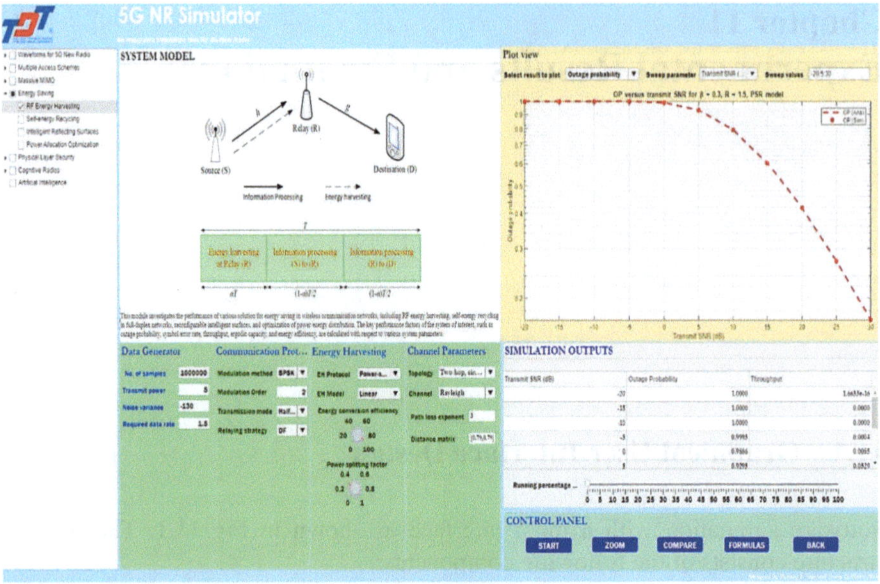

Fig. 11.1 Main user interface of the proposed 5G/6G simulation tool

11.2 Modulation Waveform Module

The input information of the modulation waveform block and the output results of the modulation waveform block module are illustrated in Fig. 11.2. Some simulation results are depicted in Fig. 11.3.

11.3 Multiple Access Module

This section presents the simulation results and interface of the multiple access technique. The simulation results of BER for NOMA are depicted in Fig. 11.4. The simulation results for SCMA are illustrated in Fig. 11.5.

For the SCMA module, the input data consists of parameters similar to the NOMA module, but in the initial version only the downlink channel and the default selected modulation scheme are available. Additionally, users can input the number of SCMA signals in one frame and the maximum bit errors before stopping the simulation. This module is programmed according to the flowchart in Fig. 3.8. After inputting the data, the program scans each SNR value and performs data generation, SCMA encoding, noise generation, and addition to the signal, SCMA decoding, and counts the number

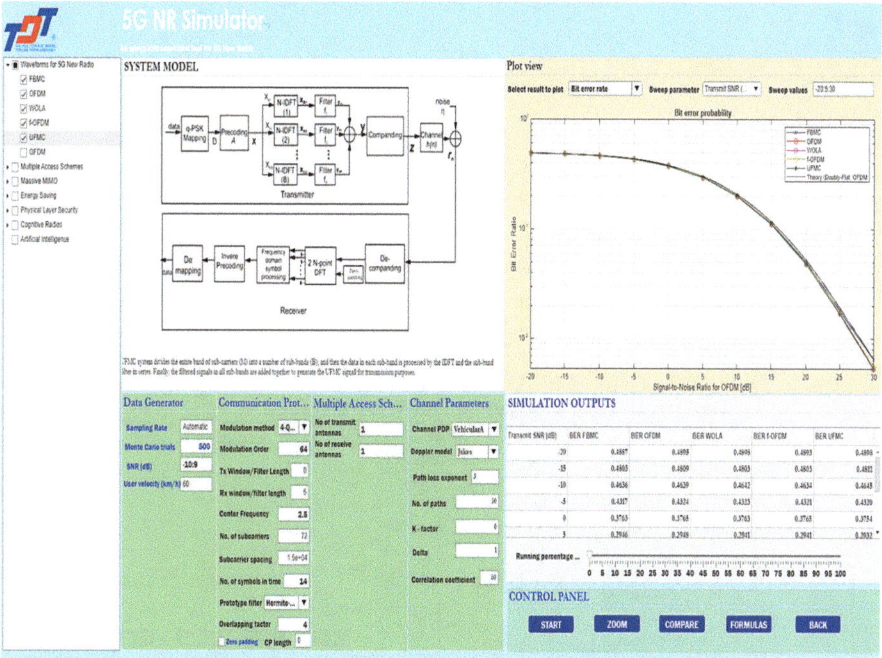

Fig. 11.2 User interface for modulation waveform module

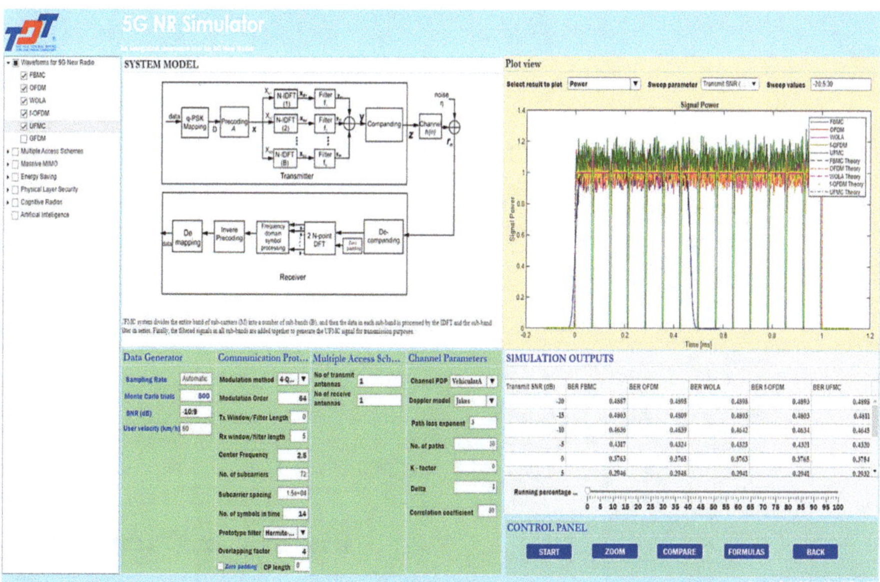

Fig. 11.3 Resulted graph of the signal power for modulation waveform's module

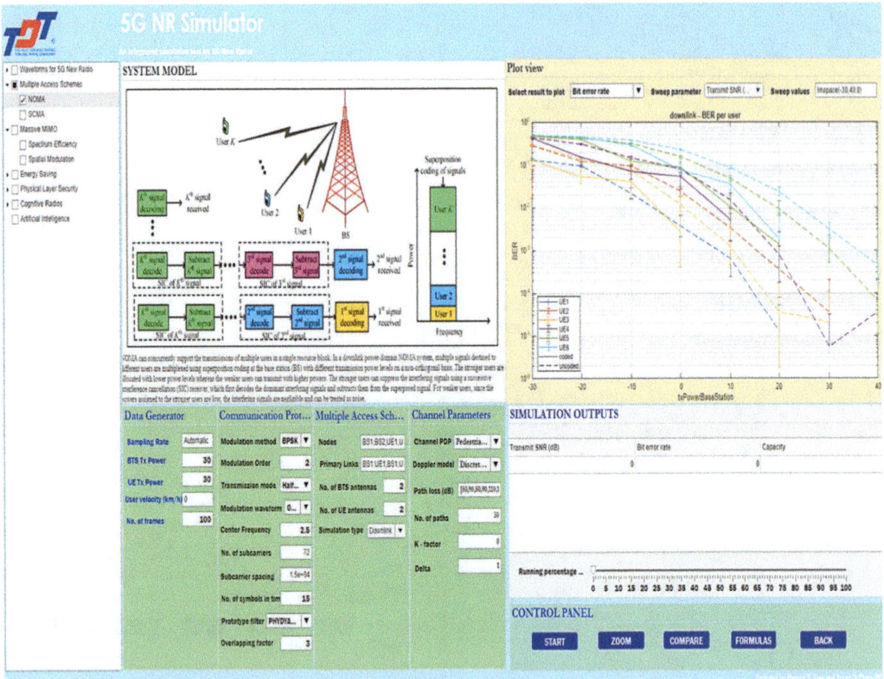

Fig. 11.4 User interface for NOMA simulation

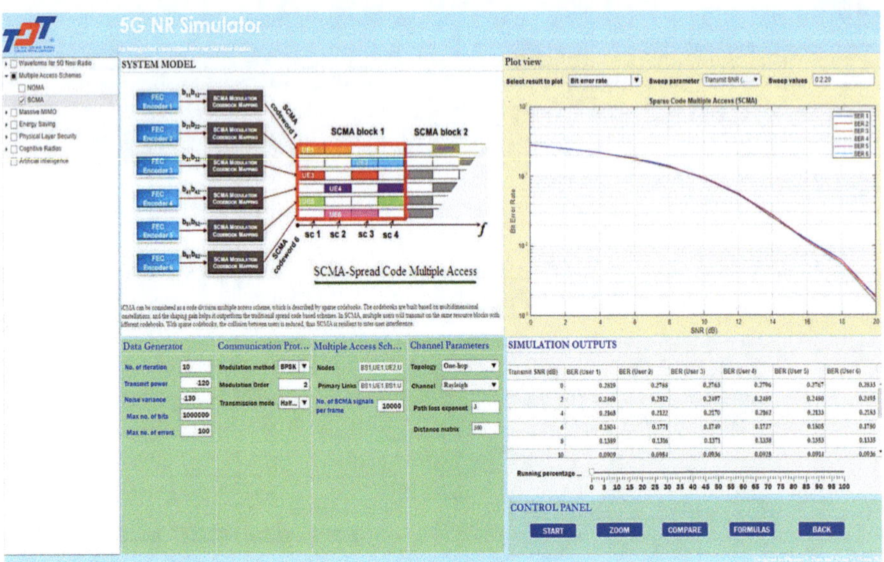

Fig. 11.5 User interface for SCMA simulation

of bit errors. The program runs until the number of transmitted bits or the number of bit errors exceeds the predefined threshold. At that point, the program updates the BER value and outputs it to a graph.

11.4 Massive MIMO Module

The designed interface for massive MIMO module is illustrated in Figs. 11.6 and 11.7. Figure 11.6 presents the simulation results of optimized spectral efficiency performance in massive MIMO systems. Figure 11.7 depicts the simulation results of BER for SM and DSM systems.

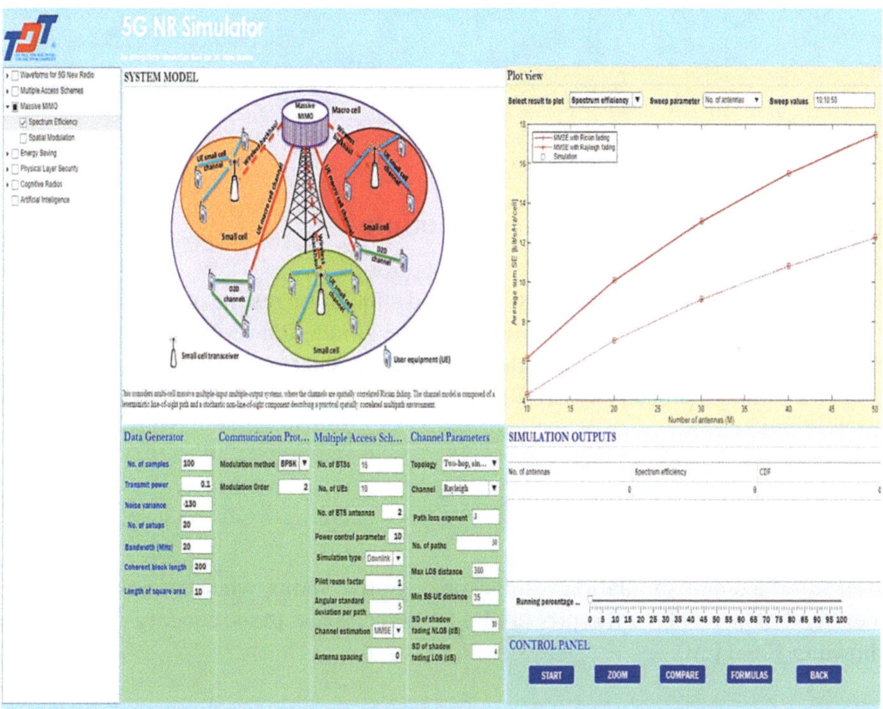

Fig. 11.6 User interface for massive MIMO simulation

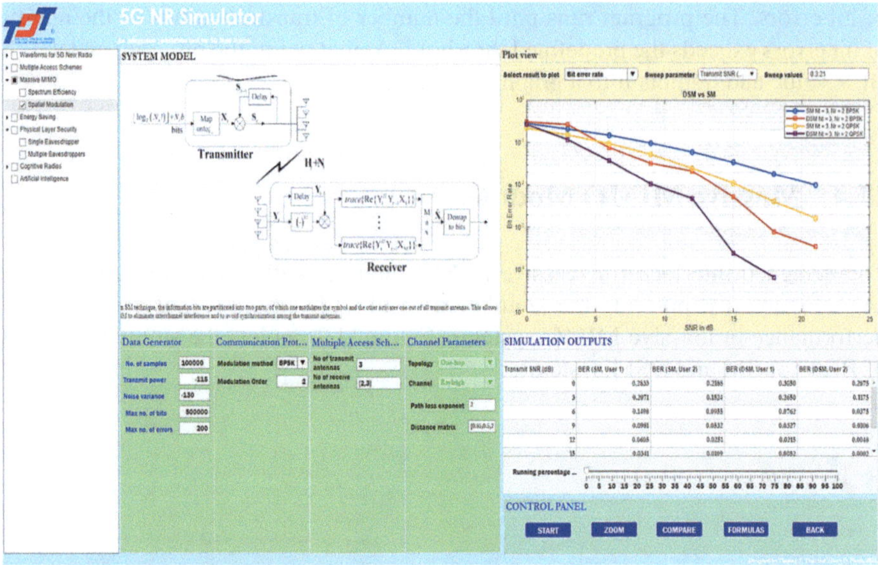

Fig. 11.7 User interface for SM simulation

11.5 Physical Layer Security Module

Figures 11.8 and 11.9 depict the interface and simulation results of the physical layer security module, where Fig. 11.8 illustrates the simulation results of eavesdropping probability (EP) in the bidirectional relay network, and Fig. 11.9 shows the 3D graph of secrecy outage probability (SOP) across energy harvesting parameters.

11.6 Cognitive Radio Block

User interface and simulation results for cognitive radio, where the output data is presented as a graph of the probability of detection at the secondary receiver, are shown in Fig. 11.10.

11.7 Energy Saving Module

The simulation results of energy harvesting using the power-splitting method are depicted in Fig. 11.11. The graph shown in the figure is OP graph plotted against the SNR ratio at the transmitter. Additionally, users can choose to plot the system throughput graph instead of OP, or plot based on other parameters such as power-splitting ratio, time, or desired data rate.

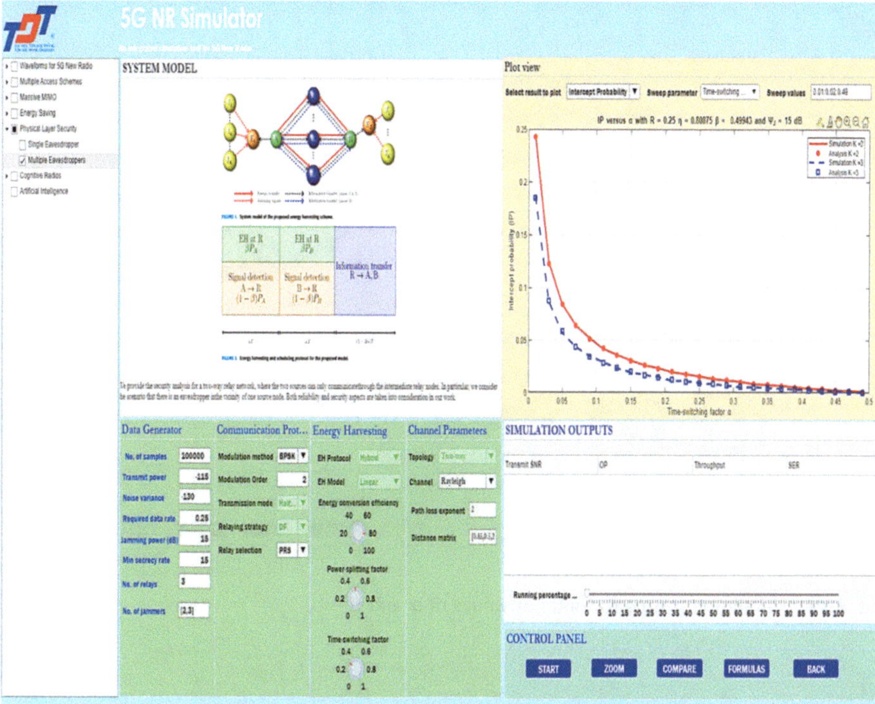

Fig. 11.8 User interface for physical layer security simulation

Figure 11.12 presents the simulation interface for energy reuse techniques in cooperative relay networks. The output graph in the figure compares OP of the system when using two different relay selection methods under two different channel distribution probability conditions. Users can also plot based on parameters such as the number of relays, required data rate, power-splitting ratio, or parameters related to channel gain distribution interference.

The simulation results of the intelligent reflector solution are presented in Fig. 11.13. The graph describes the variation of OP with SNR ratio at the transmitter for both relay node selection algorithms FSR and PSR. Users can select graph types such as OP, SEP, or system throughput. Scanning parameters can include transmit SNR ratio, number of reflector elements, or interference level.

The simulation results of power allocation optimization are presented in Fig. 11.14. This is a graph of energy efficiency plotted against the maximum transmit power level of the HRN, comparing the optimal method with reference methods. Additionally, users can observe the variation of optimal EE values with iteration count (to observe convergence of proposed algorithms), number of users, or noise power. There are two groups of simulated algorithms, one applied to AF relay and one applied to DF relay.

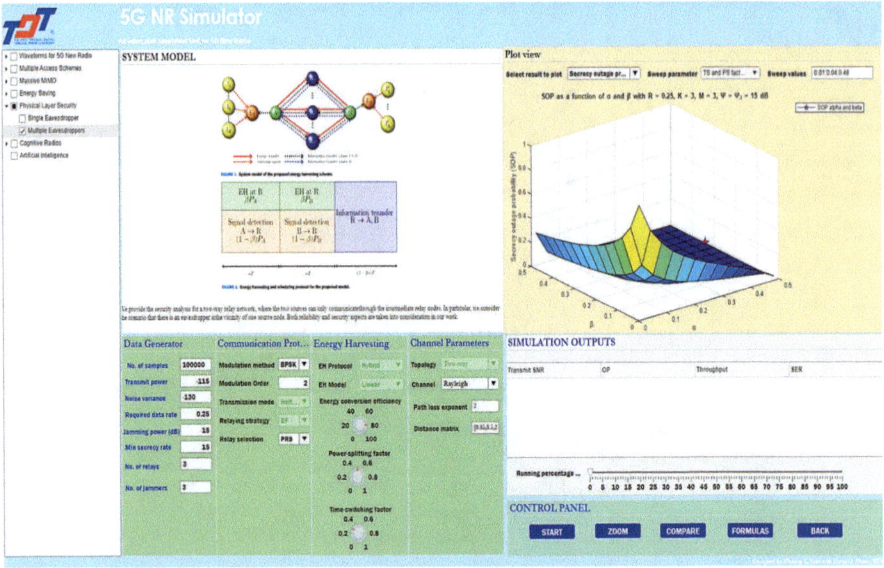

Fig. 11.9 Simulation results for optimal values of α and β

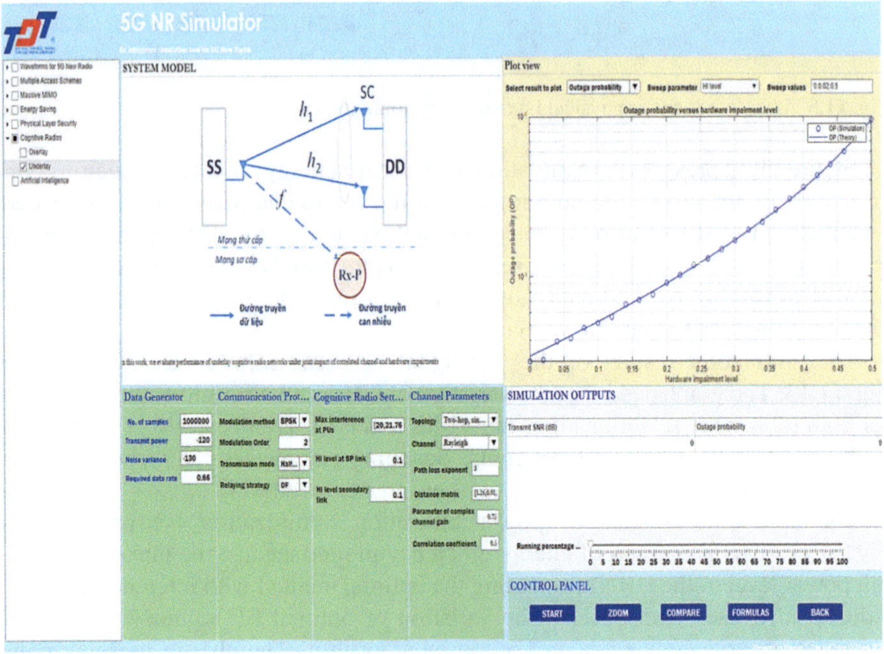

Fig. 11.10 User interface for cognitive radio simulation

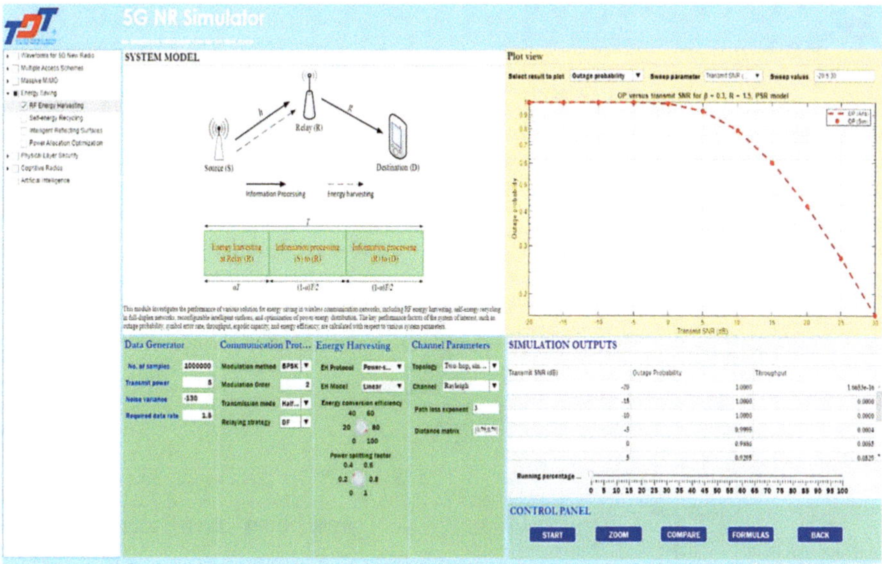

Fig. 11.11 User interface for RF energy harvesting simulation

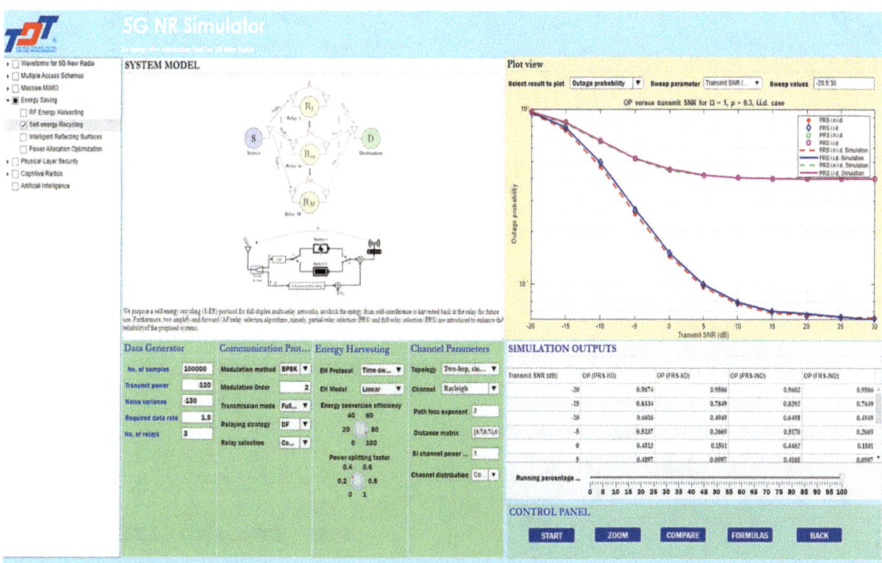

Fig. 11.12 User interface for self-energy recycling simulation

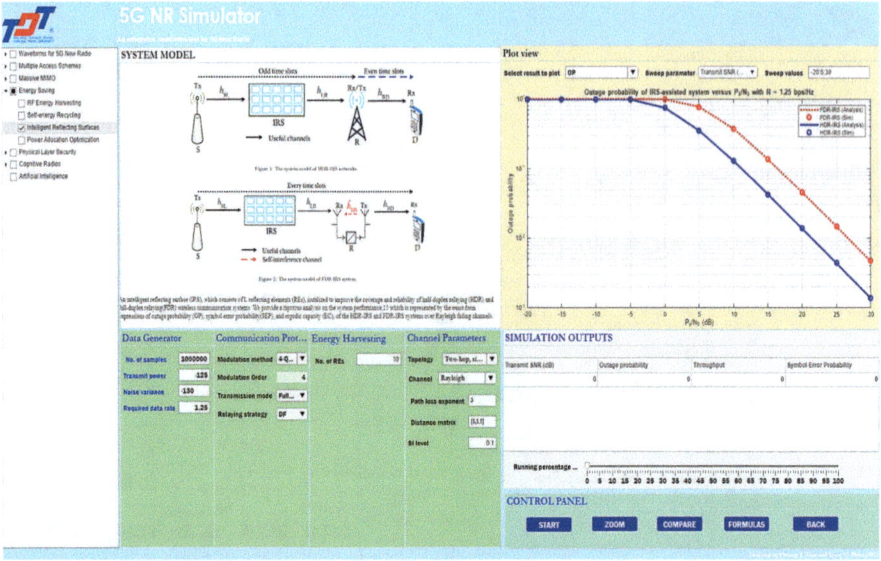

Fig. 11.13 User interface for simulation of IRS-assisted systems

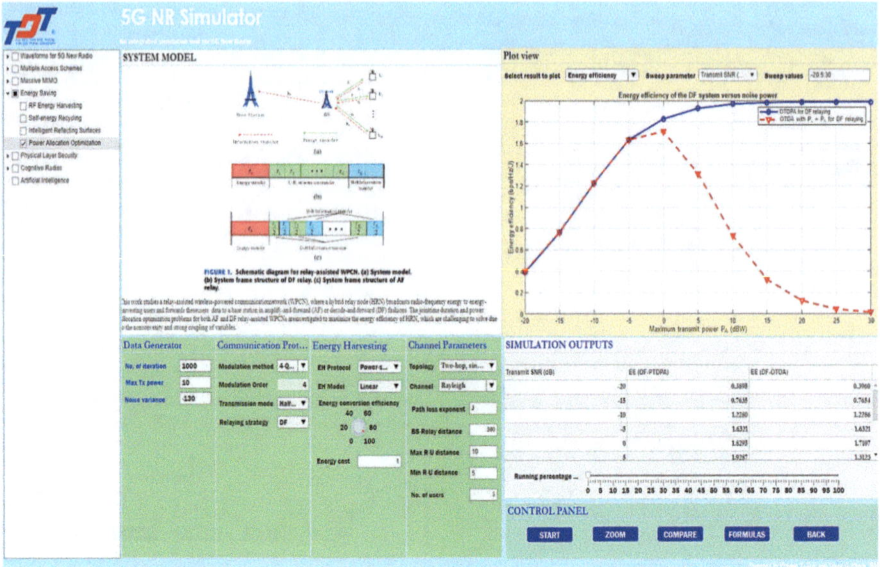

Fig. 11.14 User interface for power allocation optimization in cellular networks

11.8 Evaluation of User Interface Design

The graphical user interface has demonstrated logical consistency, including coherence shown in the design's use of blocks, icons, and unified menus for all modules, as well as the rationality in the control algorithm altering the interface content according to the selected simulation module and options regarding protocols and methods within each module.

The aesthetic appearance of the interface is evident in its relatively beautiful, clear, and harmonious layout, which impresses users through visually appearing images.

The user-friendly nature of the interface is demonstrated by providing visually intuitive instructions as well as text-based guidance. Additionally, all simulation modules display default input parameters on the interface, making it easy for users to understand how to input and output data.

The completeness is reflected in the interface's provision of relatively comprehensive advanced techniques as well as the latest research findings applied in 5G/6G networks.

Regarding processing speed, the application runs responsively on computers with MATLAB installed. Loading graph data and switching interfaces based on input parameters are relatively fast and do not experience lag.

11.9 Chapter Summary

The research has essentially achieved the main objectives outlined in Chap. 1. Table 11.1 summarizes the results of accomplishing the stated objectives of the project.

Table 11.1 Summary of the results achieved in fulfilling the objectives of the project

Objectives	Content	Results	Evaluation
Objectives 1	Establishing the general structure, components of the simulation tool	Achieved	Good
Objectives 2	Building modules to simulate modulation techniques	Achieved	Average
Objectives 3	Building modules to simulate multiple access techniques	Achieved	Good
Objectives 4	Building modules to simulate Massive MIMO techniques	Achieved	Average
Objectives 5	Developing simulation tools to evaluate energy saving solutions	Achieved	Good
Objectives 6	Building modules to simulate physical layer security techniques	Achieved	Good
Objectives 7	Writing interfaces for simulation modules	Achieved	Excellent

Chapter 12
Conclusion

Simulating 5G/6G communication signals is a crucial and essential testing and verification platform for 5G/6G technology. With key core devices and 5G base stations, the entire system software is its backbone component. This book introduces a method for designing system software to simulate 5G/6G advanced technology to support technical design of 5G/6G communication networks. High accuracy in representing channels, designs, and network orientations, and simulation tools are the best solution to address research issues in 5G/6G. As researchers, we all recognize its importance. Engineers in all aspects of research use 5G/6G simulation tools.

This book illustrates how to develop a simulation tool for learners and students to research, study, and understand 5G/6G networks at the physical layer, namely, the wireless access network. In this book, seven simulation modules are proposed, including MIMO techniques, modulation techniques, multiple access techniques, energy saving techniques, physical layer security techniques, cognitive radio networks, and artificial intelligence in wireless communications. The tool processes these technical groups with a user-friendly graphical interface for convenient observation by users. Additionally, we have summarized the operating principles of wireless information systems applying NOMA, MIMO, physical layer security, modulation, and energy harvesting techniques. Many simulation results have been presented to evaluate the system performance in terms of BER and throughput. The book can be considered as a reference source for students in related fields of telecommunications when exploring simulation techniques using this comprehensive tool. In particular, MIMO technology is leveraged to increase data rates for both fixed and mobile wireless systems and low-cost design, and electrical awareness is essential for devices using this technology, especially for UEs. Physical layer security for wireless networks does not use encryption algorithms that are widely researched worldwide. This simulation tool can be used independently or in combination with security methods using traditional encryption algorithms at higher layers to increase security levels or used to transmit secret parameters in security systems using encryption algorithms. The modulation tool provides comprehensive information at the interface level, updating on digital

P. T. Tran et al., *Developing MATLAB Comprehensive Simulation for 5G/6G Wireless Communications*, Signals and Communication Technology,
https://doi.org/10.1007/978-981-95-1117-4_12

modulation techniques in 5G digital communication systems. The energy saving tool helps us understand energy harvesting techniques, node selection techniques in energy harvesting networks, the benefit of using zero-energy devices in wireless networks, and intuitively analyzes the phenomenon of energy saving models through a graphical interface. The cognitive radio module helps the learners and researchers understand the concept of spectrum sensing and sharing in wireless networks, and investigate various communication schemes to enhance spectrum efficiency in cognitive radio networks. Last but not least, artificial intelligence (AI) is currently attracting the attention of all scientific communities. Therefore, the introduction of AI application modules in this simulation tool is irresistible and will provide a useful framework for students and researchers in the field of wireless communications.

Index

P. T. Tran et al., *Developing MATLAB Comprehensive Simulation for 5G/6G Wireless Communications*, Signals and Communication Technology,
https://doi.org/10.1007/978-981-95-1117-4

MIX
Papier aus verantwortungsvollen Quellen
Paper from responsible sources
FSC® C105338

If you have any concerns about our products,
you can contact us on
ProductSafety@springernature.com

In case Publisher is established outside the EU,
the EU authorized representative is:
Springer Nature Customer Service Center GmbH
Europaplatz 3, 69115 Heidelberg, Germany

Printed by Libri Plureos GmbH
in Hamburg, Germany